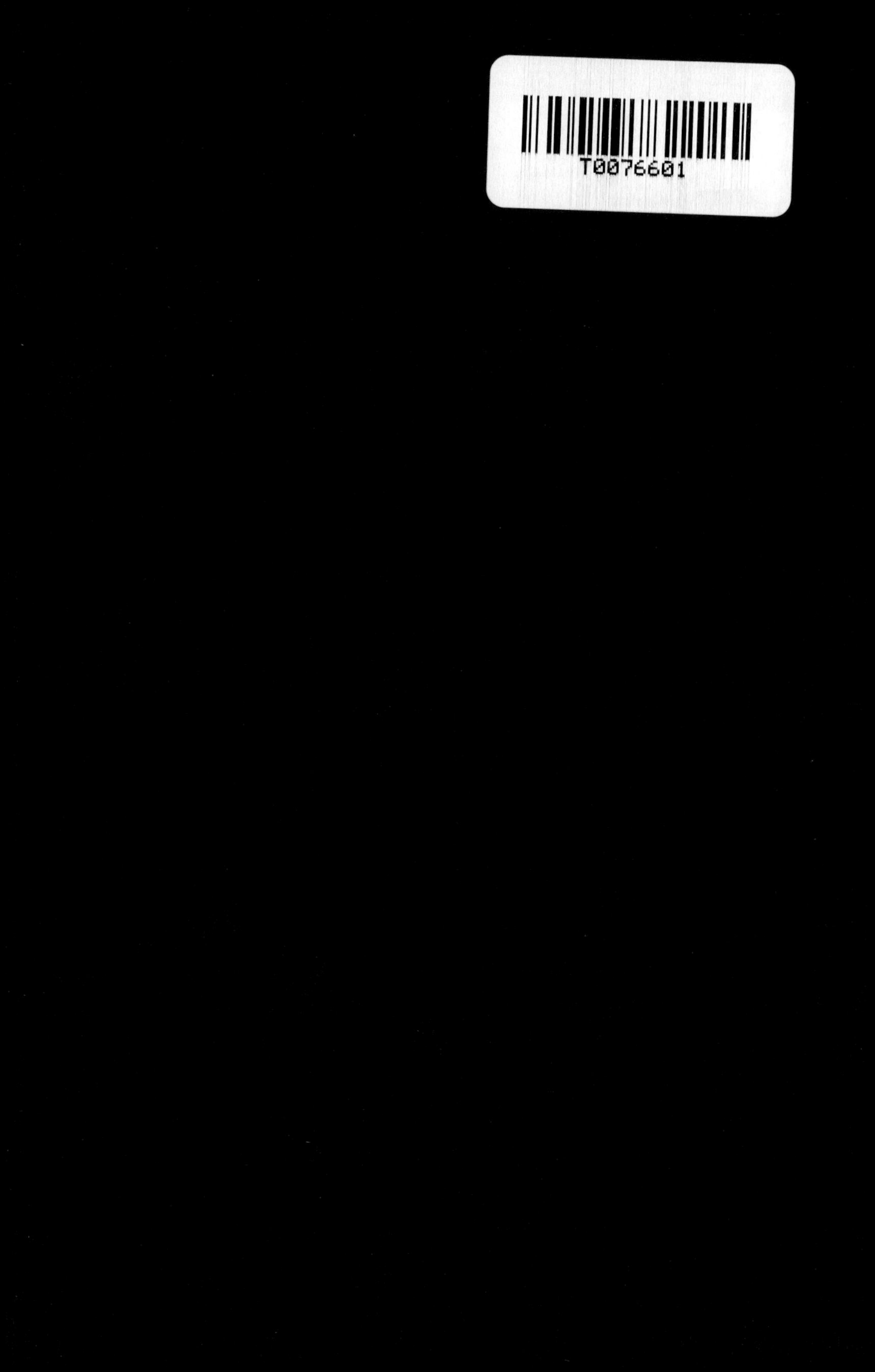
T0076601

Of Human Born

Of Human Born

Fetal Lives, 1800–1950

Caroline Arni

Translated by Kate Sturge

ZONE BOOKS · NEW YORK

2024

ZONE BOOKS
633 Vanderbilt Street
Brooklyn, NY 11218

Printed in the United States of America
Distributed by Princeton University Press,
Princeton, New Jersey, and Woodstock, United Kingdom

Originally published as *Pränatale Zeiten: Das Ungeborene und die Humanwissenschaften (1800–1950)* (Berlin: Schwabe, 2018). The translation of this work was funded by Geisteswissenschaften International — Translation Funding for Work in the Humanities and Social Sciences from Germany, a joint initiative of the Fritz Thyssen Foundation, the German Federal Foreign Office, the collecting society VG WORT and the Börsenverein des Deutschen Buchhandels (German Publishers & Booksellers Association).

Library of Congress Cataloging-in-Publication Data
Names: Arni, Caroline, author. | Sturge, Kate, translator.
Title: Of human born : fetal lives, 1800–1950 / Caroline Arni ; translated by Kate Sturge.
Other titles: Pränatale Zeiten. English
Description: New York : Zone Books, 2024. | Translation of Pränatale Zeiten. Das Ungeborene und die Humanwissenschaften, 1800–1950, | Includes bibliographical references and index. | Summary: "This book digs into the rich and mostly unexplored history of how the human sciences approached the unborn in terms of 'fetal life' by extending their gaze and research to what they called 'the period before birth.'" — Provided by publisher.
Identifiers: LCCN 2023030111 (print) | LCCN 2023030112 (ebook) | ISBN 9781942130895 (cloth) | ISBN 9781942130901 (ebook)
Subjects: LCSH: Prenatal diagnosis — History — 19th century. | Prenatal diagnosis — History — 20th century. | Heredity — History — 19th century. | Heredity — History — 20th century. | BISAC: SCIENCE / Life Sciences / Developmental Biology | MEDICAL / Epidemiology
Classification: LCC RG626 .A7613 2024 (print) | LCC RG626 (ebook) | DDC 618.3/209034 — dc23/eng/20231018
LC record available at https://lccn.loc.gov/2023030111

The ocean is used up now,

it's red and flows in their veins.

Below our feet there's now the ground,

and if we want to go to sea,

we'll have to build ourselves a boat.

For Nora, for Julien

Contents

Preface

It was quite by chance that, nearly twenty years ago, I stumbled upon a remarkable episode in the history of medicine. In the 1880s, French psychiatrists observed that Parisian children who had been conceived during the six months from fall 1870 to spring 1871, a period marked by military siege and political revolt, showed a disproportionate incidence of physical anomalies and behavioral problems. The psychiatrists wondered whether, alongside malnutrition and excessive alcohol consumption, the war trauma suffered by pregnant women might have adversely affected a whole generation of children. Just as badly as nutritional deficiencies and toxic substances, they reflected, perhaps the emotional shocks inflicted on pregnant women shaped "fetal life."

This episode seemed remarkable to me as a late nineteenth-century conjecture that pointed simultaneously backward and forward in time. On the one hand, the speculation resonates with the very ancient notion that maternal experiences imprint themselves on the form and disposition of the child-to-be like a seal in soft wax. On the other, it evokes present-day research anchored in the supposition that prenatal events set the course for a whole life, especially in terms of health and sickness.

This, then, is an episode from the 1880s that sounds an echo with the far more distant past and anticipates something that was then still the future. We might initially see that as a delightful quirk of history. But the historian is required to discover more than echo and anticipation in an event. Indeed, she needs to distrust that very figure and ask herself instead what it is about an event that makes it historically peculiar. The challenge is as follows: How can we relate what French psychiatrists in the late nineteenth century called "psychical influence" to the belief in the plastic force of pregnancy that has spanned many centuries, while simultaneously grasping its historical specificity?[1] That question raises another question in all its historicity: What were nineteenth-century psychiatrists dealing with when they sought the origins of a generation's singularities in its fetal life? This in turn confronts us with the more fundamental query: What actually *was* the unborn in nineteenth-century science?

The book that has emerged out of these questions tells two stories. First, it examines the historical foundations of what today is known as "fetal programming" or the "developmental origins of health and disease." Contrary to what is often suggested, present-day research on this theme is not a simple return to a premodern concept. Rather, in the course of the nineteenth and early twentieth century, the ancient notion of maternal impression was rearticulated in medical research as prenatal influence, something conceptualized initially in physiological terms, later endocrinologically.

Second, the book shows how the beginning of human life became an object of research in the emerging human sciences — specifically, physiology, psychology, and medicine — during the nineteenth century. When the life sciences were describing the unborn in terms of development, the human sciences were describing it in terms of the time before birth. The time spent in the womb became the "prenatal" phase in a human being's life course, a temporally defined segment from which life and biography spring inasmuch as it lays the foundations for everything to follow. Accordingly, birth itself also became an object of attention. As the threshold at which the life of the fetus ceases and the life of the child begins, birth posed above all one urgent question: When does the mind begin functioning in the human organism — before or after birth?

These two histories are entangled in a research practice of observation and experimentation about which little has yet been written, one that attended to the beginnings of the human subject and the shaping of the child by studying fetal life. How the unborn lives, what that life is capable of, what conditions constrain it, and what effects are produced when those conditions change — all these issues were investigated *in vivo* in chicks, animal fetuses, pregnant women, aborted and miscarried human fetuses, premature babies, and newborns. In this research practice, the unborn made its appearance as a "fetal life" about which nothing was self-evident and everything remained to be fathomed: How, in a manner peculiar to itself, does it take place; what is the nature of its relationship to the pregnant woman's body; and why is it able to transmute experiences of the mother into properties of the child?

This exploration of fetal life, also called "intrauterine life," was both a fetal ecology and a contribution to the ontogeny of the child, but it was something else as well. When physicians and psychiatrists turned their attention to prenatal influences, physiologists and psychologists to the beginnings of the human subject, they were always, and fundamentally, negotiating the origination of the human being in the mother's womb.

Some comments on terminology are in order here. In the sources I investigate, the terms *embryo* and *fetus* are sometimes used synonymously, sometimes one encompasses the other, and sometimes we find the classification current today, which distinguishes between the embryo, still in the process of taking shape, and the organically fully formed fetus. It is in this third sense that I myself use the two terms, along with the compound *embryofetal* when the distinction is not relevant. When it is not the physical form, but "life" that is at stake, the sources generally qualify that life as "fetal"; I follow this practice. The notion of "the unborn," drawn not from the sources but from previous research on the history of pregnancy, permits me to ask with an open mind how the unborn was imagined — without simply assuming that its constitution as embryo or fetus, self-evident and essential as it may seem to us, says everything there is to say.

Finally, a note from the translator on quotations from sources not originally written in English. Unless otherwise attributed, translations from French and German sources are her own; the English editions of French publications have been used unless they were incomplete or otherwise unsuitable.

PART ONE

Situating Fetal Life

CHAPTER ONE

Paris 1870/71: Trauma in the Womb

A child was conceived in Paris in May 1871 who would attract the attention of illustrious medical men thirteen years later. This girl, psychiatrist Charles Féré reported in 1884, had a "generally satisfactory constitution" and her skull was regular in form, though her face was somewhat disfigured by the scar of a cleft lip.[1] However, Féré continued, the child was tormented by an eyelid tic, could speak only with difficulty, and occasionally suffered from bedwetting. She read very badly and could hardly write; furthermore, she was drowsy, gloomy in temperament, and plagued by attacks of vertigo that made her drop things on the floor.

Féré ruled out family disposition as an explanation of these behavioral problems, for neither the family of the father, a respected lawyer, nor that of the mother showed any history of nervous illness, and the girl's three older siblings had no symptoms of the kind. Everything suggested that the girl should be considered one of those children known in common parlance as the "children of the siege" (*enfants du Siège*) or the "children of the Commune" (*enfants de la Commune*). She had been conceived during the tumultuous days of the Paris Commune, more precisely the early hours of May 2, 1871. Just half an hour later, troops of the National Guard had burst into the lawyer's apartment, whereupon his sensitive wife, terror-stricken, immediately fell to vomiting and did not regain her usual state of mind until several days later. The family then left Paris and the pregnancy proceeded without further complications. It was fair to surmise, Féré concluded, that the daughter's irregularities had resulted from the "shock to the mind" — the *choc moral* — suffered by her future mother in the hours of conception, when the flood of political events surged into her apartment in the shape of the National Guard.[2]

Charles Féré, recounter of this noteworthy case, was one of a handful of French psychiatrists associated with Jean-Martin Charcot who, in the 1880s, took an interest in the cohort of children conceived and born between fall 1870 and spring 1871, during the German siege of Paris and the subsequent revolutionary events of the Paris Commune. The turmoil of that *année terrible* imposed tremendous stress on the local population. In September 1870, despite the defeat of its troops, the new French Third Republic refused to accept the peace terms for ending the Franco-Prussian War. The German forces responded by besieging Paris, starving the city into capitulation. From the end of the siege in January 1871, a political struggle set the royalist majority in the French National Assembly, which concluded peace with Germany, against radical republicans in Paris, who feared a restoration of the monarchy. In March, the city rose up against the national government by electing revolutionary republicans and socialists to govern in what would prove to be a short-lived "Commune." On May 21, 1871, national troops entered Paris. In the ensuing "bloody week," twenty thousand defenders of the Commune were killed.

Now, a decade later in the 1880s, it was said that the children conceived during those months of terror showed "developmental disturbances" (*troubles d'évolution*) with disproportionate frequency.[3] According to his colleagues, Désiré-Magloire Bourneville of the Hôpital Bicêtre, founder of one of the world's first child psychiatry departments, had also observed the phenomenon; at least, his photographic collection on childhood mental illness included several portraits of children of the *année terrible* (Figure 1).[4]

The definitive portrayal, however, was that presented by the psychiatrist and criminologist Henri Legrand du Saulle in a much-noted

Figure 1. Photographic portrait of a young boy (born June 1871), taken at the Bicêtre hospital. Désiré-Magloire Bourneville, 1880–81.

1884 lecture at the Parisian Hôpital de la Salpêtrière. His discussion of the "influence of political events" on "physical and intellectual anomalies in the children conceived during the siege of Paris" established both the phenomenon and the label *enfants du Siège* in psychiatric debate as an umbrella description for the children of the siege and the Commune.[5] During his work at the Bicêtre and in the psychiatric infirmary of the Parisian police prefecture, Legrand du Saulle had gained access to empirically utilizable data that he summarized in his lecture. Of ninety-two children in the cohort in question, he reported, sixty-four had shown "physical or mental deformities"; specifically, thirty-five displayed "tubercular abscesses, a sloping forehead, a squint, epilepsy, hearing loss, stuttering, hemiplegia, clubfoot, incontinence, or rickets," twenty-one suffered from "intellectual abnormalities" such as "reduced mental faculty, dejection, apathy, inattentiveness, semi-imbecility, and idiocy." Eight children were "egoistic, perverted, immoderate, malicious, brutal . . . and obscene"; the remaining twenty-eight were feeble, but not conspicuously malformed.[6]

Just a few years later, Charcot himself noted in the clinical history of a man who since childhood had been sickly, nervous, and impaired in mobility that this patient, "born on April 13, 1871 (after the siege)" was "as one says, a 'Child of the Siege.'" The reference to the *enfants du Siège* here appears to be entirely taken for granted, requiring no further explanation.[7]

All these physicians drew a connection between political events and childhood "anomalies" that was already commonplace in the streets of Paris at the time. Pediatricians, said Legrand du Saulle in his lecture, had long been consulted by mothers whose approach to etiology was brisk: "Monsieur, what do you expect? He's a child of the Siege." That, Legrand du Saulle commented, was "exact," and it was "true."[8] But what had given rise to this connection, a link that came to light in the throwaway comment of a mother and prompted the psychiatrist to consider the date of birth in children with problems?

Legrand du Saulle had more detail to offer. During the siege of

Paris, "the innermost conditions of fetal life" had been altered in these children. Even more precisely: "Before its birth, the fetus was subject to disastrous influences, and their pathogenic effect went as far as injuries to the brain."[9] This, then, was how Legrand du Saulle understood the bond between political event and childhood anomaly: as the pathological effect of a prenatal occurrence that had intervened in fetal life.

What exactly was it that had exerted such an influence? Two factors seemed so obvious that the psychiatrist had only to sketch out familiar vignettes: "What happens during the siege? The man leaves the apartment to get drunk while the woman stays at home, hungry; soon she dips her bread in wine to ease the absence of victuals, and so she slips gradually into . . . the habits of the alcoholic."[10] But alongside malnutrition and alcoholism, Legrand du Saulle insisted, a third factor must be taken into account, namely "the state of mind in which we all found ourselves." For even if Parisians had put on a brave face, they had in fact suffered a mental trauma, a *traumatisme moral*.[11]

By referring to "trauma," Legrand du Saulle mooted a connection sure to pique the interest of the audience gathered at the Salpêtrière, which was made up of colleagues and students of Charcot.[12] Although at this time the term *trauma* in medical discourse still mainly followed the etymological tradition to describe the bodily "wound,"[13] since the 1870s it had increasingly been applied to psychological phenomena as well.[14] This expansion of the term's scope was spearheaded by the school of thought around Charcot, which psychologized the notion of trauma partly in the context of research on hysteria, but partly also in response to the events of 1870 and 1871.[15] By adding the adjective *moral*, and thus bringing mental aspects into play, Legrand du Saulle articulated just that change.

Féré, a student and close collaborator of Charcot's and later of Legrand du Saulle's, picked up this thread when, citing Legrand du Saulle's lecture on the *enfants du Siège*, he told the story of the unnamed lawyer's daughter. In Féré's view, her case was informative because it enabled the mother's traumatization to be traced back to a precisely recalled mental shock. The Charcot School saw the shock as the link that connected psychological traumatization with later symptoms.[16] Moreover, in this case there was no need to worry about a possible amalgamation of different causes. Legrand du Saulle's statistics had covered mainly children from the "poor classes," who were particularly hard hit by malnutrition and alcoholism. It was reasonably safe to assume, Féré argued, that in the case of the lawyer's daughter — the scion of "a different milieu" — the anomalies had been caused by the shock-induced "psychological state" (*état psychique*) of the expectant mother alone.[17] One could therefore contemplate here, in an isolated form, the "least well-known" yet "most interesting" factor: "psychical influence" (*influence psychique*).[18] Little wonder that Féré, who soon after the publication of his case history became the first secretary of the Société de psychologie physiologique, founded in 1885, and later vice president of the Société de biologie, would devote himself to this topic for the next fifteen years.[19]

This book is an attempt to understand the object of a curiosity that made a childhood anomaly in the present point to fetal life in the past. How could an event that took place in 1871 still be taking place in a child who blinked too frequently and irregularly in 1884, who was lost for words and could not find her way out of her dreams, who could ward off her body's floundering only by dropping objects on the floor? Starting from this question, we find that nineteenth-century talk of fetal life was not as self-explanatory as it may seem today. Rather, it reveals the historically particular and entirely novel manner in which the human sciences between the late eighteenth and the early twentieth century constituted the unborn and conceptualized the time before birth.

In the next two chapters, I introduce the theoretical and methodological orientations of this book, which combine the history of science with historical ontology. Presenting an overview of its argument, I show how *Of Human Born* proceeds by situating the story of the *enfants du Siège* within a larger history of the unborn and the human sciences.

CHAPTER TWO

Children, Fetal Life, and the Prenatal

An Event in the History of Knowledge

Charles Féré's speculations on the consequences of shock to the pregnant woman's mind are much more than an anecdote from the history of medicine. His notion of psychical influence brought together a new understanding of the unborn in the human sciences with a contemporaneous interest in the transmission of traits across generations — an interest that was, in turn, inextricable from a political concern with societal continuity. This is why the *enfants du Siège*, the children conceived or born during the terrifying year of 1870/71, form the crux of my study, as an event in the history of knowledge where such connections come into view and from which they can be explored. The children supply me with the guiding thread for an investigation in which I assume that everything about this nexus is historically specific.

That does not mean it is disconnected from what went before and what was still to come. References to pregnant women who are shaken by events and give birth to perplexing children evoke a very ancient idea: that like generates like. Thus, the ancients thought, pregnant women should contemplate sculptures and avoid funerals in order to ensure that beautiful and cheerful children would be born.[1] Equally, such comments foreshadow a present-day conviction: that the mother-to-be's lifestyle and circumstances lay the foundations for the prospective child's health and must be taken into account as a third factor, alongside genetics and upbringing, in the emergence of individual characteristics.

This postulate is currently enjoying enormous popularity, driven largely by the epigenetic turn in biology. Breaking open the always artificial dualism of "nature" and "nurture," epigenetics finds that environmentally conditioned changes in the organism have the capacity not merely to regulate the activity of genetic material, but to alter the material itself.[2] As the embryonic or fetal organism is considered particularly receptive to environmental influences, being still in the process of development, this paradigm shift adds momentum to the medically oriented research field of the "developmental origins of health and disease" (DOHaD), a field that asks how physical and psychological dispositions are "programmed" in the womb.[3] When it comes to influences on the development of the embryofetal organism, the question also arises of whether such effects will be perpetuated in the organism's offspring.[4] This interface between epigenetics and DOHaD is thought to herald a new epoch by bringing together the research objects of heredity and development, which in the twentieth century were largely insulated from one another by the disciplinary wall between genetics and embryology.

It is no surprise that such research found its way rapidly into an optimization-hungry literature of advice and self-help that judges everything the pregnant woman does, or fails to do, by the well-being of her prospective child. A popular science manual published some years ago promises to explain "how the nine months before birth shape the rest of our lives," giving both practical tips and an accessible outline of scientific research at the intersection of DOHaD and epigenetics.[5] But of the many factors potentially shaping the future child, the one that remains most fascinating today is the psychological factor, and in particular the question of how the effects of traumatic experiences — whether war, natural disasters, or genocide — can persist across generations by acting on the pregnant organism or the parental genes.[6]

A special challenge for the historian arises from the fact that Féré's hypothesis of psychological influence can at once be linked to a present-day research trend and ranged within a *longue durée* going back to antiquity. This configuration immediately suggests two possible historiographical approaches, both of which I intend to avoid.

First, the current fascination with the concept of prenatal influence makes it tempting to interpret related ideas from the past as chiefly a prehistory of the present. In this frame, today's interest would be a translation of past intuitions into positive knowledge about biological processes. All previous peaks in interest would appear as a kind of misguided scrabbling and stammering in the drawn-out process of discovering the facts. Yet this narrative of knowledge gradually becoming correct tacitly assumes that when people talked about prenatal influence at different points in history, they were always talking about the same thing. It erases the query: What exactly was it that they were talking about?

Second, the regular resurgence of the idea of influence across so many centuries seems to imply there is something unchanging at work, namely, the question of how to explain the characteristics of a human being. Yet that would be to say both too little and too much: too little, because every era formulates the question differently; too much, because the object of the investigation is defined as being an answer and, as such, one variant within a series of answers. A history of the same question, then, is unable to ask: What exactly was the question?

Both these ways of relating Féré's hypothesis of psychical influence to its precursors and descendants assume there is something always the same, whether fact or question. If I wish to avoid both approaches, that is not because they are wrong in themselves, but because they shackle the work of historicization. Only when we allow for the possibility that it is not always the same thing that is meant, or the same question that is answered, does the historical specificity of Féré's hypothesis come into view. This specificity is what interests me here, and it is what can give historical depth to the study of a present-day concern by sharpening our sense of conceptual alterity. Against this backdrop, let me return to the knot of events out of which I began to disentangle my guiding thread.

Henri Legrand du Saulle was not an embryologist, neither was he a gynecologist. He was a psychiatrist, specializing in criminology, who encountered children in the infirmary of the police prefecture. This was where everyone was brought who had "lost their senses" on the streets of Paris—*tout ce qu'il y a de délirant* is how Legrand du Saulle put it in his lecture on the children of the siege—and who were then committed to the Salpêtrière if they were girls or the Bicêtre if they were boys.[7]

Among these were some children of the 1870/71 cohort. Legrand du Saulle extracted them from the series of the children who were out of their mind, because they belonged just as much to another series: the sequence of violent events that had commenced in 1789 and "inflamed" the political arena once again in 1871.[8] Legrand du Saulle was interested in the event, and he was interested in the child as such. That was no coincidence. In the nineteenth century, science was turning ever more intensely to the nonadult, setting up children's wards in asylums, turning physicians into pediatricians, and encouraging psychologists to observe their own sons' and daughters' everyday lives. Psychologists, in particular, began to build a doctrine of development that bestowed on children the promise—and shadow—of their future, and on adults a historical depth that could now be plumbed through the figure of childhood.[9] At the same time, the "event" as such took its place within a new vision, the passing of time as a historical process. The event was the meeting-point of before and after, creating a course of things, as Reinhart Koselleck has said, by binding together the past as "experience" and the future as "expectation."[10]

Like Legrand du Saulle, Féré was neither a gynecologist nor an embryologist. He completed his doctorate in neurology under Jean-Martin Charcot's direction in 1882, and from 1887 would care for mentally ill patients at the Bicêtre, interested in precisely how physical factors relate to psychological ones and how disease spreads not only within populations, but from generation to generation. How do anomalies and pathologies wend their way through time? How do they persist, changed or unchanged, even when the bodies and biographies they mark have vanished and new ones have emerged that contain, in some way or another, what they received from their originators?

Engrossed by such questions, Féré, like his peers, named this intermeshing of disease and genealogy "degeneration." His experimentally meticulous attention was directed especially to the differing ways in which disease lodges itself in the processes of reproduction. Whether a future child can be affected by a sensitive mother's feelings to such an extent that they leave physical and psychological traces, as he suspected in the case of the lawyer's daughter, was a scientific enigma that perfectly bundled Féré's many and varied interests.

When they listened to what Parisian mothers told them of children whose steps and words escaped the norm, who could not control themselves, whose impulses, movements, and feelings slipped from their grasp, neither Legrand du Saulle nor Féré was concerned with the begetting and birthing of human beings. What the two psychiatrists saw when they counted and described such children was "fetal life" — as if the child were a transparent surface through which shimmered what had previously been. My guiding thread, then, is one twined out of event, child, and anomaly. It leads me to a place where the three came together as observers of childhood irregularities asked about the events of pregnancy and thus gave significance to the time before birth. There it was that, toward the end of the nineteenth century, the word *prenatal* arose.

In this book, I consider not the mothers — who probably wondered why anyone would bother to puzzle over the peculiarities of the terrible year's children — but those who wanted to know. That means we must think of this place as an "epistemic space," the term used by Staffan Müller-Wille and Hans-Jörg Rheinberger in their account of how the concept of heredity arose in the modern period. It was fed by sources that, heterogeneous in themselves, together "created a set of coordinates that made it possible to conceive of reproduction no longer only as the personalized and individual generation of offspring, but also as the transmission and redistribution of a more or less atomized biological substance."[11] Proposing as I do to treat the "prenatal" as an epistemic space as well, two aspects of this explanation are important for my study: what it seeks to explain, and how.

First, what does it seek to explain? In the course of the nineteenth century, the life sciences replaced the ancient idea of the "generation" of human beings by human beings (the Latin term being *generatio*) with the notion of a "reproduction" of the species. The making of children by parents was now seen less as the work of the parents than as that of the human species. The species continued its existence by remaining itself through the *transmission* of characteristics, while also giving rise to multifarious individuals through the *distribution* of characteristics.[12] As the idea of a generational connection through the transmission of biological material became more firmly delineated, a new problem took shape: Where should the contingent influences on the organism developing from that material be situated within the events of generation? At the end of the century, Féré would encapsulate the dilemma in his conceptual distinction between "true heredity" (*hérédité vraie*) and pathologically inflected "accidents of pregnancy" (*accidents de la gestation*).[13]

Historians have regarded distinctions of this kind around the turn of the twentieth century as marking a process by which "heredity" split off from "development."[14] It would be too hasty, though, to conclude that those two matters had previously been unified. What interested Féré was not simply that rule-based heredity and contingent influences could be distinguished—that very distinction had been in common use since the beginning of the nineteenth century.[15] At stake, rather, was exactly how the two factors relate to one another and together form a generative context.[16] If transgenerational continuity ever was restricted entirely to the transmission and distribution of genetic material, as has been said with regard to the twentieth century, then the distinction between two modes of that continuity—transmission and influence—had certainly been debated much earlier. The more exclusively the concept of heredity was applied to transmission, the more urgent was the need to find a conceptual framework for influence as well.

It was in response to that need that, around 1900, the term *prenatal*, generally used as an adjective, became current; the related terms *antenatal* and the German *vorgeburtlich* also established themselves. Over the course of the twentieth century, this terminology would here and there, dispersed across different disciplines, configure the very same research questions that are addressed today in the research field named DOHaD, where they are once again placed in the context of hereditary processes, this time under the rubric of epigenetics.

Any study of "the prenatal" requires a definition of its form. The prenatal is not a theory, an object, a discourse, or a discipline. It is an enabling of statements, just as Rheinberger and Müller-Wille set out for the case of heredity. The notion of the epistemic space that they apply draws attention to the ways in which different types of things enter into relationships with one another and different types of questions fuse or branch apart, without prejudging the logic of that process as the consolidation of a theory, the discovery of an object, the invention of a discourse, or the establishment of a discipline — in short, for the purposes of my argument, without regarding the emergence of the concept of the prenatal as the vanishing point of a development *toward* something. Instead, this book attends to how, in both objects and questions, the old was continued and the new was created; it tries to give describable form to a particular constellation of historical continuity and discontinuity.[17] To that end, I use the figure of "space," but I should point out right away that this does not refer to a locality antecedent to what occurs within it. This space is opened up *by* something occurring.

In an epistemic space, that something is the making of knowledge. I do not mean a knowledge that portrays or fails to portray what is given outside it, nor a knowledge that has nothing to do with its referent and is defined only through its relationship to other knowledge; neither do I mean a knowledge that adds a particular freight of meaning to something already given. The knowledge I am interested in is one that makes what it asks about through the questions it poses and that is therefore always more *doing* than product or achievement. Such knowledge becomes concrete in what Hans-Jörg Rheinberger calls "epistemic things." These are the things toward which the effort of knowing is directed. They may be objects in the narrow sense, but also structures, reactions, or functions arising through the material orders of wanting-to-know — experiment and observation.[18] In such "things," curiosity fuses with instrument, hand, and object. This, Rheinberger and Michael Hagner point out, subverts the notion of purity that tries to demarcate science from its organization, idea from materiality, yet "finds no correspondence in the process of making science."[19]

Knowledge made concrete in epistemic things is thus primarily science as praxis: an activity that is not determined either by a natural object or by its results, but that has to be inferred from its procedures, since it is through those procedures that it constitutes the things of which it treats.[20] The encounter between the children of 1870/71 and the Parisian psychiatrists harbored a thing of just this kind. When they described the children as embodying an entanglement of present anomaly and past event, the psychiatrists were speaking of what it is that can turn a present event into a future anomaly: fetal life.

Following the tracks of the *enfants du Siège* in pursuit of this epistemic thing, I do not wish to argue that the ties between child, event, and anomaly merely opened up to scientific curiosity a fetal life that was always already there. I am interested in how scientific curiosity configured fetal life in such a way that event and anomaly could be interwoven within the child, by exploring development, time, and transmission. Only by characterizing the child as development, the event as historical time, and the anomaly as a part of generational processes could the scientific gaze pass through the *enfants du Siège* to see fetal life.

This combination interlocks a history of psychological influence with another history. A fetal life constituted by development, time, and transmission was exactly what, in the nineteenth-century sciences of the human being, described the unborn. At the same time, these three things were what systematically inserted the unborn into the emergence of the human sciences as such. To study fetal life was also to raise the question of how the human subject began. As a result, my guiding thread leads me from the auditorium of the Salpêtrière into the laboratories, labor wards, and libraries where human physiologists, psychophysiologists, and psychologists investigated pregnant and fetal bodies.[21]

Starting my path through archives and libraries with the 1884 appearance of the *enfants du Siège* in the Salpêtrière's lecture hall, I tracked, on the one hand, the research to which they inspired Féré, which would become crucial to an agenda of prenatality lasting well into the early twentieth century. On the other, my search for the preconditions of Féré's work led me to the diversity of ways that fetal life was studied. I discuss these for the period between the late eighteenth and the mid-twentieth century because it is within this temporal purview—expanding in concentric circles from the 1880s—that the historical specificity of "fetal life" can best be illuminated.

There are three important consequences of my decision to mirror the genesis of my study in its written representation by consistently relating the story of the unborn in the human sciences to the story of the *enfants du Siège*. The first of these consequences is my focus on French sources. They take pride of place due to my starting point in the Parisian "year of terror" (*année terrible*) and because the making of fetal life was driven by an experimentally oriented physiology that arose particularly early and vigorously in France.[22] However, reciprocal reception and shared interests link the French studies to those of German-speaking and some English-speaking scholars and scientists, requiring me to compile a transnational corpus of sources.

My portrayal's passage through time, and this is the second consequence, also transgresses borders to an extent since it crosscuts the chronological order. I am confronted again and again with what Ludwik Fleck called the "genesis of scientific facts."[23] Those facts include the distinction between maternal and fetal blood; the transmissibility of one pathogenic agent and the nontransmissibility of another; the fetus's sensitivity to pain. Looking at such facts, I am concerned less with the logic of their emergence as captured in dates and protagonists than with how, through their study, fetal life—and through fetal life, the unborn of the human sciences—was constituted. This emphasis guides my account's movement across time, when and where it commences, and how it flashes back or hastens ahead.

Finally, the book crisscrosses a landscape of scientific fields and subjects that was just beginning to take shape in the nineteenth century. This procedure reveals the edges of the space that was staked off during the nineteenth century by addressing the time before birth. Even less than the study of embryogenesis, which would eventually achieve some identity as "embryology" but continued to be divided in complicated ways, did the study of fetal life coalesce into a discipline of its own.[24] It remained an interest distributed across medicine, physiology, and psychology.

The dispersal of material about fetal life is as historically specific, and contingent, as the story of the children who came to be known as the *enfants du Siège*. But contingency is not the same as coincidence—and this is how my attempt to contextualize one episode became a book that aims to describe the unborn in the sciences of the human being.

The Book in Chapters

The project of excavating a historically specific constitution of the unborn out of Féré's conjectures on psychological influence initially takes us back into the eighteenth century. It was then that a momentous shift, described in the pioneering work of Barbara Duden, turned a woman's being with child into the process of embryofetal development, and thus the unborn into a "biological, objective fact."[25] The "objective" component in this triad indicates that a somatic certainty of being pregnant was replaced by a scientific knowledge about pregnancy; the "biological" component shows how the object of that objectifying knowledge was constituted: as an organism. The unborn, previously said to form a unit with the nourishing body of the mother—growing in her womb by breathing, eating, sensing, and dreaming with her—now became the object of research practices that used human and animal embryos and fetuses to study everything that characterizes a living being in the biological sense: anatomical form, physiological function, and the transformative process of development.[26] As Lorna Weir remarks, this modern, biological reconfiguration of the unborn has been noted now and again, but not studied historically.[27] Although Duden, in particular, has examined the crucial role played by anatomical visualizations of embryonic development, and though we have rich research on embryology's work of "producing development," as Nick Hopwood puts it, very little has been written about the distinctively physiological investigation of the unborn in the nineteenth century.[28] It is this that I discuss in Part 2, "Living Beings."

In that part, my first step is to unpack a question that the unborn posed once it was configured biologically (Chapter 4): When a being is located in the body of another but cannot, if it is to become an individual organism, live the life of that other, how does it live? How does it form its blood, how does it sustain itself, how does it move, how does it feel? This was answered first in topological terms: the fetus lives *in* the mother, just as every other organism lives in a milieu without which it cannot feed, breathe, or move. The fetus's life, thus, is "uterine" or "intrauterine."

Yet even while the life in the womb was being defined spatially, research activity began to address its specific temporality. Fetal life was interesting as a developmental process in which vital functions took shape. In a second step, I show how this physiological perspective on the unborn kept vividly present something that anatomical work on the dead object could afford to neglect: the body of the pregnant woman, without whose labor of sustenance the fetus cannot be alive. The biological objectification of the unborn thus gave rise not only to an individualized embryofetal organism, but also a maternal-fetal relationship (Chapter 5).

Starting from this assumption, as I explain in a third step, the traditional puzzle of congenital illnesses and malformations could be teased out in new ways—no longer as an expression of experiences that have befallen mother and child jointly, but as a consequence of pathogenic influences of the maternal environment on embryofetal development (Chapter 6). In the late nineteenth century, this relationship seemed to find its exemplary embodiment in the *enfants du Siège*, so much so that they prompted Féré to shift his work on hereditary genealogies of disease into a new, experimental direction by carrying out trials on developmental anomalies. He combined the science of malformations, the "teratology" already established by naturalists, with the interest of midwives and pediatricians in the pathological unit of the pregnant woman and the unborn. At the beginning of the twentieth century, this combination gave rise to the program of prenatal pathology.

It was now possible to rethink the pregnant woman's capacity to mentally shape the unborn as a form of influence. For hundreds of years, women and scholars had explained children's abnormalities by that capacity, which was the object of the doctrine of maternal imagination so fiercely debated in the eighteenth century. Féré himself portrayed his hypothesis of psychical influence as being an echo of that idea. In Part 3, "Inner Life," my starting point is work on fetal life that was oriented on psychological questions. I first trace how, in the early nineteenth century, a visual theory of maternal impression gave way to an emotional theory, modeled by psychophysiology, of the influence of maternal feeling (Chapter 7).

As I then show, there was speculation in this setting about whether the fetus was an organism capable of sentiment and of reacting to its mother's feelings (Chapter 8). This is where the second strand of Féré's research on psychical influence comes into play: his experiments on fetal movement, performed on pregnant hysteria patients. In his experimental setup, the physiologically constituted mother-fetus relationship took a psychological turn. But the unsettling question of the unborn's interiority, which had already arisen in the early nineteenth century, was still not resolved. The psyche having come to be seen as one organic function among many, a need arose to seek the point in time when its development began. In the 1920s, this sparked a psychoanalytic dispute on birth trauma that — symptomatically — failed to find an answer to the question.

Retracing my steps back to psychological influence, I show how, in the early twentieth century, the study of hormones made it possible to grasp materially the impact of maternal stress (Chapter 9). This did not merely fill an existing lacuna. Rather, the endocrinological perspective reconfigured yet again, in yet another historically specific way, the potency of maternal mental life.

In the course of the nineteenth century, the adjective *prenatal* in all its versions almost imperceptibly edged alongside, and increasingly overtook, the adjective *intrauterine*. An axiom formulated by the educationist Gabriel Compayré in 1893 indicates what was at stake: anyone wishing to understand the child (and in the child, the human being) must necessarily expand their gaze to encompass the "obscure history of the nine months of gestation."[29] In the concept of the prenatal, the unborn was not simply defined in time. The time named in that concept has a special, historical quality. It is productive time, which passes by making something subsequent emerge out of something antecedent. It is this form of temporality that Michel Foucault meant when, in his history of the human sciences, he observed that in the nineteenth century everything empirical — thus also, or especially, "man" — came into the world as something historical.[30] The invention of the word *prenatal* was nothing other than a conceptual codification of the manner in which the nineteenth-century human sciences constituted the unborn as a "before birth" that, whether in the mode of precondition or impression, creates the conditions for an "after birth."

In the nineteenth century, this modality tied the time of the individual to the time of society. The connection came into sharp focus with the mass pregnancy accident that occurred in Paris in the six months between fall 1870 and spring 1871. In the concluding Part 4, "Politics of the Unborn," I immerse myself one more time in the story of the *enfants du Siège* (Chapter 10). This is because those children occasioned not only scientific discussion, but also a debate on the politics of memory, ignited by the question of what inheritance of the past was really embodied by the children of the year of terror. Legrand du Saulle's assertion in 1884 that the events resulted in a "degenerative," pathologically deformed link between the generations was politically explosive, given that the French defeat by Prussia and the bloody suppression of the Paris revolt needed to be assimilated into the self-image of the Third Republic. This coupling of scientific and political issues was not a factor intruding from the world outside the clinics and laboratories. It had its origins where psychiatrists listened to mothers defining their children by the historical event that had crashed into the continuing succession of generations — the chain that gave permanence to the imagined community of the nation. Thinking about the "obscure history" of the months in the womb raised fears that such continuity could be breached. Explorations of fetal life were thus also about the time of society.

That brings the story of the *enfants du Siège* to a conclusion. There is still something to be added to the history of the unborn and the human sciences, however. In this book, I often speak of research practices — of what scientists set out to do and what they did to animal fetuses and to human women and newborns. To understand this doing, we also need to investigate the relationship that arose from scientists laying their hands on their objects (Chapter 11). That relationship interests me as a social relation between researchers and objects, one no more self-evident than is the constitution of the unborn as fetal life.[31] Certainly, it gave rise to all sorts of research objects, but in the sciences themselves, it also thwarted the perfect transformation of the "child-to-be" into a biological object. The subjects of research, human scientists, were still forced to acknowledge that their object — the unborn — was a human child, and thus something that was in principle their mirror image.

In the coming chapter, I flesh out that conclusion by unfurling my research perspective from the findings of my study. These findings are grounded in the decision to do history of science as a historical ontology—as a history that shows how, in the making of epistemic things, entities are also created from which the world is composed. In the history of the unborn and the embryo, interest in this perspective has been heightened by historians such as Duden.[32] As my research proceeded, I realized the importance, too, of recent discussions in anthropology that challenge us to reflect fundamentally on what an ethnographic (or in my case, historical) investigation is actually about. Studies of present-day reproductive practices have been groundbreaking in these debates.[33] The next chapter therefore anticipates in four steps how, in the epistemic space of the prenatal, birth as a threshold was simultaneously abolished and confirmed; what that meant for the definition of the unborn; how ontological questions were turned into epistemological ones; and how this turn remained an ontological practice caught up in uncertainty. Uncertainty, because what the unborn actually *is* could not be conclusively decided through what was *known about* fetal life.

CHAPTER THREE

The Unborn and the Human Sciences

Continuity through Development

There are many ways to tell the history of prenatal life. One could start from the adjective *prenatal*, letting the whole story branch out from the observation that this term became established in scientific talk about the unborn at the turn of the nineteenth to the twentieth century. That route would lead to a narrative of the "invention" of life before birth. My study supplies building blocks for such an account, showing that despite their changing denominations — some names being based on the entity (the fetus) and others on its locality (the uterus) — nineteenth-century physiological, psychological, and pathological investigations of the unborn always revolved around the same interest. Whether "fetal physiology," "physiology of the embryo," "uterine psychology," "psychology of the fetus," or "fetal pathology," all were concerned with what comes before birth.

My own interest, however, goes beyond the history of an invention. When it looks at historical events, a narrative of that kind is bound to see only innovations — and those innovations can reveal themselves as such only teleologically, as viewed from the endpoint of the concept's invention. Historiographically, the approach is unrewarding if only because it must regard everything that happened before the term's coinage as a mere "prehistory" of what was historically new: reading prenatality as part of the continuum of a biographical and clinical life history, or, as the educationist and philosopher Bernard Perez so vividly put it in 1882, making legible the very "first page in the book" of a person's life.[1]

It is certainly true, as I will show, that nineteenth-century science's interest in the unborn produced a birth-traversing continuity that was described by the concept of prenatality. However, this continuity was not new. Before the nineteenth century, too, the human being had originated in the unborn, birth had been one moment within its becoming, and the process of becoming had been under threat. In Aristotelian doctrine, the germ becomes first a nutritive soul like the soul of plants; then a sensitive soul, thus an animal; last, it is imbued with the rational soul, stemming from the divine, that defines the human being.[2] Physicians, theologians, and legal scholars in premodern Europe had construed this theorem as implying the presence of a future human being in the mother's womb. They did not dispute the ensoulment of the unborn but disagreed stubbornly on the exact time when it occurred.[3] In the experience of pregnant women, the continuity that would carry the child-to-be over the threshold of birth commenced with the first sensation of the child stirring in the womb, "quickening." In the idea of animation in the sense of coming to life, *animatio*, that experience coincided with the scholars' postulate of ensoulment.[4] Finally, the notion that, far from being always securely enclosed in the mother, the unborn is at risk is among those topoi around the formation of human beings that have spanned the ages.[5] What, then, changed in nineteenth-century science?

Crucial to the constitution of the unborn as something prenatal was the fact that continuity could now be construed as a temporal relationship. The term *prenatal* institutes a continuum by separating "before" from "after" in time while simultaneously locking them together. This particular continuity across birth springs from a notion of development that had come to prevail at the turn of the eighteenth into the nineteenth century. It marked the end of a long and vehement dispute between the advocates of preformation, who thought of the ovum or sperm as an already-formed human body that only needed to grow, and the advocates of epigenesis, who argued that an organism arises out of initially unstructured matter, passing through different bodily forms. The work of anatomists on embryos resolved the controversy in the early nineteenth century, in favor of a new, teleological version of epigenesis.[6]

This gave a new shape to the time in which a human being emerges, as becomes clear in the contrast between preformation and epigenesis, and thus between growth and development: each is a process in time, but time is different in each case. Growth as *evolutio* in the sense of "unfolding," writes Georges Canguilhem, is an "extension along the three dimensions of space," which, despite representing "a succession, an ordered chronological series," nevertheless remains external to time in that it mediates between successive "states of the organic form that are distinct, but not unlike."[7] For development, in contrast, time is "operative time," to quote Canguilhem; it is what "has moulded" the structure of the organism, to quote François Jacob.[8] Development is a process of change in a structure during which different forms emerge, fade, and succeed each other until the organism has attained a definitive arrangement of limbs and organs.[9] What makes an emerging being identical with itself across the passing of time is not its form, but the continuing change it undergoes—in other words, a historical process.[10] It is no coincidence that in German, such development was first known as *Entwicklungsgeschichte*, "developmental history." In this paradigm, time is not the organism's riverbed; the organism's time is itself the river.[11]

When nineteenth-century physiology, medicine, and soon psychology turned to the time before birth, they were not discovering a beginning in pregnancy as something inaccessible to the soma of women, the doctrines of medicine, and the dogmas of religion. Rather, they construed the beginning in a different way, not by supposing and experiencing a moment of animation that attests to a "coming" child in the pregnant woman's body but by observing a story of development that transforms a germ into an embryo, an embryo into a fetus, a fetus into a newborn, and a newborn into a child. Time and space exchanged places. Whereas previously a child had announced itself in the "enclosure" of a female body, as Duden puts it, and the location in the mother's body defined a time of anticipation, now a child took shape through development, and time defined intrauterine events as a phase before birth.[12] In one mode, space qualifies time; in the other, time qualifies space.

Both versions of the beginning — when something interior is exteriorized and when a before is discharged into an after — involve a continuity across birth. In the second mode, though, what carries the unborn over the threshold of birth is not a childness acquired during pregnancy, but a continuum of development. This changes the event of birth itself. As the moment in which the fetus is transformed into an infant, it is just one of many events in a process that constantly pushes before into after by making a germ into an organism, an organism into a subject.

Certainly, this process could be divided into phases or stages, bringing about different entities and the corresponding research fields: the embryo and embryology, the fetus and fetal physiology, the child and developmental psychology. But these divisions were permeable. If the embryo is connected with the fetus through development, and the fetus with the child as well, then one conditions the other not simply by preceding it temporally, but by enabling it logically, and is therefore a part of it. Only an organism that has organs can live, and only a living organism can respond to its life. Development as a continuum means that morphogenesis is also functional genesis, which is also psychogenesis. In short, the chronological sequence is actually a logical relation.[13]

Once an entity is separated out from this continuum, it becomes a liminal being — it is something by dint of no longer or not yet being something else, and as such it always points to that something else. The embryo is no longer germ and not yet fetus, the fetus is no longer embryo and not yet nursling, the nursling is no longer fetus and not yet child.

For this reason, physiologists of fetal life rarely insisted on their own, separate discipline or even subdiscipline, arguing instead for embryology to extend to physiological matters and joining forces with child psychology, whose gaze lingered ever longer on prenatal physiology as it became ever more obviously the gaze of developmental psychology.

Development in the Human and Life Sciences

In the nineteenth century, the concept of development brought together medical, physiological, and psychological perspectives on the unborn — but it also tied the new sciences of the human being to the equally new sciences of life. It did so by creating the following state of affairs: The human being is an organism, just like every other living being. This organism is a product of development, and so is the interiority that distinguishes human beings from all other living creatures, to which they are nevertheless related through the shared historicity of everything that lives. This point is worth exploring because it means that the unborn's *biological* objectification was precisely what inscribed it into the *human* sciences.

The first step is to look at the eighteenth-century preconditions for the nineteenth-century human sciences. In order to become the object of empirical knowledge at all, "man," previously divided into an earthly body and an immortal soul, first of all had to be brought down to the earth of experiential facts. This was the project of the "science of man" (*science de l'homme*), which challenged metaphysics for the title of "science of sciences" in the second half of the eighteenth century.[14] It defined the human being as a composite of mind and body, a bundle of "relations between the physical and the moral" — thus the title of Pierre-Jean-Georges Cabanis's 1802 best-seller.[15] What had hitherto been a soul with transcendental origins, which took up residence in the earthly body only to move out again one day, was now defined by the science of man as a function of the organism named "psyche," the central organ of which was the brain. On this basis, the human being was declared to be a natural being.

At the same time, the *science de l'homme* accorded man an

exceptional status among his natural peers. Like all other beings, man develops a physical body, but unlike all other beings, that process continues into the formation of a mind. That was the feat accomplished by the science of man: man shared his corporeality with animals and plants, but his interiority distinguished him from them and made him the "life-form of the organism at the highest stage of development."[16] Intense attention thus now turned to the child. In 1801, for example, the Parisian anthropological society invited submissions for a competition to describe, on the basis of "daily observations of one or several infants," the "sequence in which the physical, intellectual, and moral capacities develop."[17]

During the nineteenth century, this embracive agenda of knowledge about the human being, covering both body and soul, would soon separate out into specializations. More precisely, the claims of the science of man to be the single, unified science almost immediately ceded to the disciplinary distinction between "life" and "man" that was taking shape at the same time. Groups of disciplines clustered around these two objects but remained connected like communicating vessels.

The life sciences were built on a changed view of the corporeality shared by all living beings. Until the eighteenth century, the visible structures of living creatures had been observed and used as the basis for an ordering of beings: which ones grip with hands and which with claws, which locomote and which put down roots, which see with eyes, which feel with antennae, and so forth. Now it was their *invisible* organization that came into play — the complex of organs and functions out of which life emerges. "Resemblances in depth," writes Jacob, were now more important than "superficial differences."[18] In this way, man became the same as animals and plants because, like them, he existed through organization.

There was more to this similarity, however. The nineteenth-century sciences of life also linked human beings to animals and plants through the postulate of a "single history" shared by everything that lives, the unbroken, successive genesis of natural beings.[19] When it became conceivable that whole species transform themselves by splitting and branching off over a sequence of generations, Charles Darwin was able to conceive of species change.[20] The changing form of the embryo during development had helped to train attention on that profound process of transformation, but at the same time it became clear that, unlike embryos, species change in an undirected process, eventually to be named "evolution."[21] Having toppled humanity from its transcendental throne, then reinstalled it by new means at the apex of natural beings, the principle of development thus culminated in an evolution of species that made the human a provisional outcome among the array of genealogically related species. Meanwhile, the interiority that distinguished humanity also acquired its own sciences. It was explored by a whole range of disciplines whose object was man as a living, speaking, working individual and as the subject of knowledge, to cite Foucault's definition of the human sciences.[22] No longer doubled as soul and body in heaven and on earth, man now, as an animal gifted with reason, lived halved: in a nature given to him and a culture that he carved out for himself from nature.

This intensified the striving for knowledge in both the human sciences and the life sciences. Life and the human being together formed a reversible figure, for biological research was entirely permeated by the human sciences. Even when hen's eggs and rabbits were laid out on laboratory tables, when science's objects were forms and functions abstracted from species, the search for knowledge was always "and self-evidently" (as Philipp Sarasin writes of anatomy and physiology) concerned with the human being.[23] For my own purposes, let me add that even when the shared historicity of species was at stake, the quest for knowledge was always directed to the issue of how that extraordinary species had arisen whose young come into the world so much less finished — so much less similar to the adult — than those of other species and need so much more time to achieve an adult form. It was babyhood, regarded as a developmental extension, that "made man what he is," as the philosopher John Fiske would conclude at the end of the nineteenth century.[24] Not for nothing did Darwin, while working through his zoological material, take daily notes on his children from their birth onward.[25]

In technical terms, we might explain what was involved in the nineteenth-century notion of development as follows. In this period, the human being became a species whose phylogenetic origin could be identified in the fact that it continued its ontogenesis into psychogenesis.[26] In other words, man was defined as man by his changing from organism to subject. The history of science likes to organize the nineteenth century's web of disciplines into the sciences of life and the sciences of the human being, but here the objects turn out to be commingled. Crosscutting the distinction between life sciences and human sciences was a continuous labor for knowledge of development that, first, inserted knowledge of life and knowledge of the human being into a single continuum and, second, directed the vector of life toward the human being.

Now, the vector was oriented in this way because the events themselves led from the organism to the subject, whether as a directional development from the human germ to the adult person or as the lineage, arising from nondirectional development in the shape of evolution, of a species endowed with subjectivity. That did not necessarily make knowledge of development into anthropology. If subjectivity is derived from the organic just as much as the organic moves toward subjectivity, then one can steep everything human in the logic of life just as well as one can see all life as steeped in humanity's going beyond it. Put another way: culture is just as natural as nature is cultural.[27] Both anthropology and its opposite could be built out of knowledge of development, but neither was identical with it. Knowledge of development was what at once drove and subverted the nineteenth century's division of the world into nature and culture.

A Naturalistic Puzzle

The anthropologist Philippe Descola describes that division as a naturalistic ontology splitting the world into two distinct domains, one of which, nature, is the realm of autonomous laws, while the other, culture, houses the arbitrariness of human activity. This dualism did not fall from heaven as a momentous novelty with the advent of modernity. The story of its emergence reaches from concepts of nature in antiquity up to scholarly and scientific concepts of culture in the twentieth century.[28] The nineteenth-century disciplines' approach to life and the human being was just one moment in that process. Grounding their own self-definition in the discovery of how the world is constituted, these sciences also carried out what another anthropologist, Eduardo Viveiros de Castro, calls a "massive conversion of ontological into epistemological questions": they transformed inquiry about humans' and nonhumans' ways of being into inquiry about knowledge of nature and culture.[29] The structures of the world, Michael Hagner and Hans-Jörg Rheinberger argue, increasingly looked like forms of science "turned inside out."[30] In the naturalistic world thus created, the human being relates to nonhuman beings by sharing its physicality with all of them, whereas its interiority and the expressions of interiority — consciousness, subjectivity, language — distinguish it from them all.[31]

According to Descola, naturalism has "the unique characteristic of constantly giving rise to heterodox points of view that call into question the distinctions that it draws between the singularity of human interiority and the universality of the material features ascribed to all existing beings." As soon as a naturalistic distinction is drawn between human and nonhuman, a contradiction arises — inevitably, because the naturalistic ontology is rooted in the paradox that it "persists in regarding an animal either as the lowest common denominator of a universal image of humanity or else as the perfect counterexample that makes it possible to define the specific nature of that humanity."[32] It was precisely this paradox that knowledge of development fashioned in the nineteenth century by distinguishing organism from subject, life from human being, yet simultaneously also bridging the two categories through uninterrupted transformation.

Against this backdrop, the shift in the rationale for continuity across birth — from a moment of animation to a process of development — may be interpreted as an event in the "epoch-spanning historiography of the human being."[33] At stake in the earlier disputes around the moment of ensoulment, as Maaike van der Lugt points out, had been the question of becoming human: "From what point on does the embryo live? From what point on is it a human being? Where does the soul come from?"[34] If, when considering the shift from *animatio* to development, we take into account the role of developmental knowledge in definitions of humanness, it becomes clear that this issue of becoming human did not suddenly disappear.[35] It confronted any knowledge that tried to forge a connection between organism and subject by passing from morphogenesis into functional genesis, functional genesis into psychogenesis. Jean-Claude Dupont may thus be wrong to argue that the invention of an explanatory mode for life, starting in the late eighteenth century, banished the anthropological status of the unborn to the realm of the "unthought."[36] In fact, that invention created a paradoxical point of departure for ontological questions to be converted into epistemological ones: the unborn was already human inasmuch as it was a human organism, and it was not yet human inasmuch as it still needed to achieve subjectivity.

What is historically specific, then, is not only the question of how knowledge of the unborn was produced by developmental knowledge. More fundamentally, it is historically specific that such knowledge was a way of conceptualizing the becoming of human beings — by transforming the question of what the unborn is into the question of how an organism becomes a subject.

That question was not a theoretical derivation; it materialized in local practices. Above all, it was a problem imposed by the object on the subject of research. This began with the work on early embryos by anatomists who aimed to identify a human organism within a formless, entirely unchildlike mass. Look how beautiful they are! exclaimed Samuel Thomas Soemmerring's engravings in *Icones embryonum humanorum* of 1799, which initiated the systematic portrayal of embryonal forms. Soemmerring's was an idealizing acknowledgment of natural beauty, but it was also an aestheticizing way to cope with an affront.[37] How much more pressing, in that case, was the question of the unborn's interiority. It called upon physiology, which claimed to unify the physical body and interiority by exploring the doings and being of creatures from the arrangement and activities of their limbs and organs. If the unborn was a human organism, what life did it lead—the life of a plant, of an animal, of a human being? The life of a sleeping human or one who is awake?

In the search for answers to those questions, talk of similarity with plants and of a sleeping state became less and less important, and attention turned instead to the fetus's manner of living, with examinations of the bodies of animals, pregnant women, and newborns. How much oxygen does the fetus need, and how does it form its blood? Can it hear, see, touch? Is it able to sense, and if so, what? Does it have feelings, perhaps even know that it exists?[38] Through such inquiries, the ontological question of whether the unborn is a human being was transformed into the epistemological question of how the fetus lives.

But this transformation could not be fully completed. In the study of development as an uninterrupted process, binding together the physical and the psychical and traversing the frontier of birth, a question arose that was just as impossible to answer conclusively as the earlier question of when exactly animation occurs. Yes, an organism becomes a subject, but at what point?[39] This meant that the countless questions about oxygen requirements, hematopoiesis, sensing, and feeling were not only questions of how, but more specifically, of how and when. Accordingly, research on fetal life at the turn of the nineteenth to the twentieth century fed into a controversy over the point in development at which the border between physiology and psychology should be drawn: before or after birth? Any answer to that was bound to be arbitrary, and it remained caught up in an imputation. For if subjectivity arises in a continual alteration of the relation between form and function, then the human subject has its beginnings in the germ and yet cannot be there, a place where it is so completely dissimilar to itself.

The notion of development was at once an epistemological answer and an ontological question. Binding the fully formed subject to its organic beginnings, the concept was unable to say what the organism *is* that will later become a subject. The purifying endeavor to distinguish nature from culture and human from nonhuman, described by Bruno Latour for the modern sciences, was pursued as unremittingly as it was relentlessly fated to fail.[40]

Generative Relations

By forging continuity across the frontier of birth, knowledge of development produced an entity of which one could say that it was becoming a human being but could not identify the point at which it became human-like. From the perspective of this developmental continuum, birth is not an event that sunders the unborn from the born — something like the moment, in earlier notions of the child enclosed in the womb, when something internal becomes external. Nevertheless, for knowledge of fetal life, birth did remain an event that separated before from after. This was because the being whose development crosses that boundary experiences it as a momentous injunction to organic reorientation. An organism immersed in water, darkness, and warmth and supplied with nutrients and oxygen becomes an organism that must itself undertake the work of breathing and feeding in air, brightness, and cold. Fetal life ends with birth to the extent that birth is an exchange of environment. Just as when something inner becomes outer, here, too, locality — the mother's womb, the world outside — makes a difference. But locality is not simply a locale. On the circumstance that the unborn is located in the mother's body, fetal physiology built a whole relational ecology of fetal life.

In the late eighteenth and early nineteenth century, a notion of the organism had become established that, in natural philosophy, defined the individual and, in the life sciences, defined the connection between anatomical form and physiological function. Since then, the environment or milieu had been part of the story.[41] But if, as Claude

Bernard proposed in a groundbreaking definition in 1854, life consists in "the joining of one with the other" (*la réunion de l'un et de l'autre*),[42] then there is not simply an environment that is locality and an organism that settles within it. There is, first and foremost, a relation.

In the case of pregnancy, the nature of this relation reflects what social anthropologist Marilyn Strathern has identified as the "Western" form: it separates by connecting and connects by separating.[43] Anatomy's splitting up of the corporeal unity of mother and child-to-be, further pursued by physiologists as they studied fetal life, produced not only a fetal organism and a uterine milieu, but also a relation between two organisms. Fetal physiology was as much concerned with this relation as it was with fetal individuality.

This was the context of the experimental investigations of the placenta that proliferated in the nineteenth century. By studying the placenta's function, scientists made visible an organ that literally embodied what was special about the relation between fetal organism and uterine environment. As a "third party," writes Hélène Rouch, the placenta instituted a relation that performs "neither separation nor fusion."[44]

It was this relation that now defined the unborn as an organic individuality dependent on another organism — dependent not only in the way that every organism needs an environment in order to live, but in the sense that it can be generated only within that environment. In the case of pregnancy, where one organism arises out of another, the environmental relation was simultaneously a generative relation; conversely, the generative relation became an environmental relation. This reveals the historical specificity of the relation produced by the pregnant body itself. Well into the eighteenth century, that relation was thought of as an organic unity, and in the twentieth century it would become a chemical correspondence. Between those two points stood the physiological relation of the nineteenth century.[45]

However, to the degree that the nineteenth century thought of the engendering of human beings as a connected sequence of generations through which both species and communities reproduce, the influences passed on from the maternal, generating organism to the fetal, generated one were not only a relational event, but also a process of generational transmission. That made it necessary to understand how "accidents of pregnancy" relate to "true heredity," to recall Charles Féré's terms. Disastrous influence now appeared as an event that irrupted into a course of things determined by the laws of heredity and the process of development. Against that backdrop, every peculiarity that children bring with them when they leave the womb could be interpreted as an anomaly, and anomaly could be interpreted as the materialization of a reproductive process that ceaselessly concatenates events to join the past to the future. Maternal influence became the implementation of just that enchainment. In turn, the unborn, defined by the maternal-fetal relation, was not only read in relation to the pregnant woman, but was itself a "relational body" whose time merged with its mother's time.[46] This was the articulation of events that made it possible to think of them as "prenatal" and that, in the nineteenth century, bound the pregnant woman's body to a time of society.

When "enclosed beginnings" became "fetal life" in the nineteenth century, then, two contradictory visions were created: first, a birth-traversing continuity grounded in development, and second, a distinction between unborn and born that was founded in the maternal-fetal relation. It is in this motion of simultaneously abolishing and confirming birth as a threshold that we find the doubleness of the word *prenatal*, which embeds events before birth into a transnatal developmental continuum while also, as the implementation of a generative relation, distinguishing them from events after birth.

Historical Ontology of the Unborn

In this book, I wish to trace a configuration of historical continuity and discontinuity. On the one hand, in the scientific practices of the late eighteenth to the early twentieth century, no differently than before, the beginnings of the human being were sought in the unborn: birth remained an equivocal threshold-nonthreshold. On the other, it was now development and not momentary animation that overcame the boundary, which was now staked out by relation and not locality. This configuration opened up the space in which the concept of the prenatal could surface.

Defining this space and the things that research practices fashioned in it — fetal life, the uterine milieu, the maternal-fetal relation — both as an *epistemic* space and as *epistemic* things is not solely a theoretical decision. It is also a historical description. Through these epistemic things, research as practical action constituted the question of what the unborn is in relation to the born. The treatment of this question by scientific practice is historically specific, but so is the very fact that scientific practice posed the question while discounting as unreliable other ways of authenticating the coming of a child — in particular, the pregnant woman's bodily experience of the child's first stirring.[47] That does not mean the scientific way would thenceforth be the only way of defining the unborn, which has most probably been "multimodal" in every era.[48] What it does mean is that in the acquisition of knowledge about fetal life, an entity was constituted that thenceforth existed in the world in a very specific form.[49]

As a result, a description of this knowledge is also a historical ontology.[50] It was by fashioning knowledge about fetal life that the nature of the unborn was defined, by investigating development that a distinction between human and nonhuman was drawn. In this chapter, I have used the biologicization of the unborn as an epochal rupture to contrast "enclosed beginnings" with "the developmental life of the fetus." I do not ask how these different interpretations relate to the same, presupposed item — a fetus, an embryo — but try to describe what the two constitutions of the unborn are made of, out of what things and what relations of time and space, and how they thereby differently define what the unborn *is*.[51]

The sciences of the nineteenth century produced a conception of the human being's becoming no less peculiar than the conceptions described by medievalists and ethnologists when they tell of moments of animation and transmigrations of the soul. The combination of the sciences of life and the sciences of man did not simply give rise to positive knowledge about the unborn, as existing historiography suggests — not explicitly, to be sure, but by treating the nineteenth century in a strangely reductive way or even excluding it completely, as if there had been an intermediate historical realm of pure knowledge acquisition between the ontological cabinets of curiosity in unenlightened premodernity and the spectacle of reproductive technology in the present day. I counter this self-misrepresentation of modernity by taking up a symmetrical stance:[52] the same question that medievalists and ethnologists ask of what are known as "belief systems" needs to be asked of nineteenth-century research on fetal life — not what it discovered, but what it was about. Only in this way can we see how, through the reorientation of the unborn along the spatiotemporal logic of all life, there emerged a historically specific constitution of the beginning of the human being.

This makes the history of fetal life a fundamental, if neglected, element of the history of the human sciences. It was the human sciences that gave modernity the human being described by Foucault as a "strange empirico-transcendental doublet": a being that is simultaneously the subject and the object of knowledge.[53] The story of the unborn has hitherto been disregarded in the historiography of that enterprise and left to the history of the life sciences as a minor chapter on embryology. Wrongly so. Not despite the fact, but precisely *because* knowledge in the human sciences is about the generic human being, it generated "internal others":[54] women and non-white people as flawed human beings and, in a way peculiar to itself, also a fetal organism that is already a human being and not yet a human being.[55] This fetal organism was the dissimilar beginning of the human scientist himself, something different yet the same that, "before we know it," as William T. Preyer noted in 1880, turns "into a being the same as us."[56]

That was still true of the child to which Preyer referred and therefore already true of the fetus — defined by the development that does not begin with, but is continued by, the child. If anything, it was even more true of the fetus, given that the fetus's relation to the maternal organism distinguished it not only from the adult, but also from the child. By interlocking embryo with child, the investigation of fetal life transferred knowledge of life into knowledge of man and made scientific practice an ontological practice. To pick up a train of thought from Jean-Paul Galibert, its governing question ran: How is it possible that mere cells, "numerous as they may be, end up talking to me?"[57]

As this opening section has shown, my study aims to understand how political event and childhood anomaly were brought together in the epistemic thing of fetal life. The children of the year of terror, 1870/71, gave me a guiding thread that led from that connection to a historical ontology of the unborn. My decision to write the book in a way that weaves together these two lines resulted both from the book's own genesis and from the material itself. After all, a guiding thread is not only one that marks out the path we have traveled. It is also the enwoven "red thread," a strand that, Johann Wolfgang von Goethe wrote in *Elective Affinities*, "cannot be extracted without unravelling the whole rope."[58] That includes the moment when French psychiatrists took the Parisian mothers' unquestioned "Monsieur, what do you expect. . . " and turned it into a question, one that would eventually transform the certainty of mothers into the enterprise of scientists.

PART TWO

Living Beings

> But from the outset the living being is conceived of in terms of the conditions that enable it to have a history.
>
> —Michel Foucault, 1966[1]

> The placenta is also the border zone of the fetus. Beyond it is the sea: the mother's blood; then the other shore: the mother.
>
> The border zone is a zone of heavy traffic.
>
> —Hélène Rouch, 1987[2]

At first sight, it may not seem particularly remarkable that, in 1884, Charles Féré used the term *influence* when he speculated on the link between a mother's shock and her child's anomaly. Yet his choice of words was saturated with history. Before Féré and Henri Legrand du Saulle raised the topic of the *enfants du Siège*, the onetime organic unity of pregnant woman and child-to-be had become a connection of maternal and fetal organism—a change that made it possible for the events playing out between them to be a matter of "influence" at all. Crucially for this new configuration of pregnancy, the embryo or fetus was organically individual but also predicated on the pregnant woman.

That reflected a modern concept of the organism. Since the late eighteenth century, the term *organism* in the life sciences had denoted a complex of bodily structures and vital functions; in natural philosophy, it designated the individual *tout court*.[3] Like all life, this organism was inscribed with a temporal dimension, which in its case was named "development."[4]

In order to understand the fashioning of the embryofetal entity that was defined by the concepts of organism and development, and specifically to understand what made that entity innovative, the history of anatomical representations of the unborn is instructive. Barbara Duden places the beginning of that history at the threshold of the nineteenth century, when the aspiration to produce realistic illustrations gathered pace among anatomists portraying pregnancy.[5] Earlier anatomical atlases had already made the pregnant woman's "interior body" visible and shown children curled inside its envelope,[6] but those children were neither embryos nor fetuses — they were symbolic allusions to the infant-to-be and emblems of a becoming that proceeds under cover.[7] Whenever something interior and child-like featured in an image, the generative properties of the female body appeared as qualities of concealment: egg, wrapper, nest.

According to Duden, the earliest steps in depicting the entity hidden within the pregnant body in a way that would replace emblematic representations are to be found in William Hunter's *Anatomy of the Human Gravid Uterus* of 1774. Hunter shows the child inside the womb in realistic detail — wispy hair, damp skin, plump limbs. But he does so in only three of his thirty-four plates, and if his commentaries refer to the child at all, then only with respect to its position or to the umbilical cord.[8] Hunter's "monomaniacal" interest, to cite Duden, was directed at the pregnant belly in which the child was embedded.[9]

Samuel Thomas Soemmerring, who took the critical step twenty years later, did things very differently. In his *Icones embryonum humanorum* of 1799, the female torso has dropped away from the child, which is now portrayed as a chronologically ordered sequence of embryos from the third and fourth weeks to the fourth month (Figure 2). The viewer's gaze no longer rests on the generative quality of the female abdomen, but on what Soemmerring calls the "growth and development [*metamorphosis*] of the human body."[10]

Resolutely though Soemmerring pushed the embryonic body into the anatomical limelight, he wavered in the great debate over the

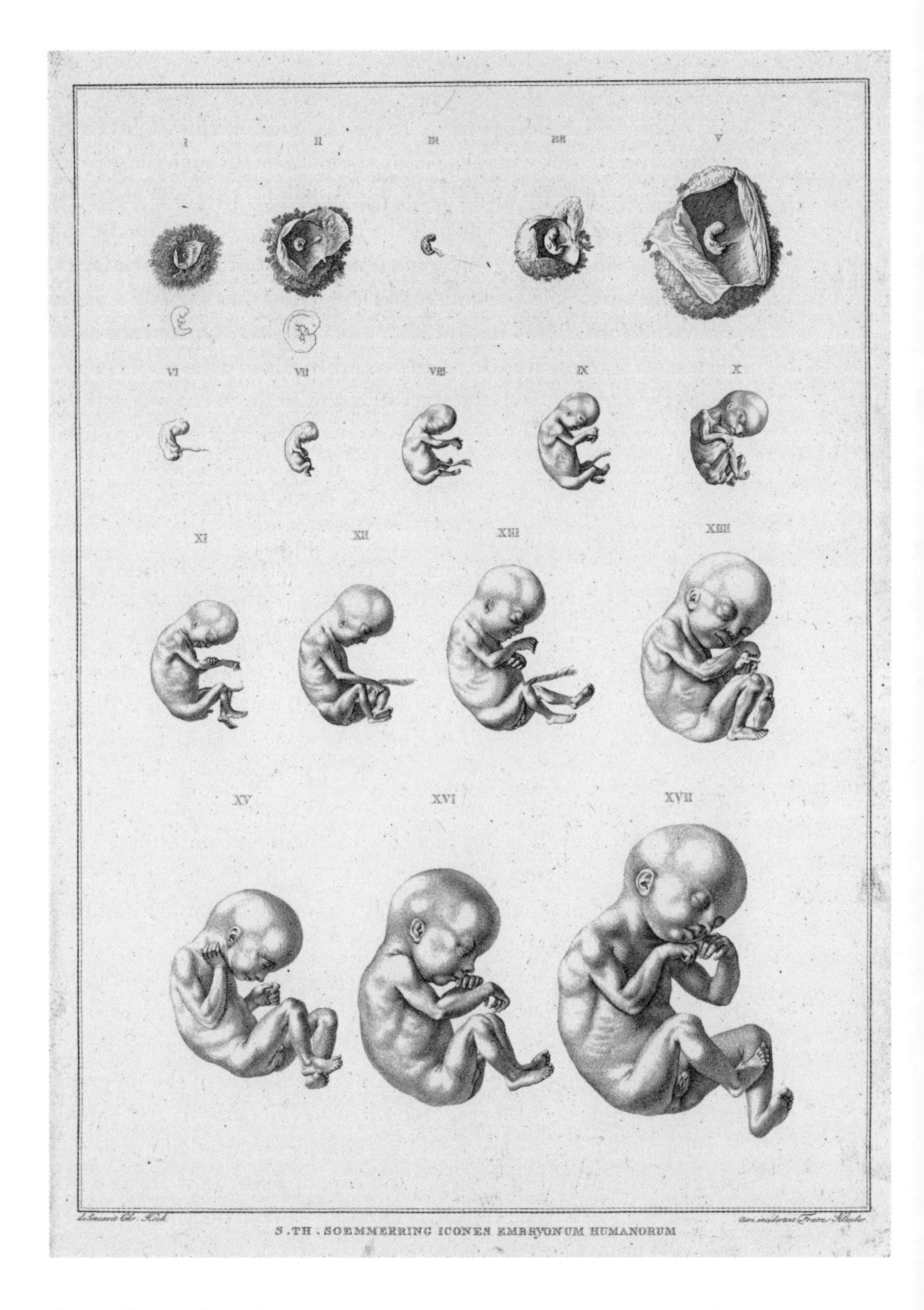

Figure 2. Plate 1 of Samuel Thomas Soemmerring's *Icones embryonum humanorum*, 1799.

logic of how such bodies came to be. At issue in this eighteenth-century dispute was whether a body comes to be through the growth of a germ already formed in the egg or seed—this was the view of the preformationists, such as the Swiss naturalist Albrecht von Haller—or whether organic form emerges out of formless matter, as submitted by the proponents of epigenesis, especially Caspar Friedrich Wolff, following ancient embryologies.[11] Only in the early nineteenth century was this controversy resolved in favor of epigenesis. In the same period, the starting point in the germ was defined more precisely. An organism, it was now thought, arises not out of entirely unstructured raw material, but out of units whose structures enable the organism to live from the very start, though they will later change significantly.[12]

In the early nineteenth century, Christian Heinrich Pander and Karl Ernst von Baer explored these structural transformations in terms of a "developmental history" or *Entwicklungsgeschichte*: a continuous train of events in which a germ becomes an organism by passing through a long series of dissimilar forms.[13] When Wilhelm His's *Anatomie menschlicher Embryonen* (Anatomy of human embryos) appeared in the 1880s, an enterprise that had begun with Soemmerring's *Icones* and been pursued in university departments of zoology and anatomy, as well as in gynecological medicine, culminated in a sequence of standardized drawings showing the development of the human embryo in tabular format.[14] It was on these foundations that developmental biology finally emerged.[15]

This was not simply a matter of turning attention to something that had always been there but had never been seen, noticed, or studied. Rather, the phenomenon itself had changed: pregnancy was no longer the coming-to-be of a child in the seclusion of the pregnant body, but a genesis of form, embodied in the embryo and named "development." To be sure, Soemmerring had accused his predecessors of a willful refusal to see,[16] and certainly the new framework depended on new techniques of looking—made possible by more powerful microscopes, measurement-based drawing procedures, the comparative analysis of species, and later also the dissection and preparation of embryos.[17] However, to apply such technologies was not simply to read knowledge off from the object. To the contrary, "the embryo" as an empirical object was the outcome of these procedures. Human and animal embryos were acquired, observed through the microscope, measured, compared and classified, sliced up, prepared as specimens, drawn, and modeled. These practices upon objects that were detached from both the pregnant woman and the embryonic membranes gave rise to an individualized entity that, also in its graphic representation, cast off the woman in order to be inserted into a developmental series where it could epitomize the process of its becoming. This is how the trompe l'oeil of the quasi-autonomous entity "embryo" took shape.

But nineteenth-century scientific engagement with the unborn went beyond the question of the development of embryonic form. François Magendie, soon to found the first journal of experimental physiology, interjected in 1817 that an anatomical description of the "growth of the organs" and its "interesting details" left much unsaid. "Let us attend to the physiological phenomena that pertain to it," he proposed.[18]

Everything about this statement is significant: that Magendie splits the anatomical from the physiological, that he locates them in a single context, and that he insists on the importance of physiology within that context. His position is firmly anchored in the modern concept of the organism, discussed above, as a "structure-function complex."[19] Rather than being something that comes about from an ensemble of visible elements (limbs, organs, their arrangement), the organism is now something that arises from the invisible ways in which those elements relate to each other "organizationally" in order to produce life. "The surface properties of a living being," wrote the historian François Jacob, "were controlled by the inside, what is visible by what is hidden." How it breathes, how it forms its blood, how it absorbs, metabolizes, and expels substances, how it stays warm, perceives, feels, and moves—everything that makes up the total organization of its life determines a being's "form, attributes and behaviour."[20] Physiology declared this to be its field of competence, and staked a double claim: first, in contradistinction to anatomy, to be a science not of form but of function, and second, to encompass both, since form is explained by function.[21]

When Magendie called for science to go deeper by studying physiology, at issue was how, in the course of nine months, a bodily form arose, but so did a life as the function of that form. This did not mark the discovery or invention of unborn life. In 1774, under the chapter heading "The Life of the Fetus" in his two-volume study of generation, *La génération*, von Haller had outlined numerous studies from the seventeenth and eighteenth centuries that addressed the embryonic body's changing consistency, its heartbeat, or its respiration. "The fetus lives early on" is how von Haller summarizes the start of his second volume.[22] This thetic statement is not the refutation of a possible opposing claim that the fetus is first dead, then alive; it refers to the special constitution of a life that forms gradually. By speaking of states or modalities of life, von Haller makes life not something that is a given, a property of the emergent being, but something that, in its specificity, defines the embryo or fetus.

In the decades that followed, this became a historically specific challenge: how to understand the emergence of the various functions that, together, give rise to life. In the human being, those functions included the interiority that distinguishes humans from nonhuman animals. As a result, investigating physiological phenomena in the unborn also raised the question of how the human child comes to be.

The organic individuality of the unborn thus arose not only out of its body, but also out of its being alive. This did not, though, immediately entail a fetus that, detached from all relationships, had solidified into a fetish of "life" as such, like the one that would take the political and media stage so insistently in the second half of the twentieth century.[23] Organic individuality was not coterminous with the free-floating independence that the anatomical image and embryological knowledge suggested when they deleted the body of the pregnant woman. In the study of fetal life, woman remained present, albeit as just an organism herself.[24]

This brings us back to William Hunter's interest in the pregnant belly. In 1774, when he demonstrated the separate systems of blood circulation, Hunter—a kind of cartographer among anatomists, "lovingly recording all the details of this terrain" of flesh and tissue, as Ludmilla Jordanova describes him—performed more than a separation between the fetal and maternal body.[25] By injecting red, blue, and yellow wax into blood vessels and tissue, he (or, as an unresolved dispute between the two protagonists suggests, his brother John Hunter together with a certain Dr. Mackenzie) discovered the precise points where one begins and the other ends, but what he explored was a region that simultaneously divides *and* connects two bodies as organisms. That is the source of the "almost oppressive intimacy" between mother and child exuded by the drawings in Hunter's *Anatomy of the Human Gravid Uterus.*[26]

In the physiology of fetal life, the pregnant woman as a maternal organism appeared in a dual guise. On the one hand, and only marginally, physiology addressed what midwives and physicians had traditionally seen as a sign of pregnancy and Magendie now, logically enough, regarded as changes in the "mother's functions": uterine tissue becoming spongy, the eyelids bluish, the mood melancholy.[27] On the other, and more importantly, the maternal organism appeared as that which sustained fetal life and thus as its precondition—the source of oxygen, water, nourishment, warmth.

From the outset, the modern concept of the organism was inextricable from the notion of an environment or milieu of this kind. Initially, in the eighteenth century, the relation between beings and their surroundings was explained rather loosely, "in a roundabout way" as part of the organism's "need or a desire" vis-à-vis its setting, but in the nineteenth century, "environment" came to include "all the external variables to which the living being was subject" and which it affected in its turn.[28] Unlike research on embryonic formation, therefore, research on fetal life could not ignore the pregnant body. It was forced, in Magendie's formulation, to consider the "relation between the mother's functions and those of the fetus."[29] What anatomical studies of placental vessels had carefully teased apart, physiology reconnected, this time in the maternal-fetal relationship that would come to be embodied by the placenta.

Such was the conceptual groundwork for the notion of psychical influence on fetal life, the influence that, in 1884, Féré believed was capable of turning a shock suffered by the woman into an anomaly suffered by the child. What was new about this hypothesis was not that it postulated a shaping power of the maternal mind, but the way in which that power became manifest, as an interaction between the maternal and the fetal organism. In the following chapters, I explore in more detail these preconditions for Féré's concern with the children of the Paris siege.

CHAPTER FOUR

The Life of the Fetus

The Embryo's Special Physiology

In 1883, William T. Preyer's study *Specielle Physiologie des Embryo* (The special physiology of the embryo) appeared in Leipzig. Preyer had worked in Paris in the mid-1860s with the leading physiologist of the day, Claude Bernard, and by the time of the publication had been directing the Institute of Physiology at Jena for many years. He presented his book as the founding act of a new scientific subdiscipline, of which there had previously only been occasional talk: what he called "physiological embryology," "biochemical and physiological embryognosy," "functional embryogenesis" (*functionelle Entwicklungsgeschichte*), or simply "physiology of the embryo."[1] Preyer's book quickly found a broad and lasting audience. The first German edition of 1883 was followed by a second one in 1885 and a French translation in 1887, which was immediately cited by Charles Féré.[2] The study would appear in thirty-three editions up to 1976.

An earlier call for embryology to attend more closely to physiological questions had come from Wilhelm His, a Swiss professor of anatomy in Basel and later in Leipzig. In influential studies published from the late 1860s to the 1880s, His argued that a comparative view of species should inform not only descriptions of how the human form takes shape, but also investigations of which physical and chemical processes drive that process and why. Quite in line with this desideratum, embryology came to accentuate biochemistry and experimentation.[3]

This was not exactly Preyer's agenda, however. Embryology, as he saw it, continued to focus on form and thus on the early stages of embryonic development, whereas he was interested in function and a more broadly conceived arc of development. Preyer's "special physiology" aims to "trace the functions one by one, from that stage of embryonic development where they cannot yet be discerned up to their transformation by birth" — from the completion of germ layer formation up to the commencement of an extrauterine metabolism.[4]

Between these points, Preyer builds his research field out of two deficits. For embryology, he remarks, morphological knowledge comes before physiological knowledge (no life without a body), yet the organism's development is based on the formation of functions just as much as on that of organs (no body without life). Physiology, for its part, has devoted itself entirely to the "functions of born humans and animals," ignoring the processes by which those functions emerge. And "just as one can only understand the organ, the tissue, and the cell when its genesis has been investigated, the function too can only be understood by way of its own history."[5] What really requires study, therefore, are the functions of the embryo and the history of function. In practice, that meant four hundred pages of findings and questions on the circulation of blood, respiration and nutrition, metabolism, secretions, heat generation, motility, sensibility, and growth.

Preyer's audacious contribution made an impact immediately in manuals of physiology and medicine. They now began, mostly citing Preyer, to give accounts of knowledge on the physiology of the embryo or fetus that were not only more detailed than before, but also more systematic. One example is the handbook of physiology published by Henri Beaunis, who had taught at the medical faculty in Strasbourg until the Franco-Prussian War and would later become the first director of the Sorbonne's psychophysiological laboratory. In the new, 1888 edition of his 1876 manual, he added a reference to Preyer's book and even created a separate chapter, "Physiology of the Embryo and the Fetus," to expand his remarks on the subject, which in the previous editions had been dealt with briefly as part of the chapter "Physiology of the Organism According to Age." The authors of the "Embryo" article in the authoritative medical dictionary of the day, as well, drew mainly on Preyer's new study for all their comments on physiology.[6]

But the impression of an opening shot is deceptive. Preyer was already able to furnish his work with a thirty-page bibliography, and his achievement was in fact less to have discovered a new object or founded a new field than to have systematized a formidable existing tradition of research. Not for nothing does he apologize in the introduction for any gaps in his reading: "In particular, despite many years of collecting, I have not anywhere near succeeded in bringing together all the references to vital phenomena, that is, physiological functions of the unborn human being and animal, which are scattered throughout the scientific literature in physiology, gynecology, anatomy, zoology, embryology, and agriculture."[7]

It was from this research praxis — dispersed not only across scientific domains, but also across the length of the nineteenth century — that Preyer sifted out the logic of an object capable of sustaining a research field of its own. This chapter argues that what paved the way for his fanfare, proving it to be just as much a finale as an overture, was a physiological particularization of the unborn through the study of fetal life. This notion of a life peculiar to the fetus must first of all be stripped of the self-evidence it seems to possess today, which I do by returning to the late eighteenth and especially the early nineteenth century. There, the thing named "fetal life" can be historically contextualized in two different ways: first, in relation to previous descriptions of the unborn and second, in relation to the history of physiology, which defined itself as the science of life and was in fact a science of the human being.

Vegetative Life and the Human Child

The idea of an independent essentiality of the human being as yet unborn was not a novel product of the separation of the embryo or fetus from the pregnant woman in anatomy and physiology. Ancient theories of generation, which retained their authority well into the eighteenth century, had held that a soul of divine origin took up residence in the embryonic body before birth. There was much debate as to how this actually took place, at which point during pregnancy it occurred, and what it meant for the legal and theological status of the unborn. What was undisputed was the fact of *animatio*—a moment of ensoulment that the erudite man posited and that the pregnant woman felt, with the child's first motion, as "quickening."[8]

Up to the eighteenth century, the idea of an individuality bestowed by *animatio* was combined with that of a unity of mother and fetus bestowed by shared experience. This was put succinctly by the philosopher Nicolas Malebranche in his *Recherche de la vérité* (Search after truth) of 1688:

> Infants in their mothers' womb, whose bodies are not yet fully formed and who are, by themselves, in the most extreme state of weakness and need that can be conceived, must also be united with their mother in the closest imaginable way. And although their soul be separated from their mother's, their body is not at all detached from hers, and we should therefore conclude that they have the same sensations and passions, i.e., that exactly the same thoughts are excited in their souls upon the occasion of the motions produced in her body.[9]

Because the soul experiences the world through the body, on this account, organic community is also a community of experience. When a child stirs, that does not prove its autonomous corporeality—quite the contrary: "Thus, children see what their mothers see, hear the same cries, receive the same impressions from objects, and are aroused by the same passions." Over many lines, Malebranche tries to put into words the closeness of the connection between mother and child, an endeavor that concludes in the notion of a total organic unity: "For basically the body of the child is but a part of the mother's body, the blood and spirits are common to both."[10]

When he described the connection, or even the union, of mother and child, Malebranche was far more interested in capturing the all-pervasiveness of that link than in discovering how living beings function by analyzing their blood and spirits. But starting in the seventeenth century, and more so in the eighteenth century, talk of the "same body" was taken seriously and assessed as a statement about vascular connections, interwoven tissues, and the composition of substances. Attention turned to organic individuality, and Hippocrates' tenet that the fetus breathes through the mother's mouth was soon rejected, while objections to the presumption of a shared bloodstream proliferated.[11]

The decisive step—decisive, that is, for both anatomy and physiology—was taken by William Hunter in 1774, when he presented the idea of uteroplacental circulation and thus framed the mother-child tie as a link between two individual organisms. "The human placenta," writes Hunter, "is composed of two distinct parts, though blended together, viz. an umbilical, which may be considered as a part of the foetus, and an uterine, which belongs to the mother." Although here, too, the parts are "blended together," just as Malebranche had argued, this entails not a shared bodily experience, but a double organic individuality. Each part "has its peculiar system of arteries and veins, and its peculiar circulation, receiving blood by its arteries, and returning it by its veins."[12] In its corporeality, through which the child had previously merged with the experience of the mother, the fetus now acquired an entitative validity previously granted by *animatio*. The fact that the fetus has its own vascular system and forms its blood itself endows it with independent essentiality.

Not every doubt had been dispelled by the beginning of the nineteenth century.[13] In the enduring controversies over shared or separated blood, however, the organic individuality of the unborn did gain ground. This shift arose in part from continuing efforts to explore the mechanisms of procreation, or, as it was traditionally known, generation. At the same time, it also formed part of the sea change that was the transition from a naturalist classification of living beings to a science of life that revolved around the question of how "life" actually works. In that process, physiology found itself faced with a new question of its own: How does the unborn relate to the born, given that it is alive under such different spatiotemporal conditions? What now had to be described was *how* the unborn lives. Organic individuality was not simply an idea—it was the outcome of investigating this question.

A breakthrough came in 1800, from somebody so skilled in the art of innovation that his oeuvre, written over the course of just a few years, would earn him fame as a founding figure in modern anatomy, physiology, and medicine. In his *Recherches physiologiques sur la vie et la mort* (translated in 1809 as *Physiological Researches on Life and Death*), Xavier Bichat interrupts his discussion to assert in a footnote that the "theory of the foetus" is "yet very obscure."[14] The note is found in the chapter entitled "Of the Mode of Organic Life in the Foetus," where he expounds his theory of the two lives, an "organic" or "vegetative" life and an "animal" life. By vegetative life, Bichat means the functions clustering about the heart, consisting of lungs, spleen, kidneys, liver, stomach, intestines, and pancreas; animal life is the complex of sensory and locomotive activity, organized by the brain, that comprises nerves, eyes, ears, skin, muscles, and skeleton and finds expression in relationships with the world. It is from this vantage point that Bichat shone his various searchlights onto the obscurity of fetal theory.

First among them is this: "The moment the foetus begins to exist, is almost the same in which it is conceived; but this existence, whose sphere is daily enlarged, is not the same [as] that it shall enjoy when it has seen the light." What needs to be described, therefore, is a life that was there from the beginning but that differs from the life after birth. Bichat calls it the "existence" of the fetus; the German translator chooses the term *Leben* ("life").[15] Contrary to what is commonly assumed, writes Bichat, this should not be compared with a "profound sleep"; rather, "its existence is the same as that of the vegetable." The fetus exists by forming blood, having it circulate, and absorbing nutrients from the maternal organism.[16] Such vegetative life is not simply a suspension of animal life—that would be sleep. In it, instead, animal life is "entirely extinguished, or rather has never commenced." In the section heading introducing the topic, Bichat makes the point baldly: it is "null."[17] The sensations of the fetus are "feeble" or almost nonexistent, its brain "in readiness," its motions "involuntary"—for although these movements are occasionally violent, they indicate nothing more than a sympathy with the affections of the internal organs, a point to which Bichat dedicates a whole section of the chapter.[18]

This description of the fetus's vegetative life proves its worth by supplying a clear response to a very consequential question. When, during a difficult birth, the "cruel alternative" arises of sacrificing the fetus or exposing the mother to "almost certain death," the former must be accepted—because the destruction of the fetus affects "a living being" but not "an animated being" (in French, Bichat's distinction is between *être vivant* and *être animé*).[19] Only after birth does it enter a new "mode of existence." That does not mean simply that its life is transformed, but that it is doubled. The vegetative life becomes an "internal" life and its functions "attain perfection all at once": the infant immediately begins to breathe, digest, excrete. And the new exteriority of the animal (or "external") life draws it out of "the nought in which it was buried" into activity, shaped through "a sort of *education*." In a "slow and gradual progress," the senses become active and endow the infant with the pleasures of dance, music, adornment; a brain that cognizes, remembers, imagines, and judges; a voice that can become word; and a locomotive apparatus that can overcome its initial tendency to "stagger, reel and fall."[20]

Describing the unborn as an organism has two implications for Bichat. First, he finds a continuity of existence across the threshold of birth, bestowed by organic life, since the life of the unborn begins almost at the moment of conception. Second, birth forms a distinct caesura, bestowed by the mode of that life. In Bichat's work, this distinction does not arise from an interest in fetal theory alone. The divide between born and unborn paradigmatically names the divide between vegetative and animal life. This is why Bichat emphasizes the frontier so frequently and with such rhetorical brio when he describes the vegetative life of the fetus as an absence of animal life, for example, or the functions of animal life as emerging from zero. His very first sentence on life after birth conjures a child who, just one page later, is embedded into the author's human community and is capable of becoming a dancer, musician, cook, or philosopher.

The logic of this break is an entirely physiological one. Birth is a process that transposes life's ways of functioning as it adapts to changed conditions. Bichat's descriptions of fetal functions—how the fetus already forms blood but as yet senses almost nothing—are therefore always interleaved with equally detailed descriptions of the specific conditions in the womb. Those conditions are characterized by uniformity and the absence of stimuli. The surrounding amniotic fluid retains the same temperature for all nine months; the walls of the uterus always remain mobile and exclude all light, all smells, all sounds. Only with birth does the child experience difference and stimulus as it is discharged from an imperturbable womb into an environment that is light and dark, warm and cold, noisy and quiet, and which it now experiences as its exterior.

By making a living being into an animated being, the physiological caesura of birth additionally marks the momentous ontological change of becoming-child, something that resonates terminologically with the ancient *animatio*. Bichat's unborn is novel, but, situated at an epistemological hinge point, it also harks back in time. In pre-biological notions of generation, Barbara Duden points out, the child-to-be was "a-topical," "latent," or "*non-dum*"—not-yet.[21] It had no place; it was not yet as it would *be*; and the manner in which it would *be* was both there and not-there, immersed in nothingness. Now, though, all of these things are qualified physiologically. The external world into which the child comes at birth is the precondition for animal life; the future that the fetus lacks is a mode of life; the fetus can be both there and not-yet-there because the continuity of life connects it to what went before.

Physiological animation, then, is not the becoming-human found in the older idea of *animatio*, which took place as the entry of a "spirit" peculiar to humankind and was betokened by the child's first movement. To rule out any confusion between the two concepts, Bichat devotes a whole section to showing that fetal motion is involuntary.[22] Physiological animation, he shows, sets off an animal life that, in the human being, gives rise to what is human. To be sure, this step creates a kind of scientific Rubin's vase. Since the physiological study of animation addresses the becoming of a child, knowledge of fetal life will also have an anthropological aspect; and since becoming-human arises out of fetal life, it is also a matter of physiology. In this construction, physiology is anthropology, and vice versa. This is why Bichat's "yet very obscure" theory of the fetus could expect illumination from "animals somewhat similar in their organization" as it tried to discover the origin of the process that is capable of producing philosophers, cooks, and dancers.[23]

This anthropological element in the description of fetal life was brought to the fore by the physician Christian Friedrich Nasse when his 1824 paper on the animation (*Beseelung*) of the child inaugurated his new journal of anthropology. Nasse felt obliged to add a refreshing drop of ideality to what he perceived as Bichat's French materialism. On one point, Nasse writes, Bichat was mistaken: the reason why the law rightly protected the fetus's life was "not because of what the fetus is, but because of what it can become." Aside from this issue of potentiality, however, Nasse fully accedes to Bichat's fetal theory: "The fetus is without animal or human animation [*unbeseelt*]; it has only the life of a plant."[24]

That does not mean it *is* a plant, for "by virtue of this life, and receiving directions for development on the soil of the mother's body, it develops into a human form." Having assured the congruence between the unborn and the born in this way, Nasse is able to insist on what is different about the fetus's life:

> But that it puts down roots like a plant; that, like a plant, its parts develop successively; that it sends them outward like a plant; that, like a plant, it extends its extremities just as a plant bears the sexual parts upward; that the mixture

> of its fluid and solid parts still fluctuates between those of the animal and the plant body; that, finally, in its development, it is tied to precise temporal orders just as is the plant — all this accords completely with the kind of life we . . . must ascribe to it.[25]

The fetus, then, nourishes and forms itself just like a plant, and when it moves, that is "a mere product of the concatenation of bodily factors," in no way testifying to sentience (*Empfindung*) or will, especially as its "sensory organs" are barely formed. Even the presence of a brain is not a sufficient condition for "mental activity" (*psychische Thätigkeit*).[26] All that changes with birth. Then, the "human plant" experiences "the entry of air into the lungs," and by willfully taking its first breath, it becomes an "animated [*beseeltes*] being."[27]

As emphatically as Nasse names the point in time when animation takes place, the manner by which it occurs remains unclear to him. Does animation happen "all at once during birth," or as "a gradual process in the fetal state"? The problem is so thorny that Nasse can resolve it only through the "mystery" of the first human's creation: the Bible tells him that God breathed the breath of life into the clod of earth. Although he secures the border between human plant and animated being by means of something "unfathomable," Nasse is not resorting here to the metaphysical soul. His anthropology is indebted to a psychophysiological view of the soul, on the basis of which, at the beginning of the text, he defines just what a human being is, namely, an animal animated by sentience and will that differs from other animals only in possessing "self-perception."[28] Based on this, he soon turns from the manner of animation being unfathomable to its being something not *yet* fathomed. The question, he argues, needs to be further pursued "by means of science," and specifically on the basis of "experiential data."[29]

Bichat, with his theory of the two lives, had drawn the line between a plantlike unborn and a humanlike born both firmly and rather blithely. Nasse complicated the matter by asking *how.* This made Bichat's distinction precarious. If the mind both becomes gradually and occurs momentarily, because physiology begins with the fetus and then transforms at birth, what then actually distinguishes the born from the unborn in anthropological terms? Nasse hoped empiricism would help him to answer that question, yet empiricism was the very source of the confusion: the more clearly fetal life became function and function became the object of experiment, the more the categorial unity of Bichat's mode of existence splintered into an ensemble of functions all in the process of emerging.

The Experimental Life of Functions

In the first half of the nineteenth century, physiology's jurisdiction on function was what grounded its claim to be autonomous from anatomy. That claim was underpinned by giving function precendence over form. Physiology was no longer to be a mere facet of anatomy. If anything, anatomy should be regarded as part of physiology, because far from life ensuing from the arrangement of organs and limbs, the arrangement of organs and limbs ensued from life as the overall function of the organism. What physiology's claim obscured was the fact that its method of choice, the experiment, had already been widely applied by "old" anatomy, even on living animals.[30]

The key protagonist of this self-assured physiology was François Magendie. With a healthy dose of understatement, Magendie liked to describe himself as a "ragpicker" of the empirical findings to be gleaned from the realm of theoretical speculation.[31] His approach earned him the chair of medicine—now renamed "chair of experimental physiology"—at the Collège de France in 1831. Magendie believed that Bichat had collected far too little such material. He set aside Bichat's doctrine of the two lives and radically replaced vitalist ideas with the concept of function: the organism lives not because an internal vital force of some kind or another wants it to, but because it is organized in such a way that life must necessarily ensue from it. In 1816 and 1817, Magendie set out his stall in his influential two-volume *Précis élémentaire de physiologie* (Elementary manual of physiology), which was also translated into German (as *Handbuch der Physiologie*, 1834).

Fetal life is a systemic element of this opus in two respects. First, it appears in the descriptions of individual functions. Magendie was interested in how functions change over the course of the various phases of life—among which he counted life's emergence. Generally using just a few lines, he outlines the formation of each of the organic foundations in the fetus, always adding the question of how it is activated. In the fetus, he writes, the inner ear forms early, but the middle and outer ear are not yet operational, so that the baby is born without sensitivity to noise.[32] The olfactory system is barely developed at birth, present in only a rudimentary form. The fetal gustatory system is for the most part already formed, though it is not known whether the fetus uses it before birth; most certainly, newborns can be seen to respond to the distinction between sweet and salty.[33] Vision, like hearing, forms only after birth.[34] It is, Magendie adds, improbable that the fetus has a sense of touch, but its movements suggest that it can sense the processes occurring in its organs and thus react to irregularities in the mother's circulation and breathing.[35] Magendie anchors all these deliberations in brief but insistent references to observations and experiments on newborn babies, as well as to investigations of fetal movement.

Second, Magendie's *Précis* addresses fetal life as a domain of the physiology of generation, made up of processes ranging from the secretion of sperm to lactation. It is here that we find the program of a specifically physiological view on the unborn, cited in the introduction to this section of my book, that leaves the study of the organs' growth to anatomy and instead seeks the "physiological phenomena that pertain to it."[36] Entirely unclear, Magendie writes here, is how life exists in the egg, where vessels, nerves, and circulation are still absent, or why the heart becomes active before the nervous system has formed.[37] The physiological processes in the physically still incomplete embryo, he continues, are largely unknown, though one must assume a kind of circulation of the blood. The case is different for the fetus, in which the majority of the organs have already developed. There, functions can be investigated—especially the circulation, to which Magendie devotes several pages.[38] Much nevertheless remains to be discovered. Regarding nutrition, Magendie mentions all sorts of speculations in the literature. The only thing he believes to be certain is that the placenta draws nutritious substance out of the maternal organism, though that substance's composition and manner of operation is unknown.[39]

For Magendie, unlike for his predecessor Bichat, the life of the unborn was not defined by a modality. Instead, he saw a multiplicity of functions whose formation could be described, and the genesis of an organism from the ovum, via the embryo, to the fetus. This frayed the edges of the line that Bichat had drawn so neatly. Fetal life was now a phase in the organism's coming-to-be; on the other hand, what for Bichat had been animation at the moment of birth now became the activation of each individual function at a different moment in time. This ramification became increasingly clear as physiology moved function to center stage and experimentation made it visible.

Müller, considered the founder, with Magendie, of physiology as an independent discipline, was another researcher dedicated to this undertaking.[40] An adept of both physiology and comparative anatomy, Müller criticized Magendie's work for its excessive fragmentation of the organism into individual functions, but he frequently invoked the French physiologist when it came to experimentation and vivisection.[41] Müller first made his mark in the annals of physiology as a student of medicine, not yet twenty years old, with an experiment on fetal life. His prizewinning findings came out in Latin in 1823, and a year later Müller published a sixty-page German disquisition titled "Zur Physiologie des Fötus" (On the physiology of the fetus) in Nasse's *Zeitschrift für Anthropologie*.[42]

In this paper, Müller, like Magendie, expressly demarcates his own project from the anatomical question of embryonic formation. Referring early in his treatise to research on "the history of formation" of brain, skull, spine, intestines, lungs, heart, teeth, and bones, he specifically excludes all of these from his own domain of investigation, on the grounds that they are "more closely affiliated with morphology than with physiology."[43] Again like Magendie, Müller begins at an early stage of development, first setting out the physiology of the egg and the fetal membranes, with comments on the corpus luteum, yolk sac, allantois, and amnion.[44] This is followed by descriptions of the amniotic fluid, lung, skin, and secretions.[45] The focus of the study, though, is on proving that the fetus obtains oxygen not from the amniotic fluid, but from blood that circulates in the placenta and umbilical cord.[46]

To this end, Müller describes a series of cruel experiments mostly carried out on living animals; he mentions rabbits with particular frequency, but also used sheep and cats. With enormous attention to detail, Müller recalls how, "on the twenty-seventh day of February, 1821," "in the company of several of my friends," he opened a sheep's abdomen and raised the curtain on a vivisectional spectacle: "During the incisions into it and into the membranes, the fetus moved vigorously, and after the *liquor amnii* [amniotic fluid] had drained off, it threw itself to and fro and attempted to rise up."[47] The presence of witnesses was invaluable when, occasionally, a fetus "continued to move so vigorously that it had to be held fast by two people."[48]

Previously, on February 6, 1821, a rabbit fetus had also been exposed to observation. Detached from the maternal organism and membranes and placed in a glass container, it was gradually deprived of oxygen. "Anxious and creeping," the creature threw itself about, "tried to lean on its feet, opened its mouth, and perpetually thrust its head against the glass wall of the receptacle." It fell unconscious after a quarter of an hour, but then, after oxygen was restored, it "revived," "yawned," and "contracted its abdominal walls, almost as if it were making convulsive respiratory movements." Finally, "laid upon the table, it was no longer heeded, because we moved on to a different experiment."[49]

In Müller's experiments on living animal fetuses, the darkness of Bichat's "obscure" fetal theory became a theatrical scene of medical students holding down live animals, incisions releasing a flow of amniotic fluid, glass containers shaken by convulsions. Here we see a callously down-to-earth zest for sounding out the frontiers of research practice, as physiological will blazed the trail toward an empiricism that was capable of making visible the invisibility of function.[50] That included physical measurements and the investigation of chemical changes, but also, and most importantly, the experimental determination of causal conditions and the anatomical localization of functions.[51] With his work on oxygen requirements, Müller advanced an investigation of fetal life that would reveal again and again the very multifarious bundle of active and nonactive functions that Magendie had assembled in his *Précis*.

Historical Physiology

Fetal life now appeared in two different guises: as a phase and as a bundle of functions. The challenge to which that gave rise deserves further attention, and the comments of Ignaz Döllinger are particularly illuminating in this respect. Döllinger was appointed to a professorship at Würzburg University in 1803, at almost the same time as his philosopher ally F. W. J. Schelling. He made some sweeping changes there: "Why give lectures? Bring some animal here and dissect it—and then do the same with other ones."[52] This injunction stood at the start of his collaboration with Karl Ernst von Baer, who had come to Würzburg in 1815 to learn comparative anatomy from him, and with Christian Heinrich Pander, who joined the two men in 1816. Under Döllinger's direction and in collaboration with the engraver Eduard d'Alton, Pander set about meticulously reconstructing the transformation of embryonal form, especially in the chick. This endeavor, later continued by von Baer in Königsberg, was what founded embryology as the science of "developmental history."[53] Döllinger had set himself an even more ambitious goal, however. Out of the newfound embryological knowledge of formation, he aimed to build a physiology of the whole of life, grounded in *Naturphilosophie*.

Preparing *Grundzüge der Physiologie* (Outline of physiology, published posthumously in 1842), Döllinger tackled this task by organizing the work as two related parts. "The Emergence of Human

Individual Life" and "The Progress of Individual Life" are distinct, but both are governed by transformation. Life is not simply a series of events from conception to death. Rather, an "incessant activity and creation" happens "in the interior of the body," so that "conception can constantly renew itself and what has been produced will pass through certain stages of development."[54] This process of emerging, continuing, renewing, and passing-through follows a regular course, and its rule is that of concatenation in time: "For everything that appears as a part of the life process at a certain time has its peculiarity only by virtue of what preceded it and thus prepared it and gave rise to it, and its value consists only in its contribution to the establishment of a developmental stage to come." In this sense, knowledge of an individual human existence "cannot truly be other than historical." The aim must therefore be "to ascertain for each separate phenomenon in the course of life its historical place, and its relationship to what was before it and what will come after it, and thus also its significance in the total process of life."[55]

When he describes life as a historical process, Döllinger is stressing not the open-endedness of events, but the fact that they enable a living being to remain identical to itself across its many changes.[56] It is this historicity as a property of the organism that underpins Döllinger's division of his *Grundzüge* into two parts. The first, "Emergence," describes germinal life: the human ovum before and after fertilization in its vesicular form and its membranes. Then comes "the issue . . . of the real individual out of the germ," the genesis of the embryonal form as it passes through fibrocartilage, tubes, radii, salts, and fluid. A closing comment headed "Metamorphosis" argues that although life is always process, there are two modes of processuality. One is the "constant continuation" of life, which is "tied to nerves, muscles, bones, and the mass of the blood"; all this renews itself ceaselessly. The other is a "history divided into periods." This ensues from the transformations called forth by changes in "exteriority," which instigate "alternating metamorphoses of the blood's course, the digestive canal, the generative organs, and the senses."[57]

The second part of the study, "Progress," abandons the chronological organization to portray, in three sections, the cell system, the bone system, and the blood system, though the explanations of the associated functions do include remarks on their emergence. The section "History of the Bloodstream," for example, successively presents the "embryonic bloodstream," the "fetal bloodstream," and the "bloodstream of the born."[58]

In the introduction to that section, we find the relationship of born and unborn described, following the second mode of processuality, as a relationship between two physiological "epochs." As the heart cavities multiply, as vascular trunks converge and branches ramify, the "development of the individual organism" proceeds as a self-organized separation from a "whole." That whole becomes an "exteriority" of the "part," which "separates" and "orients itself" vis-à-vis the exteriority. Like everything else, the human being arises from the entirety of the Earth, which utilizes the "individuals already present" as its "organs" to the same end. The human being that is not yet separated therefore exists as a part of the "maternal body that generates it." It is from this body that it must detach itself, an act for which it prepares in the progress of gestational life; later, it will have to defend the independence it has achieved against the "cosmic forces." Here, the criterion of exteriority defines two main periods in the life of the human being. The first comprises "gestational life [*Fruchtleben*] before birth," which then splits into a "double existence": "one in the egg, and the other in the womb." The second is the life that begins with birth and continues until death. This distinction "also lays down the two main epochs of the formation of the bloodstream."[59]

Döllinger's *Grundzüge* appeared as an unfinished work in 1842, soon after his death. According to the editor, gloom and misfortunes had put a premature end to Döllinger's writing. Having never quite recovered from a bout of ptomaine poisoning after dissecting a child's cadaver in 1827, Döllinger caught cholera in 1836, and he also became increasingly exasperated by the careless physiological research of his contemporaries. All this, writes the editor, soured his appetite for continued scientific work.[60] An 1841 obituary interpreted Döllinger's sudden lapse into silence as a reaction to the appearance of competing publications — Müller's *Handbuch der Physiologie* (Handbook of physiology), especially, had eclipsed his work. Historian of science Owsei Temkin, in contrast, blames the difficulties raised by the book itself. He finds a fatal flaw in the part I have just discussed, concerning the main periods of life and the epochs of bloodstream formation, since the periodization of life and the discussion of the cardiovascular system cannot be really bracketed together systematically.[61]

The interesting point for my purposes, though, is not the success or failure of Döllinger's endeavor, but its productivity. What Döllinger wanted was precisely what Temkin sees as doomed to failure: an interlocking depiction of life's functions and its ceaseless change. This brought into play two competing distinctions, the difference between embryonic formation and the progress of life commencing in the fetus; and the difference between unborn life and born life. Müller resolved this more pragmatically in the second part of his *Handbuch.* Although his chapter "On Development" also reaches from the ovum to developments after birth, it presents not a constantly iterated conception, but a succession of periods of life in which the changes after birth are less fundamental than those in the life of the fetus.[62]

In the somewhat impenetrable thicket of ideas with which his study abruptly closes, Döllinger was trying to capture the dual location of fetal life in physiology: as a history of functions and as a phase of life. This was a radical articulation of developmental historicity as an organizational principle. In it, Döllinger's physiology rose to the challenge he had posed to his students, that of understanding by way of embryonic development how an organism remains self-identical by changing.[63] The whole of Döllinger's historical physiology revolves around such continuity. What characterizes life from beginning to end, he argues, is the drive to autonomy, but in order to achieve and maintain that autonomy, life has to change.

Being able to understand continuity through change was crucial for a reason that may not be immediately obvious to us today. It promised to prove something far from self-evident: that, as Döllinger himself put it, the human embryo does not pass through animal forms in the course of its formation, but "is at its very first appearance really and immediately a human being."[64] The same task of demonstrating ontological continuity arose with regard to function. Dissimilar as the initial ensemble of layered membranes — the chorion, amnion, and yolk sac — is to the fully formed child, just as dissimilar is the initial fetal pulsing to the later workings of the "spirit" in which the "perfection" of the human organism will ultimately manifest.[65]

The alterity of fetal life, which Bichat described as being like a plant and Magendie divided into asynchronously activated functions, posed a special difficulty with regard to the "spirit" or "mind." It was strangely banal, yet important: true, the mental was just one function among others, yet this particular function had a critical significance. After all, according to Döllinger, the "ultimate task of all physiology" is to describe human nature in its specificity, and that specificity lies in mental activity, which was once seen as the soul and is now considered a function of the organism.[66]

A Psychophysiological Prelude

At this point, it is worth looking in more detail at what happened when, in February 1821, Müller began to expose the living fetuses of animals. The young Müller's experiments on fetal oxygen supply may have been a little casual, but they were not random. His description of his experimental scenery features an important secondary arena, the capacity for sensation. Müller touches on sentience, which had been pivotal for evaluating fetal life in the writings of Bichat and Nasse, when he recounts how the animal fetuses, picked out of the drained-off mother animals and placed in glass containers, squirmed in fear. He makes no further comment on the point — his main interest was in oxygen requirements, and in his experimental setup this behavior was nothing more than an indicator of how much oxygen was too little, sufficient, or too much. But Müller did not neglect to apply the "stimulus of a knife" to an exposed rabbit fetus, whereupon the fetus made "a kind of sound."[67]

This experimentalization of fetal sentience by stimulating the live creature supplemented the inspection of the sensory organs in dead bodies, as carried out by Magendie on the inner ear. In the decades that followed, it would be repeated many times — sometimes systematically, when physiologists scratched at vivisected animal fetuses, but often unsystematically, when obstetricians pinched the toes of aborted human ones.

The fact that Müller provoked fetal sensation in passing, but nevertheless intentionally, is not particularly surprising. A year before the results of his experiments on oxygen supply appeared in print, he had presented a thesis for his doctoral examination tellingly entitled "Psychologus nemo, nisi Physiologus" (Not a psychologist unless also a physiologist). Müller, who had gone to school in the then-French city of Coblenz, was alluding to an assertion made by the *médecin-philosophe* Pierre-Jean-Georges Cabanis in 1802: "The physical and the mental natures [*le physique et le moral*] thus merge at their source, or, to put it more accurately, the mental is only the physical nature considered from certain particular points of view."[68]

As Claudia Honegger notes, at the core of Cabanis's postulate is sentience or *sensibilité physique*, a feature highly valued by both science and society in general at the time.[69] It is by means of physical sensibility that organs translate impressions into motion, and out of the organism's resulting perception of something external or alien, a self arises. Cabanis elevates this sensibility to the most general principle of life, which simultaneously explains the human being. The mind, spirit, or psyche is nothing other than a motion of the brain, caused by impressions that may be immediate or mediated by the nerves.[70] This is the context in which we should view Müller's scratching the skin of a rabbit fetus. He says no more about it than he does about the fetus's fearfulness. Yet perhaps, when wielding the knife, he was thinking of what he had read about fetal life in Cabanis's best-selling *Rapports du physique et du moral* (translated in 1802 as *On the Relations between the Physical and Moral Aspects of Man*).

In that work, Cabanis initially follows Bichat by arguing that the faint movements of the fetus in its mother's belly are no more than a "prelude" to animal life, which begins only when the organism starts to nourish itself, that is, upon birth.[71] He then adds, however, that because the "principle" of such movement is the same as in the born infant, we may assume that the fetus has "already received the first impressions," out of which the "idea of resistance," a notion of "foreign bodies," and finally an awareness of the self or "*moi*" take shape. Cabanis reaches this conclusion by interpreting the conditionality of fetal life in a way diametrically opposed to Bichat's reading. As the enclosed fetus moves, he writes, it feels restriction and constraints, to which it reacts with needs and desires. It already has a sense, however "badly conceived and vague," of well-being and malaise, and tries to prolong the former, eliminate the latter. In fetal life, which is surely just the overture to another life, "the child's brain has already perceived and willed." He has "some feeble ideas; and their recurrence or their turning into habit has produced in him inclinations." This is the "first origin," the point of departure that must be sought by anyone who wishes to understand the "intellectual operations" of the human being.[72]

At the start of the nineteenth century, then, Cabanis postulated a connection between the physical and the mental, physis and psyche, within the fetus's life of sensation. The physiologists who followed him derived from this a continuum of the mental that was not interrupted by birth—it had its organs and origin in fetal life. Some decades later, this would feed into Döllinger's notion of developmental history, and origin became beginning: the beginning of a process in which the being's continuity was conferred by change.

Such deliberations also gave rise to the studies on brain development that started to appear in greater frequency from 1810.[73] But these were anatomical investigations and had no need to concern themselves with the activities of the mind. The physiology of fetal life, in contrast, was preoccupied for the whole nineteenth century with the problem of whether the human being already perceives and wills in the mother's womb (I return to this in Chapter 8). Müller himself, after his work on fetal oxygen requirements, pursued neither route. He remained—"with childlike joy," he wrote to Döllinger's student von Baer in 1828—attached to *Entwicklungsgeschichte* (developmental history), though he now studied the eggs of insects. As he put it in the letter to von Baer, whom he saw as an expert in the field: "Now I am hatching something too."[74]

Individuality in the Sleep of the Soul

If everything psychical arises from something physiological, and the origin of the human mind is to be sought in the fetus, then the question is no longer whether mental activity starts before birth, but at what point. This is exactly what Karl Friedrich Burdach asks in his 1828 handbook of physiology, in the chapter on life in the womb: "But when does the psychical life of the embryo begin?"[75]

Initially, and with some élan, Burdach insists on the "continuity of life in the begetter and the begotten, as regards the soul just as the body."[76] In this short formula we find, first, a model of procreation that sees the emergence of an individual as a realization of the species. When human beings procreate, the "general existence" of life continues unbroken in the "particular, concrete" begotten, as Burdach explains a little later in his text.[77] Second, the formula stresses the connection between body and mind. Matters of the mind are not excepted from the continuity of life, so procreation itself is already the "awakening of an individuality that takes effect in self-formation and the sense of self, and develops from an imperceptible starting point."[78] A beginning of mental life in the embryo must therefore be sought at the point where the organism begins to take shape:

> Nowhere do we see a hiatus in life, a leap in the course of development, and so we must assume that the germ of the sense of self, that dynamic unifying point of life, is given from the moment of fertilization; that it begins, in an

> unrecognizably minimal form, as the ovule of psychical life, the origin of which we can see only with the telescope of reason, just as only the microscope can make the first trace of the organic body physically visible; that the sense of self is in a process of latent life, just like the material existence of the egg, which has no particular organs and no perceptible functions but nevertheless truly lives.[79]

The mind, too, has its ovule, its egg, in which it is enclosed from the very beginning as a self-consciousness that cannot be seen for anatomical purposes with the microscope but only, for the purposes of fathoming life, with a telescope. Two aspects of this are key. Mental life fuses with a concept of individuality—already highly charged with *Naturphilosophie*—that is defined by self-formation and self-consciousness. And the question of beginnings is organized by a dual continuity: life leaves no gaps, and development makes no leaps. As a result, the question of how mental life begins cannot be answered by naming a moment within life in the womb. What Burdach calls the "latent" is not Nasse's idea of potentiality: Burdach specifically rejects Nasse's belief that animation begins with the first breath.[80] Instead, the key point is graduality, confounding Bichat's notion of birth as a caesura. The latent is not what will be, but what has always been becoming—something that "develops from an imperceptible point in a gradual series of stages."[81]

This paves the way for inquiry into "different psychical states in gestational life."[82] For mammals, before they are "unveiled" by birth, the mind is no more than a "muffled brooding . . . because the substances in the egg satisfy the needs of the embryo and are presented to it, as to a plant, without its own free action." Barely measurable pulsations allow us to infer a very slight brain activity, but the "state of the sense organs" and the lack of sensory stimuli suggest that the senses have not yet been activated: "The fingers are curled, the eyes closed, the ears sealed with a membranous substance, the nose filled with mucus." Bathed in an ever-constant fluid without smell or taste, light or sound, the life in the womb receives stimuli only from movement, temperature, and pressure.[83] Accordingly, it is without consciousness.

This state of brooding or incubation has a name: "We now denominate this state as *sleep*, and thus see human life in the womb to be a sleeping life."[84] In such life, however, the absence of the signs of mental life is offset by the presence of its component elements. Although the fetus has no awareness of itself, it does have organic "self-determination and self-preservation," as expressed in its active "appropriation" of substances.[85] And even though the mind is still incubating, a point in time, the "fifth month," can be named at which, with the completion of organ formation, a "balance among the various organs" is reached and a sensation of "pleasure and displeasure," directed at a central point, becomes possible, marking the start of "the psychical life as something real."[86]

Over three decades, through the detour of vegetative life, the unborn had become a sleeping human being, a guise that Bichat had refused to acknowledge because he needed the unborn to stake out the distinction between two modalities of life. Now, in place of a vegetative life defined by the absence of animation, there is a fetal life that ramifies into a history of functions and a phase of development. This grounds a continuity of being: the fetal functions, in Magendie's sense, that are not active dissolve into the continuity of life (which knows no gaps) and what for Nasse was a problem of the beginning dissolves into the continuity of development (which makes no leaps). What really makes a human being is now thought to have always been there: an organically given individuality, which reaches fruition in the form of a psychical function.

Traversing Birth

From this viewpoint, Preyer's *Specielle Physiologie des Embryo* of 1883 seems to be less an innovation than the systematization of a research agenda that had its conceptual roots in the early nineteenth century. The program grew from the conjunction of the question of fetal life, which arose when the idea of a joint corporeality of the pregnant woman and the unborn collapsed, and the postulate that life emerged in a developmental history, which arose from investigating embryonic formation. That interface constituted fetal life as an epoch or period, embodied in an organic individuality that accorded entitative legitimacy not to a child-to-be, but to an unborn. The unity of such life immediately fractured into a multiplicity of functions that, despite forming a single organic context, were embedded in different, nonsynchronous times: some already are, others are not yet; some are earlier, others later, some completed, others incomplete. On the other hand, fetal life acquired unity through the principle of development that designated it as a biographical epoch. Fetal life meant the genesis of functions, and the genesis of functions meant developmental life.

This was bioscience. But from the outset, the description of fetal life also carried an anthropological question: how the arrival of a human child takes place. It is no coincidence that "sensation" absorbed the ontological dimension of that question and converted it into an epistemological praxis. The fact that sensation is expressed in motion ties it to the ancient idea of *animatio* that previously authenticated the existence of a child in the mother's body; this is one reason why discussions of the fetus's sentience are always accompanied by interpretations of fetal movement, as prefigured in the Aristotelian tradition.[87] At the same time, sensation is an organic faculty that enters physiology as a relay between body and soul and the functioning of which is, in principle, accessible to empirical investigation. With the transformation of child-to-be into fetal life, the process of becoming a child became the genesis of a psychophysiological function, and the moment of ensoulment, *animatio*, became a developmental history of the mind. Or, put the other way around: as *animatio* lost its special significance because the human was now one animal among many, the unborn became developmental life.

In the course of the nineteenth century, questions such as these generated the numerous studies that Preyer collated in his *Specielle Physiologie des Embryo*. He was able to insert them into an ensemble that portrayed fetal life as a connected set of emerging functions—blood circulation, respiration, nutrition, excretion, thermogenesis, motility, sensibility, and growth line up as chapters in Preyer's book. So extensive were the available findings, especially on blood circulation and oxygen supply, that Preyer could even define standards on certain aspects, such as the "normal heart rate."[88]

At the same time, Preyer described fetal life as a phase of development by showing how it was connected to a before, in the germ, and an after, in the newborn. I will look at this portrayal not in Preyer's comments on individual functions, but in his depiction of birth. In the early stages of the fashioning of fetal life, there had still been struggles over what birth meant—caesura or not?—but Preyer now sees it as a relationship of continuity and discontinuity. First and foremost, birth is a switch between opposites:

> Before, for a long time it [i.e., the human being] is in a fluid, swimming and closed off from the atmosphere so that the entry of air is enough to kill it; after, it can withdraw from atmospheric air only for a few moments without risking death. Before, nourishment is passed to it effortlessly through the umbilical cord directly into the blood mass; after, food must be taken in incomparably more slowly and laboriously through the mouth, stomach, and gut. Before, it abides in uninterrupted darkness; after, in the light of the world; before, alone in soundless quiet, after, in noisy society; before, in always the same warmth, after, in colder air with fluctuating temperatures. Before, it moves only involuntarily as if asleep, finding insurmountable resistance everywhere; after, freely, without the barriers of the uterus wall.[89]

The organism completes this transition—from liquid to air, ease to travail, darkness to light, warmth to cold, constriction to freedom—through "changes" in its "interior." It must now breathe, eat, look, listen, regulate its temperature itself, and its motion is liberated.[90] Birth, in Preyer's narrative, marks a physiological breach between born and unborn, as the conditions of life reverse and trigger functional adaptation.

However, this rift is bridged by the organism that undergoes it.

Whereas the change in the "nearest surroundings" instigated by birth is "major and sudden," the change inside the organism is gradual and nuanced.[91] For one thing, the ground is prepared: the organism that is born has already formed its ability to reorganize its life. Before the very first breath there is an oxygen supply, before the first feeding there is nutrient intake, the fetus's auditory nerve is excitable before the newborn can hear. Furthermore, the transition does not take place all at once: the limbs move before birth, respiration begins at the moment of birth, digestion starts when the infant first feeds, the senses develop as its life proceeds and as intellect is achieved by the child. It is in this staggered form that fetal life becomes the life of a child, before, during, and after birth.

Crossing the discontinuity of physiological adaptation at birth, which makes the fetus into a child all at once, there is thus a continuity of development that makes the fetus into a child gradually. Preyer had already hinted at this relation of continuity and discontinuity in 1882, when he announced the publication of his *Specielle Physiologie des Embryo* in a book titled *Die Seele des Kindes* (Figure 3; translated in 1893 as *The Mind of the Child*). The rationale he gives for treating fetal life and the infant mind in two separate books is purely pragmatic:

> I proposed to myself a number of years ago, the task of studying the child, both before birth and in the period immediately following, from the physiological point of view, with the object of arriving at an explanation of the origin of the separate vital processes. It was soon apparent to me that a division of the work would be advantageous to its prosecution. For life in the embryo is so essentially different a thing from life beyond it, that a separation must make it easier for both the investigator to do his work and for the reader to follow the exposition of its results.[92]

It is no coincidence that sensation is what connects the two books. Apart from a few pages on growth, *Specielle Physiologie des Embryo* ends with a chapter headed "Embryonal Sensibility," while a chapter on "Development of the Senses" opens *The Mind of the Child* and begins to present "continuity in the capacity for sensation" on the

DIE SEELE DES KINDES.

BEOBACHTUNGEN

ÜBER DIE

GEISTIGE ENTWICKELUNG DES MENSCHEN

IN DEN ERSTEN LEBENSJAHREN

VON

W. PREYER,

ORDENTLICHEM PROFESSOR DER PHYSIOLOGIE AN DER UNIVERSITÄT UND DIRECTOR DES PHYSIOLOGISCHEN INSTITUTS ZU JENA, DOCTOR DER MEDICIN UND PHILOSOPHIE.

LEIPZIG.

TH. GRIEBEN'S VERLAG (L. FERNAU).

1882.

SPECIELLE

PHYSIOLOGIE DES EMBRYO.

UNTERSUCHUNGEN UEBER DIE LEBENSERSCHEINUNGEN VOR DER GEBURT

VON

W. PREYER,

PROFESSOR DER PHYSIOLOGIE AN DER UNIVERSITÄT JENA.

MIT 9 LITHOGRAPHIRTEN TAFELN UND HOLZSCHNITTEN IM TEXT.

LEIPZIG,

TH. GRIEBEN'S VERLAG (L. FERNAU).

1885.

Figure 3. Title pages of William T. Preyer's books *Die Seele des Kindes* (1882) and *Specielle Physiologie des Embryo* (first published 1883).

very first pages.[93] Sensation links the continuity of development to the continuum of physiological and psychological processes. More than anything else, this connection between Preyer's two books suggests that the path from Bichat's caesura between vegetative and animal life, via Magendie's activation of functions and Burdach's notion of a sense of self given from the very start, culminated in Preyer's definition of developmental life through the relation of continuity and discontinuity. Tellingly, Preyer picks out the fetal capacity for sensation from the crowd of puzzles that remain unsolved. That capacity is still underresearched, he argues, similarly to nutrition but unlike circulation and respiration.[94]

Preyer's own special interest here is in fetal feelings of pleasure and unpleasure, which he considers indistinct but not implausible, since babies express them immediately after birth.[95] The capacity to distinguish pleasure from unpleasure, he adds a little more boldly one page later, is "prenatal and inherited, and innate in the proper sense of the word." The same is true of feelings of hunger and muscle feelings. Pain, in contrast, is felt "without doubt only to a minor degree, even by the mature fetus" — the newborn reacts hardly at all to slight irritations of the skin, though noticeably to stronger irritations.[96] Even this slight sensitivity to pain, he continues, indicates that a wakeful state may be possible at least from the sixth month, since from then pain can waken the fetus, "like a hibernating animal or a nursling fast asleep in a quiet, dark room."[97] The sensitivity of the skin, on which this phenomenon depends, must already be present "long before the embryos are viable on their own," which is why they "markedly often react vigorously to painful interventions, especially strong electrical traumatic, chemical, and thermal ones (cooling or warming), through all sorts of initially disordered, then ordered reflexes." And the only reason they can react that way is that the capacity for movement arises before sensibility.[98]

Preyer here explores developmental life in great detail as the genesis of particular functions in their interaction, but he also situates that complex in relation to something else. In a passing yet enormously significant comment, he happens across the word *prenatal*. It serves him as a way to distinguish between two forms of continuity in sensation that is not interrupted by birth: the innate, as that which carries the organism over the threshold of birth, divides into a continuity bestowed by the germ and one instituted by development. The former is hereditary innateness, the latter prenatal innateness. In a different formulation: the former is heredity, and the latter is the stage of development that has been reached at birth.

It is very evident that Preyer saw the adjective *prenatal* as a means to raise the profile of the latter form of continuity vis-à-vis the former. The occurrence and usage of the word at this point reflects what Frederick B. Churchill has called a "watershed" in the study of intergenerational relations in the 1880s.[99] At the time, biology's new findings narrowed the broad concept of inheritance down to the more specific *Vererbung* or hereditary transmission, as a transgenerational continuity that is bound to a carrier substance and follows particular laws. Whereas the innate was previously the inherited — that which comes down to the child from its parents either through transmission or through influence — this complex now separated out into the prenatally influenced and the inherited. The organism draws its capacities from a continuity within which two modalities can be distinguished: one that is given by the germ and described as heredity, and another that is given by development and described as prenatally acquired. Preyer makes that distinction only perfunctorily in the passage I have cited, though he discusses it in more depth in *The Mind of the Child*. What is important here is that, for the first time, being before birth made sense in a complementary relationship to heredity, and that it managed to fulfill the task of conceptualizing continuity across birth while also identifying developmental life up to birth as a single unit.

In 1931, the embryologist and historian of science Joseph Needham would credit Preyer with having written what was still the only book on embryonic physiology.[100] Another ten years on, the reviewer of a new English book on fetal physiology remarked that no further efforts of the kind had since been made. This, he argued, was surely because embryological research was still concerned primarily with form, not with function.[101]

However, that Preyer's claims for fetal physiology as a distinct field remained a one-off cannot be laid at the door of embryology. Rather, it reflects the fact that developmental life as an object was scattered across different research fields and formulations of the problem. In the twentieth century, the continuity of the organism was assured by genetics, which was initially criticized by developmental researchers on the revealing grounds that it simply rehashed the old idea of preformation.[102] Meanwhile, "development" was distributed between embryology, interested in form, and psychology, interested in the child. Between them, both logically and temporally, stood the object of fetal life, with its inherent ambivalence: birth, as discontinuity, demarcated the fetus from the child and classed it with the embryo, whereas the continuity of development connected it to both. This is why Preyer's appeal did not result in the establishment of a discipline. Nevertheless, his studies paved the way for embryology and child psychology to expand their scope by looking at fetal life.

CHAPTER FIVE

Pregnancy as Relation

Placental Barrier and Substance Transmission

In 1825, the journal *Annales des sciences naturelles* published a brief but important note on the blood of the fetus in vertebrates by the Swiss physiologist Jean-Louis Prévost. Later to become Stendhal's physician in Geneva, Prévost was first and foremost a proponent of experimental medicine. He investigated blood, urine, and muscular contraction, and his paper on the process of fertilization won prizes in Paris.[1] In the 1825 letter to the *Annales*, Prévost outlines a study of something he introduces as "one of the thorny questions of physiology," which has long exercised "all physicians" and many anatomists: the still controversial issue of separate blood circulation.[2]

Working, as he often did, in tandem with the chemist Jean-Baptiste Dumas, Prévost had discovered that during the very first days of incubation, the blood corpuscles in the chick embryo differ in form and volume from those in the adult bird. If the same thing applies to the fetus of viviparous animals, his letter tentatively concludes, then "the question of blood communication between mother and fetus would be easy to resolve." And indeed, his subsequent experiment on a nanny goat shows that the volume of the blood corpuscles in the fetus is double that in the pregnant goat. "There is, thus, a material and incontestable difference between the blood of the fetus and that of the mother, a difference that can only be understood if we suppose that the embryo carries out sanguinification itself, and on its own account, using the materials supplied by the mother."[3] It is not the vessel ends but the blood itself that resolves the question: the embryo does have its own activity. At the same time, the embryo can only be active to the extent that it is connected with its mother. That is to say, wherever we find discontinuity of blood, there must be continuity of supply. Where blood divides, nourishment must connect.

This matter continued to interest Prévost in the years that followed, when he turned to the "liquid of the cotyledons." Fifteen years after his note on fetal blood, he spoke to the Society of Physics and Natural History of Geneva on his new topic in a lecture titled "Physiological and Chemical Studies on the Nourishment of the Fetus." In that paper, he reiterates that the fetal circulatory system is "entirely separate." After all, he points out, it is easily observed that circulation still continues for a while in the fetus of a just-killed pregnant ewe.[4] At issue here, however, is not blood, but a liquid to be found in both the fetal and the maternal portion of the placenta. This liquid can be weighed, its acidity and smell can be studied when it is exposed to the air and its desiccation when left there; the scientist can determine the temperature at which it coagulates, how its color changes in that process, and what happens when it is treated with cold or warmed alcohol, ether, or acetic acid. Prévost and his colleague Antoine Morin had done all those things, and the outcome of their degradations, dissolutions, and combinations of the liquid was a list, measured in grams, of proteins, phosphates, salts, fats, and more than one substance that had never previously been found in the animal organism. Together, the authors conclude, these comprise the one-eighth "solid matter" extractable from a liquid that is the same whether taken from the maternal or the fetal side of the placenta — from which they deduce that it "is transmitted [*transmise*] from one to the other without being altered."[5]

As Foster de Witt points out in his historical survey on theories of the placenta, Prévost's lecture to the Genevan naturalists was the first systematic study of the placenta's nutritive function.[6] For my purposes, though, the passage cited is important less as a moment of innovation than as an indicator of the easily overlooked fact that Prévost studied fetal supply as a *function*. His experiments on blood had taken the same approach. The two sets of research combine to form a unit, since even though one addresses separated circulatory systems and the other transmitted nutritive material, both are about one and the same organ, the placenta, whose function evidently is both to divide and to connect. This is why the blood can be materially different while the nutritive material remains the same.

Prévost's combination of research on the discontinuity and the continuity of substances resolved with great elegance a problem that had driven François Magendie almost to despair in 1817. Magendie, too, had engaged with the question of shared or separated blood. Before him, William Hunter had described how maternal and fetal vessels ramify and terminate without connecting. But this left the terrain from which the ramifications of both initially spring, a tissue they share and that embraces them all. There, the picture was not quite as clear as it may appear in retrospect. After a series of fruitless experiments "even in living animals," Magendie concluded in 1817 that there is "no direct communication" between the maternal and fetal vessels. Doubtlessly, however, "the blood of the mother," or at least "some of its elements," must somehow pass through: "probably it is deposited by the uterine vessels on the surface or in the tissue of the placenta and is absorbed by the radicles of the umbilical vein." Why else was he able to smell in an exposed dog fetus the camphor he had injected into the pregnant animal a quarter-hour earlier? There was clearly a need, said Magendie, for "new experiments."[7]

Experiments on an Organ of Relation

When Magendie called for fresh experiments, a fact became a question. Agreed, there is no direct communication through vessels, but how do substances nevertheless pass from the mother to the fetus? Notably, one reason why Magendie found "new" experiments necessary was that old experiments on the issue existed. He meant the trials in which wax (or, less often, another substance such as quicksilver) was injected into the maternal or fetal side of the placenta, or into the mother's pudendal artery in animals or the cadavers of pregnant women, in order to discover whether it ceased to flow or found its way across the placenta. Some of these experiments went back to the sixteenth century, and what they brought to light was confusingly contradictory. They had helped Hunter reach his postulate of two separate circulatory systems, but experiments of the same kind before and after Hunter's work did not serve simply to identify the courses and connections of vessels. They were also allied with two further questions: What nourishes the fetus, and how does nourishment reach it?

Discussing seventeenth- and eighteenth-century embryology, Joseph Needham identified four sets of hypotheses that responded to these two inextricably related questions. First, the fetus is nourished by menstrual blood that flows through the umbilical cord. Second, it draws nutrition through its mouth from the amniotic fluid or uterine milk. Third, its nutrition is fetal or maternal blood, menstrual blood, uterine milk, or amniotic fluid, each of which reaches it through the umbilical cord. And fourth, it absorbs nutrition through the pores of its skin.[8] Given this complex backdrop, it is not surprising that Albrecht von Haller's search for clarity in the second volume of *La génération* (Generation, 1774) seems rather circuitous. Meandering from datum, to doubt, to question, von Haller wonders whether blood or a nutritive liquor passes through the umbilical cord to the fetus, which, however, certainly always also draws nutrition from the amniotic fluid—only to arrive at the conclusion that in any case there must be some "free and unbroken passage" between mother and fetus.[9]

After Hunter, Needham tells us, the question of fetal supply was formulated differently. At issue was no longer whether blood, water, or milk flows through the mouth, veins, or pores, but how tissue is able to transmit matter.[10] In other words, the question of the impeded versus unimpeded flow of one liquid or another had become a question of the mechanics of transmission. This was why old experiments needed to become new.

Hardly had Magendie noted this deficit than, still in 1817, a seminal German journal of physiology published work by August Carl Mayer, then professor of anatomy and physiology in Bern, who claimed to "have found the key required to penetrate more deeply into the secrets of the fetus's life than has hitherto been granted to physiologists."[11] Mayer pondered whether veins could "suck in" matter, as had been said of the *venae lacteae* since Hippocrates and Aristotle, though disputed by others; that is, whether "drippable liquids and substances that are dissolved in them pass through into the venous system."[12] To find out, he injected dyes, salts, metal oxides, and oils into the lungs of living animals and monitored their diffusion in the blood, urine, sweat, milk, tissues, and organs. By including gravid animals in his study, Mayer expanded his scope to the placenta and fetus. He called the experiments using potassium ferrocyanide "very convincing": it had been possible to detect the chemical not only in the fetal membranes, but also in the fetus's stomach, kidneys, and urinary bladder.[13]

In his *Specielle Physiologie des Embryo*, William T. Preyer would cite Mayer as the first to prove the "crossing of a substance foreign to the organism."[14] What was special about Mayer's experiment was that he did not inject an experimental substance, simulating milk or blood, into uterine or placental vessels in order to investigate their permeability. Instead, he introduced a foreign substance into the circulation of the mother animal. In this experimental setup, the object of interest was not a flow of substances from body to body, but the transmission between two organisms with distinct circulations.

Another study, identified by Preyer as being one of "the first reliable experiments on human beings," makes the same point in its very title.[15] "Uebergang von Medicamenten in die Milch der Säugenden und in den Fötus" (Transition of medicines into the milk of the lactating woman and into the fetus), published in 1858, presented clinical observations made by the physicians Adolf S. Schauenstein and Josef S. Spaeth.[16] Schauenstein and Spaeth had been able to detect in both the amniotic fluid and the excretions of newborns the iodine with which the syphilitic mothers had been rubbed before giving birth. They conceded that experimental proof was still lacking but argued that this was exactly how medical praxis hastened ahead of physiological experimentation—after all, physicians had been deliberating for many years on the effects of medication on pregnant women. In fact, the two authors' evaluation of the clinical data was not unsystematic. They also examined a newborn that had not yet been given its mother's breast, aiming to exclude the possibility of transmission through the milk.[17]

Such experimental studies of transmission were flanked by copious descriptions of the placenta. An article from the famous *Dictionnaire des sciences médicales* of 1820, also much cited in the German-speaking world, portrays the placenta as a "fleshy, spongy, vascular mass" where, on the one hand, the "absolutely independent" circulation of the fetus's blood takes place and which, on the other, "seems to serve as an intermediary between the mother and the fetus during pregnancy."[18] As an intermediary, it establishes a "mediate circulation" and "reciprocal exchange" between the mother and the fetus.[19] Most physiologists, the article continues, are now persuaded that the placenta is "the principal means of communication between mother and child," since it transports nutrition and probably oxygen, but without creating a connection of nerves.[20]

In the *Dictionnaire*'s equally weighty successor, the *Dictionnaire encyclopédique des sciences médicales* of 1886, further points appear to have become firmly established. That the placenta supplies the fetus with oxygen is now certain, as is nutrition through "placental exchange," even if much still remains to be discovered regarding the latter.[21] It seems to be almost entirely taken for granted that the maternal and fetal organisms are not connected directly, but that "communication" takes place between them, carried out by the placenta as a "temporary organ."[22]

The interlock between the description of the placenta and the experimental exploration of its transmission function in the nineteenth century is well illustrated by Preyer's *Specielle Physiologie des Embryo*. There, Preyer calls the placenta the "nutritive organ of the fetus," which both transfers solute and easily diffusible substances from the maternal to the fetal blood and carries out "the transportation of very small particles through, for example, leukocyte migration."[23] Preyer admits that this has been conclusively demonstrated only for oxygen, but the long series of experiments that he summarizes, starting with Mayer's study, leaves him in no doubt of its validity. Countless experiments since then had introduced substances including copper, rose madder, camphor, atropine, nitrate, quinine, phosphorus, quicksilver, lead, arsenic, and sulfuric acid into a gravid animal and investigated their residue in the fetus.

Incidentally, the reverse route, the "crossing of substances from the fetus to the mother," as Preyer calls it, was studied in terms of a postulated reciprocity. In 1898, the authors of a paper on "fetal-maternal transmission" would outline several experiments on this — first and foremost their own, which had shown that substances injected into animal fetuses found their way into the mother's urine as long as "the little ones" were alive and that this process immediately ceased upon their death.[24] To be sure, the significance of fetal-maternal transmission was far less obvious than that of maternal-fetal transmission.

By the end of the nineteenth century, then, the placenta had become an organ that divides two organisms by accommodating the terminals of two circulations, and simultaneously connects two organisms by mediating between them. It embodied an entirely new constitution of the pregnant woman's connection with the unborn: through placental function, this became a relation between two organisms that, as Adolphe Pinard puts it in his handbook article on the fetus, are "completely independent in anatomical terms" but "intimately correlated in physiological terms."[25] If these authors constantly and unquestioningly spoke of "exchange" despite the doubtful presence of reciprocity, that only emphasizes the intensity of their interest in framing a connection as a relation.

Milieu as Relation

In his historical physiology, published posthumously in 1842, Ignaz Döllinger, too, deliberated on questions around fetal supply. He approached them through analogies. If nutrition and oxygen reach the fetus by the same route, Döllinger reflects, then this is the only unique aspect of the "relation of reciprocity" that arises from the placental circulation. In all other respects, the fetus relates to the pregnant woman exactly "as the human being, or any animal developing outside the womb, relates to the inhabited planet and, through the planet, to the universe."[26] Birth is the tying of new bonds: "The child, separated by birth from the mother's body — if not yet completely, because of its suckling — now, as a citizen of the Earth, enters into a relationship with external Nature similar to the relationship it has hitherto had with its mother." The only reason why these relationships are "similar," not the same, is that the unborn child is doubly enclosed, within its mother's body and within the cosmos. "In fact, for the fruit that it contained, the womb was only the mediating organ by means of which the life of the universe and that of the newly generated being entered into interaction."[27]

Döllinger thought of life not only in a radically historical way, but also in a radically environmental way. But what being environed — surrounded by something as a human being, animal, or embryo — meant for him was nothing other than being in relation to something else, whether in the womb or on the planet. The fetus is enclosed in its mother in the same way that people and animals live on the Earth: by being intertwined with her.

This was characteristic of the nineteenth-century life sciences. To live *in* something meant to live *by means of* something. In the eighteenth century, natural historians had already been intrigued by the fact that organisms live on particular soils and in particular atmospheres. It was this that explained why species breathe, eat, and move in ways that are so regular within the species but differ so markedly between species. When natural history became life science and the classification of living beings was replaced by the explanation of life processes, humans' and animals' inhabiting of natural spaces changed as well. It was now a relation, connecting organisms to what surrounds them.[28]

In Döllinger's thinking, this kind of environmental or milieu relationship is realized by the placental tissue. Like any other life, fetal life is tied to the conditions of its existence and cannot exist otherwise than through that very tie, which is described with a plethora of relational concepts: communication, exchange, interaction, interchange. This aspect takes Pinard's description of the placenta in another direction. In the placenta, two vascular systems are independent from one another and yet correlated, which is what identifies the fetus as an organism, whereas the maternal body is both organism *and* milieu. If relatedness makes the fetal organism the same as all other organisms, it is distinguished from them by *what* it is related to: it lives in a milieu that is itself an organism.

This circumstance prompted particularly intense debate among authors with an inclination to *Naturphilosophie*. In 1820, Carl Gustav Carus located the "the most important singularity of the fetus's life [*Lebenseigenthümlichkeit*]" in two properties: first, it "does not stand in free and immediate interaction with external Nature," but is buried "in the maternal organism"; second, it lacks "self-determination" because it is "still part of a larger individual organism."[29] In 1835, Gabriel Gustav Valentin defined "embryonic life" by its "lack of independence and free individuality."[30] Karl Friedrich Burdach's influential handbook of physiology, around the same time, describes a relation "between independence and interaction" that resembles the relation "between a parasite and its host organism," which are both "individuals that strive to achieve or assert a singular existence, but also form a joint whole, interact with each other, and are dependent on one another."[31] And Johannes Müller sees pregnancy as "the most intimate possible juxtaposition of two beings that are in and of themselves quite independent," though in practice it is only upon birth that the child becomes "independent" and "a body alien to the uterus."[32] What distinguishes the fetus from human beings and animals on Earth is the manner in which it is an individual: as an organism, it is part of another organism, and that marks its individuality as an unfree individuality.

Another author who employed similar semantics, but without sounding out the philosophical overtones, was Claude Bernard, one of the founding figures of experimental medicine and a former student of Magendie's. Bernard took a radically function-based view of life, and he thought of those functions as relations. He was not interested in musing on the unborn. Development, an object so difficult to capture experimentally, threw Bernard off his stride, yet he could not sidestep it entirely, given that the whole of his physiology revolved around nutrition.[33] Of interest here, though, is not exactly what made Bernard turn to the subject of the unborn, but how he stumbled over the maternal-fetal relationship and clambered back up by finding a way to conceptualize unfree individuality.

What troubled Bernard was a complication in studying glycogen production in the mammalian fetus. An "initial difficulty," he explains, arises from the fact that the fetus "remains in communication with the maternal organism through the placenta." As a result, one must assume "that the substances detectable in its tissue are not the product of a fetal activity, but come from the mother." This "objection" goes further than glycogen alone: "In fact, the mammal embryo is not free like the embryo of birds, which develops in isolation; conversely, the relationship of dependence with the maternal organism is not as close as was once believed." After many years of error, Bernard reports, we now know that mother and fetus are not nourished by the same blood in a direct vascular connection, and the "relative independence enjoyed by the fetal organism" has now been proven; between the two organisms, we find not "continuity," but a "passage" of liquid substances.[34] To encapsulate this relation, materialized in the discontinuity of vessels and the passage of substances, Bernard finally chose a concept that brought together many of the points debated at the time: "relative autonomy."[35]

The Unavailability of Fetal Existence

That Bernard fell to thinking about the maternal-fetal relationship when he encountered difficulties with investigating the mammalian fetus was more than a coincidence. It was inevitable that this relation, brought to light by numerous experimental studies of the placenta, would pose an obstacle to research — and that obstacle itself brought to light, again and again, the unfree individuality of fetal life. For Preyer writing in 1883, the numerous endeavors around this matter culminated in both a stock of knowledge and a methodological reflection.

On the first two pages of Preyer's introduction to his *Specielle Physiologie des Embryo*, he defines his program as the "physiological developmental history of the individual being." Preyer then straight away embarks on a long and convoluted discussion of the difficulties involved in this task. Thus, "as regards the *material*, one must from the outset dispense almost entirely with what is most interesting,"[36] for "dead human embryos at early stages of development" can seldom be obtained, and "the same thing applies, even more so, to living ones." Living embryos were precisely what the researcher coveted. It would be "desirable," Preyer tells us, to "observe the living fetus in its natural environment, or at least, as far as possible, to detect its expressions of life while it is still developing in the uterus." However, observation in the human being can "be executed only imperfectly because of the opacity of the uterus's walls and the parts surrounding it."[37] The pregnant body stands between the physiologist and his object as a doubly unbreakable barrier: as a milieu, it is inextricable from the object; as a living human organism, it is unavailable to the researcher for evident ethical reasons.

During the nineteenth century, this impasse produced much creativity, which Preyer outlines in his introduction. His account begins with substitute objects taken from the human body. First among these are living miscarried embryos. They are not well suited to the purpose, since not only are they scarce, but obstetric practice (which is where they are to be found) has other priorities than exploiting the medical event for research: "as a rule," one has "so much to do for the miscarrying woman that the egg is examined only when it has long since gone cold and the embryo is dead." Furthermore, "the eggs obtained in this way, through abortion — that is, a process that is not physiological but pathological — are often enough already pathological." Nevertheless, Preyer continues, one must make use of whatever is at hand, since "each observation, even the apparently least significant, about its movements if any" may "become valuable through comparison with other findings."[38]

In the same way, says Preyer, stillborn fetuses that are born too early should not be ignored, because "in the human being, such cases substitute for vivisection." Equally instructive are premature births, for "the behavior of seven-month or eight-month premature babies . . . may in many essential points be regarded as the same as that of the unborn seven-month or eight-month fetus." This "material," too, is scarce and is certainly nowhere obtainable "frequently enough for one to be able systematically and in detail to observe the babies and experiment with them, quite apart from the often insuperable difficulty of separating the infant from its mother or attendant."[39] Even after birth, the woman withholds the object of study from the researcher's grasp. As the nourishing organism becomes a caring person, the tender gesture that enfolds the child takes the place of the body that enfolded the unborn.

Preyer thus moves on to other substitute techniques. The physiologist, he notes, can have recourse to procedures with which, "as is well-known, gynecologists and midwives regularly concern themselves." These include chance actions, as "when examining the woman in labor the finger introduced happens to touch the mouth of the fetus, which sucks it."[40] More important is work "with the sense of touch and hearing." Preyer sees little that is "physiologically significant" in the "experiences" of obstetricians and midwives themselves — he omits to mention that the data on fetal heartbeat, on which he presents numerous statistics, were gathered using auscultation in maternity hospitals.[41] What he does regard as clearly open to experimental utilization is palpation of the "strange movements of fetuses in the second half of the intrauterine period." Such movements "could be made the object of more thorough studies regarding their dependence on differing states of the mother, their vivacity, and their relation to the fetal position, without any special difficulty and quite well, in some cases through meticulous observation of the *visible* elevations and dips in the mother's abdominal wall." What obstetrics saw as signs of pregnancy and of the unborn's vitality or position, the physiologist could use as "those few symptoms of a special existence of the fetus in the maternal organism that can be identified without injury to the latter."[42]

But Preyer's emphasis on visibility indicates what was still lacking. Palpating and listening went some way to fulfilling the physiologist's desire to observe the living fetus "in its natural environment," yet they also fell short. For that desire was not simply to read off information, but to be able to observe transparently, without distortion. An illumination apparatus designed by Preyer, the "embryoscope," testifies to this, even though it could only be used on an avian egg, which was anyway exempt "from any consideration for the mother animal."[43]

Preyer's comment introduces animal substitute objects as a next possible step on the path to suitable research material. Unlike that of human beings, the "natural environment" of fetal dogs, cats, rabbits, guinea pigs, and mice can be cut open — or, at least, physiology considered itself entitled to do so in the era of vivisection. Even here, though, Preyer faces problems. For one thing, mammalian fetuses cannot be obtained in the large numbers that would be required for systematic research because "in most experiments, the mother is sacrificed as well." Second, the vivisectional intervention deforms its object: since "the connection with the mother animal is no longer the same as before," the exposed mammalian fetus finds itself "in an abnormal situation." Finally, inferences drawn from the animal object regarding the human object are bound to be unreliable.[44]

Palpation, listening, or substitution by animals, the immature embryo, or the premature baby — regardless of the technique, the object always eludes capture. Much remains hidden from the hands and ears placed against the female belly; mothers and nurses restrict access, and the gaze that gains passage through the uterus wall in animals by means of light or incision becomes entangled in species distinctions. Every access and every object remains just one more approximation.

The same is true for work on the last substitute object named by Preyer, the "mature newborn." These are abundantly available in "large foundling homes and lying-in hospitals," but the labor room is not a laboratory. It lacks the "extensive apparatus" that the experimental physiologist would need in order to study such things as metabolism or neural responses. "The purely external inconvenience that just-born infants cannot be brought to the physiological institutes frequently enough, and that the institutes' instruments, in some cases difficult to transport, are not easy to bring to the lying-in hospitals, thwarts many an attempt to carry out methodical experiments on the newborn child."[45]

Something far more than an "external" circumstance highlights unavailability of a different kind from that dictated by the maternal body. If it is not the species (human versus animal) or the type of intervention (incision, gaze, hand) that is modified but the moment (unborn versus born), a new obstacle takes the place of opacity, and that is the passing of time.[46] Releasing the unborn from the body of the mother, birth also changes it in a way that removes it from the reach of fetal physiology's interrogation. It now repeats within itself the change of its environment, moving from water into air, from darkness into light, from quietness into noise, from warmth into chill.[47] This is why it is so important to catch the "just-born" in the newborn before it becomes truly "born." The physiologist's dream of amalgamating the spaces of physiological laboratory and labor room is nothing less than a fantasy of synchronizing the times of research and birth. In fact, the moment when the born child is physiologically still a fetus but organically already removed from the maternal body remains imaginary for research practice because it is just as fleeting as the maternal body is opaque.

In 1883, Preyer introduced his *Specielle Physiologie des Embryo* by painting a panoramic picture of research practice that, since the beginning of the century, had been unable to study the living human being except by dissecting the gravid animal; that turned the inside out but thereby separated what belongs together; that could feel and listen but could not see. The fact that fetal physiology had to work with objects that were not its object, and with techniques that offered no more than approximation, concerned Preyer as a problem of unavailability. That went beyond the impossibility of seeing the living fetus in its natural environment, for in the unborn, as everywhere, life happens as a function — and vital functions, unlike bodily functions, cannot be looked at, cannot be observed. Physiological observation always and necessarily tips into experimentation, and the experiment disrupts, sometimes even destroys, the organism.[48] Physiologists can test sensation by irritating the skin, identify oxygen requirements by waiting for rabbit fetuses to die under glass, detect transmission by injecting toxic substances into pregnant animals. But in the case of the unborn, any experimental disruption or destruction must encompass not only the fetal organism but the maternal one, and in the case of human beings, as Preyer laments, this means having to do without the most important material.

Due to this lack, research practice in fetal physiology was an ever-defective approximation, missing its actual object or changing it to the point of being unusable. Yet that failure was a productive one. True, the practice of fetal physiology could never approach the living human fetus directly, because the pregnant body encloses the fetus and because as soon as the unborn is disclosed, it ceases to be unborn. That very fact, however, confirmed again and again that the precondition of fetal existence is its relation to the maternal organism and to the time of development. In research practice, even (or especially) where it encountered insurmountable obstacles, the existence of the human being in the womb became fetal life, and thus something that could only be explained by the logic governing all that lives.

CHAPTER SIX

Prenatal Danger

Passage of Toxins, Pathological Unit

In the chapter of William T. Preyer's *Specielle Physiologie des Embryo* titled "Passage of Substances from the Blood of the Mother into the Embryo," stains including indigo, saffron, and potassium cyanide appear as experimental substances. The chapter also bristles with toxins and pathogens that "pass through" the placenta or, just as importantly, do not: "alkaloids of opium," "alcoholic drinks," the "variola toxin," "anthrax bacilli," the "syphilitic virus," "scarlatina toxin," "tubercular bacteria," "malaria toxin," and many more.[1] There are studies of whether "after chloroform inhalation by the woman in labor and after morphine injections, the fetus is or is not contaminated as well," or whether there is any prospect of immunizing the fetus in the womb by vaccinating the mother.[2] Questions arise around issues such as diffusion time, degrees of permeability specific to different infective substances, how a particular substance interacts with the quantity of oxygen circulating, and how its effect correlates with the organism's size.[3] Physiological experiments join forces with observations by medical practitioners, which in turn suggest "future investigations . . . using animal experiments."[4]

This pendulum movement between experiments with toxic and infectious substances and clinical observations forms part of the nineteenth-century habit of investigating the normal by means of the pathological. Describing that particular imbrication of physiology and medicine, François Jacob writes: "If knowledge of the physiological state was obviously necessary for the interpretation of pathological conditions, the study of pathological conditions also provided a precious instrument to study biological functions."[5]

Sure enough, Preyer's chapter on the passage of substances is located not in a section concerned with pathology, but in the discussion of nutrition. The reason Preyer addresses morphine, scarlatina, and syphilis there is that simultaneous contamination of the mother and the fetus may shed light on the workings of the placenta in normal physiological supply — placental function is a passage that now denotes not vascular permeability or tissue porosity, but a process of transmission. It is thus perfectly logical that the chapter is preceded by one on the "influence" of changes in maternal blood and circulation, and followed by one on the "crossing" of substances from the fetus to the mother. At stake is not only the fact that the maternal and fetal organisms are functionally connected, but also the exact nature of that connection. Why does the placenta transmit one substance and not another? How does that relate to the composition of the substance concerned, and what does it tell us about the links between the nutritive function and other functions?

Co-contamination, then, illuminated the normal physiological mediation between the maternal and the fetal organism; in terms of pathology, it explained how disease arises in the fetus. This was the basis of the notion of "influence" that obstetricians and physicians applied with such self-evidence throughout the nineteenth century. The term may seem unremarkable, but in fact it marked a momentous change. "Influence" articulated in a novel way the ancient notion that mother and unborn child form a single unit of disease.[6]

In the teachings of antiquity, the unborn was subject to dangers arising from malnutrition or illnesses that altered the fluids in the mother's body, from impacts or blows that jarred her abdomen, from a womb that was too broad or too narrow and left the generative matter too much or too little space.[7] In this view, the coming child's injuries stem from its participation in the maternal being in whose body it is enclosed. A different perspective emerges in a 1749 study by William Watson that is considered the first experimentally structured investigation on the placenta's permeability to pathogenic agents. The child before birth must, Watson writes, be regarded as an independent "Organization." Although it is nourished by the maternal fluids, when the mother falls ill with smallpox, it is affected differently and at a different point in time.[8] Again, mother and child form a pathological unit, and again the unborn partakes in the mother's experiences — but it suffers them in its very own way.

In 1827, Christoph Wilhelm Hufeland took this as grounds to call for a distinct "fetal pathology."[9] Even if the mother's blood vessels do not pass directly into those of the child, since the two are connected by the "intermediate body" of the placenta, it is nevertheless "always the blood of the mother that the child receives." What is essential is not only whether that blood is abundant, but also whether the child "receives pure or diseased and tainted blood from its mother." Hufeland's comments did not reflect the state of the art in fetal physiology as regards shared blood, but they were logical enough in inferring from the issue of co-contamination a "*pathogeny* and . . . *materia medica of the fetus*."[10] This may be read as a medical formulation of the organic individuality of fetal life, since Hufeland took it to mean that "for the physician," the human being "exists, lives, and makes claims on his attention and care already upon the first, invisible beginning of its generation" — and not only "when it has become a visible and audible member of human society" whom the "parish registers" were prepared to record.[11]

One year later, the pediatrician Charles Billard staked out the terrain of this fetal pathology. Without actually designating it as such, he insisted upon its independence as a field. Functions are interconnected in different ways in each different phase of life, Billard argues, and "the aberrations of those functions or the illnesses that result from some or other disruption occurring in the organization likewise vary according to . . . the different epochs in the life of one and the same being." He reiterates the point on the subsequent page: just as, in a state of health, "the ovum, the embryo, the fetus, the adult man . . . fulfill their functions in a particular manner," in a state of illness they "also present particular symptoms, the form and mode of which most assuredly differ according to those different phases of the organization."[12]

Each epoch of life thus has its own manner of being ill. It is not only upon birth, "as the philosophers have said," that the human being begins to encounter "the series of evils that afflict his species." The annals of art offer plentiful evidence "that the child has experienced ailments during its intrauterine life and only too often brings their cruel consequences with it as it is born." Introducing his clinical observations of newborn foundlings, Billard extracts diagnostic guidelines from the principle of a pathology divided according to the phases of life: it is only by knowing about the possibility of intrauterine disease that he is able to recognize "congenital maladies" as such.[13]

An equally definite formulation, though with less in the way of detail on fetal life, was offered by Xavier Bourgeois. In 1862, this passionate advocate of homeopathy published a treatise on "the influence of the pregnant woman's illnesses on the health and constitution of the child," discussing the transmissibility of tuberculosis, scrofula, syphilis, and various febrile diseases. Bourgeois writes: "The mother acts upon the fetus that is in her womb. It is nourished by her blood; it develops, forms its organs, grows by drawing from that blood all the nutritive elements it requires. Fastened to the mother by the most intimate of connections, it lives, so to speak, on the life of the mother, and the majority of morbid influences that cause havoc in the maternal organism have a faint or stronger resonance in the organism of the fetus."[14]

Underlying all these appeals for a form of medicine that—unlike the philosophers and the parish registers—takes account of what comes before birth was a particular notion of the unit of disease: not as a joint state of illness shared by the mother and the coming child, but as the transmission of illness from the maternal to the fetal organism. The pathological unit now involved a pathogenic influence of one upon the other, just as, physiologically, the pregnant woman and the unborn had become a functional coalition of two differently organized lives. This conception of the pathological unit, in other words, mirrored the new physiological conception of the unborn. Hufeland gave the organic individuality of the fetus a fetal pathology; Billard measured out a prebirth epoch of disease for fetal life; Bourgeois supplemented the milieu relationship with pathogenic influence.

By the last third of the nineteenth century, the pathological unit in this form had become established as a component of systematized knowledge on fetal life. In his encyclopedia article of 1878, Adolphe Pinard lists the "pathology of the fetus," alongside anatomy and physiology, as one more subfield of fetal knowledge and as a separate facet of pathology. The fetus in the uterus "may present the most varied pathological states," but even more importantly, "from an etiological point of view, one may say that during its intrauterine life, the human being is more at risk than it will be in any other epoch of its existence." Pinard draws up a catalog of hazards that constantly threaten the being even before its birth. They begin with the germ, which may carry "certain morbid affections" that will "immediately disrupt and hamper its development and in some cases destroy it," so that "the source of its life is also that of its death." Then there is the "maternal terrain," which may become sick and "furnish it with insufficient materials or even directly transmit sickness to it"; a "reservoir" in which it is contained and that may expel it before it can live "a proper life"; the "ephemeral organs" of placenta, umbilical cord, and membranes that "normally nourish it" but can also inflict malformation, mutilation, or even death; and injuries, for despite being "contained in the maternal organs," it is not always protected there.[15]

Pinard's fetal pathology makes good on hypotheses inherited from the early nineteenth century, but it also offers an initial taxonomy by distinguishing between "external" pathologies caused by contractions of the uterus, "internal" pathologies resulting from deficient nutrition or disease transmission, and malformations. As yet, Pinard concedes, this whole complex is no more than a sketch, since, the topic having been "sunk in the most complete obscurity until recent times," the "nosological framework" remains fragmentary.[16] What is already quite clear is why the fetus requires its own pathology. This is no longer only because its time constitutes a particular phase in a human life, but due to the particular features of that phase. Unlike disease in the child or the adult, disease in the fetus has "a special stamp, a characteristic physiognomy," because it suffuses a terrain "in the process of rapid evolution."[17] In short, fetal pathology is about impacts on development. These impacts mean that the unborn and the pregnant woman now suffer pathogeny in different ways.

Pinard set out a conceptual orientation for fetal pathology by bringing together the pathological unit, previously the concern of physicians, with errant development, the remit of teratology (the science of malformations) since the early nineteenth century. The figure who would put that conceptual link into empirical practice was the very Charles Féré who intervened in the debate over the *enfants du Siège* as paradigmatic cases of "disruption to development" in 1884.[18]

Experimenting with Development

In the 1883 proceedings of the Société de biologie, Féré had presented the case of a pregnant hysteria patient addicted to morphine, who had gradually reduced her drug consumption under his care. He observed that the fetus reacted to every reduction of the dose with violent motions and displayed withdrawal symptoms after birth.[19] The fact that Féré was already working on fetal pathologies may help to explain why, one year later, he paid such careful attention to Henri Legrand du Saulle's lecture on the *enfants du Siège*, immediately followed it up with his own description of the phenomenon, and began to puzzle over the momentous conditions that the drama of 1870/71 had imposed upon fetal life.

Féré, regarded by his contemporaries as an extraordinarily methodical researcher familiar with every chapter of the life sciences, approached this question from two different research perspectives.[20] On the one hand, he inserted it into his psychophysiological studies on perception and motion, to which I return in Part III of this book. On the other, it inspired him to move beyond the observations on fetal pathology he had already gathered in clinical practice and deepen them systematically by means of experimentation.[21] That is because when he thought about the *enfants du Siège*, Féré encountered a problem in attributing cause and effect, one that had already dogged his earlier observations.

Féré identified three aspects of this difficulty, which he describes in a retrospective essay. First, he explains there, it has become obvious that the same intoxicant or infectious cause can have different types of effect. For example, syphilis in the mother may result in a stillbirth or in the child's infertility or "debility." Conversely, the same effect can arise from different types of cause; for example, the effects of alcohol or lead poisoning are similar to those of syphilis. Third, experiences of physical or psychological shock may have the same effects as poisoning and infections, which they "copy" (*reproduire*) by triggering "disruptions to nutrition." In view of this complexity, it is important to discover how "a particular influence acts upon generation in line with the conditions proper to it."[22] And that can only be achieved through experimentation.

So it was that, in the 1890s, Féré set to work on hens' eggs. He shook, lacquered, and perforated them, varied their temperature and position, enveloped them in vapors of turpentine, aromatic essences, musk, mercury, phosphorus, and ammonia. He divided sets of several dozen eggs at a time into control and experimental groups, manipulating them in the incubator twenty-four or forty-eight hours and seventy-two or ninety-six hours after the beginning of incubation, and found embryos malformed, dead, alive, or normally developed. Féré called his procedure experimentation on "external influences," and described it as a continuation of research undertaken by Camille Dareste, who had visited him once a year in the laboratory and never left without giving some piece of good advice.[23]

Of Féré's contemporaries, Dareste was the greatest protagonist of the science of teratology, which since the early nineteenth century had been studying deformities as developmental anomalies. In the seventeenth and eighteenth centuries, two-headed or half-headed, many-limbed or shrunken-limbed "monsters" had been displayed in cabinets of curiosities as caprices of nature; now, difference in form was read as a deviation from normal development. Though caused by contingent circumstances, that deviation proceeded in just as regular a manner as formation itself.[24]

This conclusion had been reached not by gazing at monstrous exhibits in the cabinets, but by manipulating embryonal development. In Germany, the anatomist Johann Friedrich Meckel the Younger laid the foundations by proposing the idea of "impeded formation" (*Bildungshemmung*), while in France, Étienne Geoffroy Saint-Hilaire began to experiment systematically and in detail on "arrested development" (*arrêt de développement*).[25] After the mid-century, Dareste continued this research and gave it the name "experimental teratogeny," which Féré would later adopt to describe his experiments.[26]

Féré, however, showed "a little initiative," as he put it, and modified the experiments. He drew on Pasteur's innovation of using control groups in biological experimentation, making it possible to identify the "conditions" under which anomalies arise as opposed to merely producing as many monstrosities as possible—such was Féré's scathing account of his predecessors in teratology.[27] But Féré's sense of initiative took him beyond the use of control groups; his interest in preconditions expanded the experimental arrangement in another respect. Or rather, by expanding the experimental arrangement, he changed the question. In his second group of experiments, Féré moved from external to internal influences, introducing into the albumen substances "capable of modifying the embryo's nutrition."[28] Inspired by the cases of poisoning and infection he had witnessed in his everyday work as a physician, he applied toxins such as nicotine, strychnine, prussic acid, or alcohol and microbes such as the syphilis pathogen.[29]

The new experimental setup supplied Féré with a "certain number of general facts," to quote his summary article of 1899, which groups these into three aspects.[30] First, disruptions may provoke or suppress development; in the latter case, anomalies, delays, or a complete halt may result. Because the same influence, in the same experimental group, elicits both malformed embryos and embryos that are only rather puny, it must be concluded that the same cause has different effects and that something else determines the type of effect. Variation arises not from the *type* of influence, but—and this is the second aspect—from its intensity and duration; it also depends on the individual germ, because there are always some embryos that can resist anything. Third is something that Dareste suspected but had not been able to prove: the moment in the developmental process at which it occurs is what determines whether a disruption (*trouble*) will lead to a malformation or a pathology. "The same physical or chemical agent has a teratogenic effect when the embryo is still in the process of formation and a pathogenic effect later on. . . . The experiments showed that the same injurious influence may, depending on the period in which it takes effect, result in infertility, monstrosities, abortion, stillbirth, developmental delays, or congenital debility."[31] Féré accords different values to these findings. Duration and intensity remain somewhat elusive, but he believes that the embryo's individuality is an important discovery and regards the point in the developmental process as nothing less than a new rule of dependence on developmental status.[32] Indeed, it was only a few years later that the "critical point" in development would be discovered.[33]

Féré's insights were not the fruit of a simple accumulation of individual experiments. Importantly, he placed within a single experimental framework things that had previously been treated as categorially distinct, and this enabled him to create a crucial synthesis. On the one hand, his findings slotted teratological developmental anomalies into a general fetal pathology. If malformation is one effect among others, all of which may be caused by the same influence, then it is not itself the phenomenon that needs to be fathomed—the object of research must be the conditions under which a pernicious influence will act in one way or another. On the other, Féré now applied to congenital disorders the principle of developmental disruption that had been articulated in teratology: if such maladies, just like malformations, are a question of timing, then they too are developmental anomalies.[34] Féré explained the connection succinctly in a magazine article of 1894: "Like monstrosity, a morbid predisposition is the result of a disrupted evolution."[35] The experiments had given rise not only to a synthesis of teratology and pathology, but also to a new categorial unit. Congenital deformation and illness did not differ as effects of different causes; they were different effects of the same causes, which impacted differently at different points in time.

After Féré's death, his friends assembled an index of his extensive publications. The photograph they chose to illustrate it was an unpretentious portrait showing Féré in a lab coat, his arm resting on a table with a few open books (Figure 4).[36] This was not the genius inventor of a concept capable of synthesizing the medical gaze on the time before birth with traditional thinking on monstrosity, which had worked its way from divine will, to the whims of nature, to developmental anomalies. Féré contented himself with the conjunctive "teratogeny *and* general pathology," the title he chose for an overview published in 1899 in the prestigious venue of a Société de biologie anniversary volume.[37]

Modest though its presentation was, Féré's work was received with great enthusiasm by John William Ballantyne, a Scottish physician who, early in the twentieth century, collected everything that had been discovered over the previous hundred years regarding congenital anomalies, the passage of toxins, and malformations. The resulting two-volume work was programmatically titled *Manual of Antenatal Pathology and Hygiene.* In 1895, Ballantyne had already given Féré a prominent place in his short-lived *Teratologia: A Quarterly Journal of Antenatal Pathology*, but his manual of 1904 exalted the French physician as having succeeded in uniting teratology with fetal pathology.[38]

That synthesis offered Ballantyne the basis for a proposed "new department of medicine": antenatal pathology. He located the novelty

Figure 4. Charles Féré (1852–1907). The epigraph reads: "Every individual who works participates in social evolution, and he has the right to believe that he is working for something eternal."

of this field not in its research object, congenital anomalies and illnesses, but in its way of looking at them, its "point of view."[39] What Ballantyne meant was a rigorously scientific style, which he still felt obliged to distinguish from metaphysical interpretations, especially of malformation.[40] Most of all, Ballantyne saw antenatal pathology as a new way of approaching an old problem, since it combined into a single discipline three groups of phenomena — "monstrosities," "fœtal diseases," and "morbid predispositions" — that had hitherto been the object of separate investigations. Antenatal pathology, in sum, "is concerned with all the morbid processes which act upon the organism before birth, and with the effects which they produce by their action."[41]

This was the mark Féré had made on the writer who launched the adjective "antenatal" with such élan. Of course, Ballantyne did not invent the word; the terms "antenatal," "prenatal," and "pre-birth" (*vorgeburtlich*) had already been finding their way into physiological and medical writings for some time. Among others, Pinard, Ballantyne's much-admired contemporary, and Hufeland in the early nineteenth century had seen the advantages of these adjectives as alternatives to the entitive "fetal" and the spatial " intrauterine." As we saw in Chapter 4, Preyer used the terminology at the end of the nineteenth century to distinguish developmental from hereditary events. Ballantyne did the same — but more than that, he used "antenatal" with programmatic intent, to describe a branch of science whose raison d'être he derived from its capacity for "centralisation."[42]

At the beginning of the twentieth century, therefore, the term "prenatal" or "antenatal" was not simply a catchall, but a concept that was capable of integrating studies dispersed across numerous objects and research domains because it designated a perspective they all shared. By this point, birth had been constituted in a new way, as I have already shown: on the one hand, it was an event that drew the line between a before and an after, but on the other, it linked the before and the after within a continuum of development in which the former conditioned the latter. All Ballantyne had to do was to pick up this notion and subject it to an etiological turn. Birth, he noted, is "not a beginning," but marks "a stage in the life of the individual," connected to the subsequent stage by "projection" inasmuch as the effects of "morbid agents" on the organism *in utero* continue to impact on its "extrauterine life."[43] Within this stage, Ballantyne distinguished the maternal and paternal influence on the germ from the maternal influence on the embryo and fetus. Stages of development, and not causes or consequences, were what now organized the field of congenital diseases and deformations. For that, the label "antenatal" made obvious sense.

"When a word appears in the title of a book or paper," writes Georges Canguilhem, "it has been recognized as more than a mere metaphor by the competent scientific community,"[44] and after some delay, the concept of prenatality was finally adopted in the twentieth century in the titles of physiological texts. The book considered to be the successor to Preyer's *Specielle Physiologie des Embryo* appeared in 1940 with the subtitle *Origin and Extent of Function in Prenatal Life*; in 1946, a volume with the main title *Researches of Pre-natal Life* would follow.[45]

Pregnancy Accidents and Heredity

There is a reason why Féré did antenatal pathology yet did not invent either a name or a subject field for it. His interest in developmental troubles lay in a particular phenomenon upon which he oriented the interpretive order of all his experiments. Alongside the problem of ascribing causes to effects, he found himself faced with the further complication that all the "various agencies" imitating each other — infections, poisoning, physical and psychological shocks — themselves "copy" another process, namely, the effects of "hereditary degeneration in morbid families."[46] The observation raised the question of how developmental disruption relates to the inheritance of disease. It is within this larger frame of intergenerational pathogeny that Féré's experiments must be seen.

This was not the discovery of a new distinction. When the Société royale de médecine's competition called for work on hereditary diseases, their therapy, and their prevention in 1788 and 1790, many candidates responded by distinguishing the diseases that are the parents' legacy to their children from those caused by physiological or mechanical events during pregnancy.[47] Although the latter might, writes the physician Jacques Millot in a popular guide of 1801, be named "hereditary," this is imprecise, since they "belong to the fetus": its mother has never suffered them.[48]

At the end of the nineteenth century, thus, there was nothing novel in the distinction between a hereditary disease and a pathology acquired by the germ, embryo, or fetus due to contingent circumstances. Nevertheless, numerous questions remained unresolved, especially regarding the intergenerational unit of parent and child and thence the genealogical concatenation of units. For those interested in this wider context as opposed to individual patients, it was not easy to detect what was at work, when, and how. Féré believed that what he saw was a process he called "degeneration," in which illnesses seem to twine about a genealogical trellis, insinuate themselves into families, and burrow into lineages; to be passed on from parents to children and from grandparents to grandchildren; and to remain the same even as they emerge, decay, or change their shape. The complexity only increased as one sought the regularity assumed to be present in everything that lives. This problem was omnipresent in Féré's oeuvre.

In 1884, the same year that the *enfants du Siège* were discussed at the Salpêtrière, Féré published an acclaimed study on the dilemma in the neurological journal coedited by Jean-Martin Charcot.[49] The theme of the article, titled "La famille névropathique," was not a family with neurological problems, but the family of neurological diseases and their kinship with diseases such as arthritis, rheumatism, gout, or tuberculosis—"in a word," says Féré, "with all the degenerations."[50]

In the years that followed, he expanded the paper into a book, the first edition of which appeared in 1894 with the same title and an added subtitle, "Teratological Theory of Heredity, Morbid

Predisposition, and Degeneration."[51] Féré's revision, as he tells us in the introduction, took account not only of new publications on degeneration and heredity, but also of current research on questions around infective diseases. The latter he found particularly instructive regarding "the role of the terrain." It is there that a pathogenic factor begins to unfold its specific action when it encounters a "pathological predisposition" that is "handed down [*leguée*] to the individual by his forebears." Féré adjudges this predisposition in the terrain of the individual body to be nothing less than "the primordial cause — the cause of all causes" of disease.[52] The book-length revision was also informed by the first results of Féré's experiments in teratological pathology, whereas the original study had been based mainly on clinical observations. This shift is even clearer in the second edition of 1898, which offers rich insights into Féré's concern with the connection between developmental disruption and heredity.

Theorizing heredity, predisposition, and degeneration, as announced by the subtitle of Féré's *La famille névropathique*, was no small undertaking. Above all, it was one in which medical interests interlocked with questions in the life sciences. "The general laws that we have been able to infer from the observation of biological phenomena," Féré writes in his introduction, "can also be applied to pathology; and among them, the laws of heredity seem to be confirmed most frequently in the study of diseases."[53]

Indeed, the pathological had been crucial to research into the processes of heredity since the early nineteenth century, particularly in France.[54] It was medical professionals, and especially psychiatrists, who had revitalized an early medieval tradition and, drawing on the legal concept of inheritance, begun to speak of "hereditary" illnesses. Out of this initially metaphorical and adjectival usage, the noun *heredity* soon arose, oriented on biological processes. Whereas the earlier adjective *hereditary* had simply designated those illnesses that could be seen to pass from parents to children, the noun *heredity* described a force or mechanism by which traits are transmitted from generation to generation according to particular laws.[55] Underlining the originally medical context of this semantic transformation, around the same time the verb *degenerate* gave rise to the noun *degeneration*, which came into use in the early nineteenth century as a near-synonym of *heredity*.[56] Taken together, the two terms designated an arc of events that span generations and by means of which species—and families—form a connection enduring over time. The key question they helped to address was why some characteristics can remain the same in the continuity of procreation while others change, why some vanish and others emerge. How is it that both species and families remain self-identical across time even as they change?

The shift that gave rise to "heredity" and "degeneration" had its roots in just that inquiry. Previously, the question of what exactly connects forebears and descendants so that they share some things but not others had been posed in a very different way: How is it that children are similar or dissimilar to their parents? Well into the eighteenth century, the answer was to regard procreation, from conception to breastfeeding, as a creative act. The offspring was considered the work of the parents, the product of an always singular act of procreation under always singular circumstances, the exact nature of which produced similarity or dissimilarity.[57] Unprepossessing parents could generate a beautiful child in a particularly passionate embrace, or a clumsy act by otherwise dexterous parents could generate an eccentric human being. In the nineteenth century, this procreative action of two people became the propagation of a species; questions about the similarity or dissimilarity of parent and child became questions about the stability and variation of species characteristics, with heredity tending to describe the former.[58] The assumption that these processes must follow clear rules made them appear even harder to fathom. By the last third of the nineteenth century, a plethora of speculative notions of inheritance had arisen in response.[59]

This was the context in which Féré wrote *La famille névropathique*, and Féré himself related his work to a contemporary breakthrough. "The transmission of certain physical or moral characteristics from parents to children across a large number of generations," he notes, "has been well-known from time immemorial." Only recently, however, have theories been available that may claim "to explain the phenomena of heredity in their entirety."[60] Among these, one particularly worthy of attention is the "ingenious theory" presented by August Weismann in 1885, which seems "at first sight" to raise an "essential objection to morbid heredity." According to Weismann, writes Féré, the fertilized egg contains in its nucleus all the organism's characteristics, and through division it equips each cell with a complete sample (*échantillon*). In this way, a continuous line leads from the preceding to the subsequent generation—from the germ out of which the individual emerges to the contribution that it, in turn, makes to the generation of its offspring. If that is true, then the "personal influence of the generators on the quality of the germ plasm, which contains all the attributes of heredity, must be nil." The generators themselves would be nothing but temporary administrators of "this heritage"; the characteristics "*acquired*" by them could not be "*transmitted* through heredity."[61]

What was at stake in this theory? Among historians of science concerned with heredity, Weismann's 1885 study on "the continuity of the germ plasm" has been rated as pivotal.[62] Weismann succeeded in solving the puzzle posed by the shift of attention from the parent-child unit to transgenerational reproduction: How can a mature organism produce germ cells containing the same materials that it received from its own parents in the germ out of which it itself developed, but which it no longer *is*?[63] Weismann's response was to postulate a distinction between somatoplasm and germ plasm. The somatoplasm gives rise to the cells of the organism, while the germ plasm contains the hereditary substance, untouched by that process, which is made up of parental components and enters into a new composition through procreation. In this way, an uninterrupted line can pass through generations, the point that Weismann aspired to demonstrate.

Two conclusions could in principle be drawn from this explanation, and they formed the basis of Féré's detailed engagement with

Weismann's work. First, Weismann's distinction between somatoplasm and germ plasm presented the sharpest demarcation so far between development and transmission. What the organism receives from its forebears and will pass on to its descendants is separate from the body into which it develops.[64] Second, it surmounted the contradiction between stability and variation that had troubled the notion of heredity ever since heredity had come to denote something that is not only shared by parents and children, but also forges a chain between generations. Gazing into the cell nucleus, Weismann persuasively proposed a substance-like composite of parental components, the germ plasma, that could explain both stability (by way of transmission) and variation (by way of recombination) through the same modality.[65] Both these conclusions helped to pave the way for the genetic concept of heredity that would gain firmer contours in the early twentieth century. For the time being, though, they raised as many questions as they answered.

Féré did not query the continuity of germ substance as uncovered by Weismann, but he was not convinced that it explained everything. In particular, as a physician with an interest in physiology, he was not prepared to reduce the explanation of life to structures in cells. Féré could, he tells us laconically, easily dispense with the "details" of Weismann's theory "and how the intervention of fertilization complicates the problem."[66] More important was whether and how the germ's continuity could be reconciled with the observation that "under the persistent influence of the conditions of existence, there appear in living beings new aptitudes and characteristics that remain lastingly acquired across successive generations and the transmission of which can only be explained by hereditary influence (Darwin)." What Féré wanted to understand was the stabilization of variations that arise not from fertilization on the cell level, but from influences upon the organism. The pathological form of that process was Féré's concern in *La famille névropathique*: specifically, the "undeniable fact" that the "predisposition for nervous affections" is acquired through circumstances and perpetuated through hereditary transmission, just like the physiological characteristics that once arose "from an animal species' adaptation to a given milieu."[67] Féré now asked: If what Weismann says about the continuity of the germ plasm is true, how can a pathology burst into a family in such a way that it persists over generations?

Certainly, Weismann's germ continuity contradicted the inheritance of acquired characteristics as postulated by Jean-Baptiste Lamarck at the beginning of the nineteenth century and evoked by Féré's analogy between adaptation and the sudden appearance of disease. Reducing transmission to the germ line alone left no place for the inheritance of acquired characteristics; there was no means by which the experiences of the organism could change the structure of the germ. However, something different is at issue when, as Féré cites Weismann himself as saying, "surrounding influences" affect both the organism and the germ enclosed within it—"not only the generator, but the generator and his product at once."[68] This sufficed for Féré to sweep away the contradiction as a merely ostensible one, since the germ may receive through transmission a "sample" that remains unchanged for its whole lifetime, but that does not mean it cannot acquire pathological alterations through influence.

I have noted a Lamarckian resonance in Féré's comparison of pathogenic environmental influences on the germ with the emergence of new traits through an organism's adaptation to environmental conditions.[69] Féré himself, though, would probably have concurred with a later remark by Weismann himself: "That there are illnesses which are acquired and nevertheless 'inherited' is certain, but it has nothing to do with the LAMARCKian principle, because it concerns *infection of the germ* and not a particular alteration in the constitution of the germ itself."[70] As Rasmus Winther has shown, Weismann distinguished variations during the process of inheritance ascribable to changes in the organism from those ascribable to changes in the germ plasma. With growing conviction, he considered the latter type to be possible.[71]

It is no coincidence, then, that Féré's introduction to *La famille névropathique* does not mention Lamarck's name. Instead, his very first footnote leads us to an influential and encyclopedic study of heredity by the psychiatrist Prosper Lucas, *Traité philosophique et physiologique de l'hérédité naturelle* (Treatise on the philosophy and physiology of natural heredity) of 1847–50.[72] Lucas thought of generational transmission as a processual composite of two processes: new traits arise spontaneously through influence on the events of procreation (*innéité*) and become stabilized in the subsequent process of heredity (*hérédité*).[73]

It was this interplay that interested Féré, since when applied to pathology, it really amounted to the question of how degeneration is possible. Unsurprisingly, the next footnote in the introduction to *La famille névropathique* references Bénédict Augustin Morel's treatise on degeneration, *Traité de dégénérescence*, a no less influential synthesis of thinking on degeneration that had appeared soon after Lucas's *Traité*, in 1857. In Morel's concept of degeneration and Féré's reading of it, the interaction between innovation and stabilization is a couplet of influence and predisposition.[74] The parents' unfavorable life conditions produce a predisposition for disease in the child through influence on the germ, embryo, and fetus. Often a simple overexertion in one generation may set in motion a cascade of pathologies potentially lasting for many generations: a disease may develop in the morbidly predisposed offspring and, in turn, predispose their descendants to illnesses that are not identical to those of the second generation, but progressively more serious, and so on.[75] The linking element in this chain is not a particular disease that persists across generations, but the constantly repeated emergence of a predisposition in the child through parental disease in the intergenerational unit.

In other words, it is through influence that Féré here explains how pathology can break into a lineage and endure as degeneration: the morbid disposition can become stable or, more precisely, produce itself again and again.[76] This, in turn, accounts for the peculiarity that in a "degenerative" lineage, the pathology changes its shape, so that, for example, a nervous illness in the mother may be followed by a malformation in the child and a nutritional disease in the grandchild. Indeed, that is the scenario indicated in the title of *La famille névropathique*: the notion that the generational sequence of a human family may be overlaid by a different genealogical connection, the lineage of the maladies themselves. This seemed particularly true of "nervous diseases," the subject of the book. Such diseases, writes Féré, "are far from being passed on in the same form by heredity; rather, they succeed each other in accordance with their intimate kinship [*parenté intime*], by frequently changing within one and the same lineage."[77]

Like Lucas, then, Féré is interested in the cross-generational impact of influence on the events of procreation. But unlike Lucas, he does not call the element of stabilization in these events "heredity," or at least not heredity in the biological sense used by Weismann, which Féré defines as "true heredity" (*hérédité vraie*).[78] True heredity does exist in the case of pathologies, but there it describes how a particular phenomenon, such as restricted growth or lunacy, perpetuates itself through the changing generations.[79] It is "outside heredity," in contrast, that the "pathogeny of anomalies" takes place, arising either from the parents' disease (if degeneration is already underway) or else, originally, from toxic influences on the germ, embryo, or fetus, which Féré, focusing on the maternal organism, now calls "accidents of gestation."[80] Both cases result in "developmental defects" (*défauts de développement*), which produce disease or deformity in the child and predispose its descendants to the same problems.[81]

The experimental production of precisely such "defects" was the goal of Féré's experimentation, and summarizing this work, he stresses that the process by which they arise is one not of "transmission of acquired characteristics that have become hereditary," but "on the contrary, the dissolution of hereditary characteristics."[82]

But why was this clarification necessary in the first place? If the pathogeny of anomalies in the germ, embryo, or fetus is seen as the mechanism of degeneration, then it seems indeed to be a generational connection, just like heredity. In their different ways, both are processes of transmission and thus account for continuity — whether the continuity of traits or their progressive dissolution through disease. This goes a long way to explaining the polysemy of the term *heredity*, which can be noted quite generally for the nineteenth century and makes itself felt in Féré's work. At one point he calls an inheritance enclosed in the germ "biological" or "true heredity," at another he describes as "morbid heredity" the fact that the different diseases manifested by forebears and progeny are related genealogically.

This uncertainty was not due to a failure to distinguish between two things. The concept of heredity itself was ambivalent, because it described as a connection between generations something that had previously been the procreative action of parents. As a result, the emergence of pathologies in the unborn, which since the eighteenth century had been seen as a matter of milieu, was inserted into the same question of generational continuity that brought forth the phenomena of "heredity" and "degeneration" in the nineteenth century. This is why an instance of intoxication during pregnancy could also be regarded as a generational event.[83] And it is why Féré was interested in new findings on infective processes just as much as findings about heredity, while a physician like Bourgeois prefaced his treatise on the pathological unity of the pregnant woman and the unborn with an extensive section on Lucas's theory of inheritance.[84]

When, following Weismann, heredity alone became responsible for transmission, that did not put an end to the confusion but inaugurated a new arrangement of things.[85] Up to then, development and heredity had been at once independent and, within the overarching issue of transmission, interconnected. Féré's studies were firmly anchored in that arrangement, so he had no particular need for a novel designation for his combination of fetal pathology and teratology. From his perspective, experimenting on developmental defects meant expanding the scope of teratology by adding the questions of pathology that arose from his interest in degeneration. What Féré was looking for was new insight into a well-established complex of questions. Dareste, in particular, had already tied teratology to the issue of continuity and change in species characteristics. In the introduction to his 1876 summary, he recalls how, inspired by Darwin, he varied the environmental conditions of chick embryos in order to provoke not only serious anomalies, but also "slight anomalies, *varieties*," such as could also arise in viviparous species "through the intermediary of the maternal organism."[86]

Ballantyne, in contrast, marked a parting of the ways when he issued this appeal in his antenatal pathology somewhat later: "After many years, in which the seed has monopolised attention, a time has arrived in which our thoughts are directed to the soil."[87] Only with the crystallization of a concept of heredity that encompassed generational connection inside the germ did it become necessary to terminologically codify the issue of pathogenic influences on development as well. Thus, in the early twentieth century, even those who drew on Féré's work in France would transplant it into Ballantyne's conceptual framework. The pediatrician Émile Apert, in particular, and the obstetrician Henri Vignes, secretary to the French Eugenics Society, now spoke of the need to account for "prenatal influences" alongside heredity as such.[88]

Maternal Shock as an Emotional Toxin

The discussions on the *enfants du Siège* intensified Féré's interest in pathogeny before birth, but they did not initiate it. "For twenty-two years," he points out in *La famille névropathique* of 1898, "I have been researching facts suited to showing that developmental defects are the most efficacious individual conditions for morbid predisposition."[89] From 1884 onward, however, the children of the year of terror became a regular reference in his work.

One reason for this was that they presented an exemplary case of pregnancy accidents, sharpening the distinction from "true" heredity in that even if they were "completely comparable with the type one observes in the children of the mentally ill," as Legrand du Saulle noted, inheritance could be ruled out as a source of their anomalies because their siblings showed no similar problems.[90] A further reason was that the *enfants du Siège* embodied the heterogeneity of factors that Féré was seeking to unify in his experiments when he modeled every form of pathogeny before birth as a nutritive disruption to development. In these children, two directly physiological causes had been identified, malnutrition as a deficit of nutritive substances and alcoholism as the toxin par excellence.[91] All the more interesting to Féré was the factor of maternal trauma as psychical influence. He turns to it sooner or later in all his important publications on degeneration and developmental pathogeny, where, constantly changing his terminology, he writes sometimes of "emotional drunkenness," sometimes simply of "the passions, the emotions."[92] If it could be plausibly shown, Féré reasoned, that something felt could exert a pathogenic effect on development, this would be the extreme case to confirm the postulate of nutritive disruption, though such disruption would then have to come about indirectly.[93]

Féré was not alone in his fascination with mental factors. For the whole of the nineteenth century, and crosscutting scientific domains and fields of medical practice, the psychological component was firmly entrenched in speculations about dangerous influences during pregnancy. In his 1862 book on that subject, the physician Bourgeois discusses nonsomatic factors in a chapter titled "On the Influence of the Passions, the Mental Emotions, the Imagination."[94] In Karl Friedrich Burdach's handbook of physiology, published in 1828, a passage on the "influence of the *mother's life* on the development of the embryo" jumps from the dangers posed to the unborn by infectious diseases to, just a few lines later, endangerment by the pregnant women's anxiety during "wartime unrest."[95] In 1827, the physician Hufeland counted among the "illnesses and dangers to which the child is exposed in its mother's womb" the possibility that "in the case of neuropathic and hysteric mothers, the child may already experience convulsions in the womb." He offers an example: "A mother who had spent the second half of her pregnancy in constant distress and anxiety during wartime gave birth to a child that suffered seizures immediately, from birth onward, and on the ninth day died in convulsions." Indeed, "the whole nervous constitution of the day, at least among the higher ranks," can be explained by the fact that "this disposition is already imparted to the children in the womb."[96]

Biologists and heredity theorists, too, delved into the question. In his treatise *Hérédité psychologique* of 1873 (translated into English as *Heredity* in 1875), the philosopher and psychologist Théodule Ribot, a colleague of Charcot's and Féré's, discusses not only the inheritance of mental qualities such as instincts, passions, or imagination, but also, as a nonhereditary phenomenon, the influence of temporary mental states experienced by the parents between the moment of conception and birth. It can hardly be doubted, opines Ribot, that certain mental dispositions of the child depend on the "actual and momentary state" of the parents "at the instant of conception."[97] Just as intoxicated parents engender epileptic or "insane" children, so also even more "transitory states which exist at the moment of conception," such as passions or mental affections, "may exert a decisive influence on the nature of the being procreated."[98] And in 1895, the influential biologist Yves Delage considered it at least unrefuted that "transitory states," among which he includes a "sickly disposition" and "drunkenness" but also "very violent feelings," have consequences for the child if they occur during conception and gestation.[99]

Vague as many such suppositions remained, Féré actually had a rather precise view of the possible mechanism by which the effects could arise. "A certain number of facts," he writes in *La famille névropathique*, "indicate that it is through the mediation of the nervous system that physical and mental shocks are able to act upon the embryo and fetus." Experiences of shock are, after all, nothing other than agitations of the nerves, and agitations of the nerves can disrupt nutrition because nutrition "depends" on the nervous system. Féré concedes that, even after the research he describes, "experimentation has not yet managed to confirm the reality of that influence."[100] Yet it was probably Féré's insistence, especially in his accounts of those experiments, on the possibility of mental influence that prompted the remarkably ambivalent evaluation of his studies by the embryologist and historian of science Joseph Needham in 1931. Needham calls them unique and interesting, but also flagrantly unsystematic and tinged with magical thinking.[101]

It is true that, as an author, Féré does not make life easy for his readers. He constantly links everything to everything else, starting from scratch each time, and rarely remains faithful to a single term across more than one study. Needham's suspicion had a different source, however. Féré, he says elsewhere, tried to "give a scientific foundation" to the "popular belief" that thunder and lightning could kill the embryos of fowls.[102] In fact, that was not so much a popular belief as a caricature. Needham was reducing Féré's experiments on hens' eggs to a hypothesis of psychical influence in which he saw nothing more than the folkloristic remnants of a past, more superstitious age. He could hardly have been more derisive; after all, the embryo of a barnyard fowl has the least to fear in this respect, being enclosed in an eggshell and not in a mother. At the same time, his jibes were not entirely without foundation. Ultimately, Féré's notion of psychical influence did evoke a very long-standing idea that he was deeply concerned to prove correct—even if not necessarily for the case of poultry.

PART THREE

Inner Life

The imagination is a function without an organ.

— Georges Canguilhem, 1980[1]

And as for the infant! It seems quite unfathomable that the family members are the only ones whose attention is captivated by the gradual unfolding of the infant's senses, its will, its mind, its passions, its virtues.

— William T. Preyer, 1880[2]

In Charles Féré's conjecture that the female mind could influence the unborn organism, we find the echo of a theorem that, in numerous variations, goes back to ancient sources. It was discussed with particular enthusiasm in the seventeenth and eighteenth centuries as the doctrine of parental, but especially maternal, "impression" or "imagination," in German as *Versehen* (literally mis-looking; seeing with a shaping or negative effect) or *Einbildungskraft* (the force of imagination). In the early modern period, this meant the view that when combined with an intense emotional sensation, visual impressions on the conceiving and pregnant woman — and these included imaginary impressions, images in the mind's eye, such as a longing for something — leave their stamp on the form of the unborn and cause "monstrous" dissimilarity in the shape of birthmarks and deformities.

This position was described succinctly by the sexologist Iwan Bloch (under the pseudonym Gerhard von Welsenburg) in an 1899 historical overview of *Versehen* in this sense: "If a potent psychical impression affects a pregnant woman to such an extent that the *cause of that impression* becomes visible in the body of the child, then the woman has 'mis-looked.'"[3] For the temporally delimited instant of procreation, the same force was also attributed to the father's mind. That version of the trope invoked the ancient emphasis on conception as the crucial moment, a notion that lost some of its currency in the modern era but did not completely disappear.[4]

Féré points out the longevity of this traditional idea of mental influence and ascribes it to an intuition external to science: "The influence of the parents' mental state at the moment of conception on that of their children captivated attention before medical men began to attend to it."[5] From the rich treasury of anecdotal examples, he chooses not cases that claimed to explain physical imprints, like the ones that had featured so prominently in the seventeenth- and eighteenth-century theory of impression, but stories concerning character traits. The Greek poet Hesiod, he tells us, advised against conception after attending a funeral, since the resulting child would be melancholy in nature. Erasmus of Rotterdam explained his *folie* by the fact that he "was not the fruit of some dull conjugal love"; Laurence Sterne's Tristram Shandy attributed the untoward aspects of his character to a question posed by his mother at a highly inopportune moment. And the personality of one of Louis XIV's illegitimate children remained marked all his life by the emotional crisis that had troubled his mother, Madame de Montespan, during his conception.[6]

Féré's choice of examples, from literature and the reservoir of popular belief, and his interpretations, oriented on character traits, are helpful in historically situating his postulate of psychical influence. Anecdotal exempla "from the oral and book tradition," as Jean-Louis Fischer puts it, had always been adduced to prove the power of the parents' mental state, but they now had less and less of a role to play in underwriting the truth of the impression doctrine.[7] That teaching had experienced a boom in the sixteenth and seventeenth centuries, primarily as an explanation of deformities, and in the eighteenth century it was reinforced by the new value placed on imagination as a creative faculty of the subject. At the very same time, it also came under fire in a vehement dispute that has been called the "quarrel of imaginationism."[8] By the start of the nineteenth century, it was virtually impossible to avoid taking a stand on the issue. According to John William Ballantyne's manual of antenatal pathology, almost everyone concerned with midwifery, surgery, childhood disease, or dermatological complaints at the time had been forced to declare themselves either a believer, a skeptic, or neutral.[9]

In the course of the dispute, across the eighteenth and especially the nineteenth century, the idea of maternal impression acquired scientific respectability in a new shape: as the causal connection between the mother's feeling and the child's anomaly, whether physical or psychological. It was in this form that, in the early twentieth century, the old power of the maternal mind would find its place in prenatal pathology, which counted the mother's mental state as one factor of developmental disruption among others. This was a period when historical surveys of the theme proliferated, tackling the body of tradition by sifting out what was fantastical and securing those elements that could be reconciled with the science of the day.[10]

Féré did not go into such historical detail—he simply copied his examples from Prosper Lucas's *Traité de l'hérédité naturelle* (Treatise on natural heredity).[11] What interested him was how to *prove* psychical influence. The teratological and pathological experiments on developmental disorders that I discussed in Chapter 6 contributed to that end by revealing the manner by which pathogeny before birth occurs. But how could experiments detect whether the mother's mind was a causal factor in such events?

To answer this, Féré first needed to determine whether mental processes in the pregnant woman can make themselves felt in the womb at all. That inquiry belonged to a research field that interested Féré almost as much as the field of degeneration: psychophysiology. Turning to the unborn in psychophysiology, Féré found himself facing an additional set of problems—for a psychophysiological perspective on the unborn raised no less a question than that of mental life before birth. As I showed in Chapter 4, that issue had haunted the physiology of fetal life from the outset. During the nineteenth century, it was continually articulated and repeatedly subjected to experimentation. This research praxis configured maternal impression not only as influence, but also as communication between two subjects.

The chapters in this section of the book take the history of the imaginationist doctrine into the nineteenth century, but not in order to trace how it was passed down in medical instructional literature, writings on hygiene, "folk medicine," or guidelines on how to "make beautiful children." Revivals of the doctrine in concepts of magnetism and hypnotism, too, are treated only marginally, and I leave aside entirely the theme of maternal impression in literature. Nor do I discuss how the creative power of imagination found its way into aesthetics and gynecology.[12] Instead, I aim to uncover the historical ties that bound the "old" theory of maternal imagination to the "new" pathological and psychological interest in the prenatal. Several scholars have already indicated that such a connection existed.[13] Here, I argue that what instituted the link at the core of science was the fabrication of the maternal-fetal relationship in the practices of research.

CHAPTER SEVEN

How the Pregnant Woman Feels

Contesting Maternal Imagination

In the seventeenth and the eighteenth centuries, the power of the mother's mind was part of a psychosomatic complex that included the unborn in the body of the pregnant woman. Put schematically, and setting aside occasional variants, the logic ran as follows: the mother's sensory impressions, when accompanied by intense feelings, simultaneously direct the vital spirits in her own body and the child's to the point where the impression occurs; if this elicits a sensation (such as pain) in the woman's body, it leaves an imprint in the unfinished body of the unborn like an imprint in soft wax.[1] The challenge was to explain, whether mechanically or physiologically, exactly how the mother's agitation continues in the child's body. René Descartes suggested that images received by the woman are transformed into physical signals, but much more common was the idea that the mother's blood or nerve fluid within the child moves to the places that she has in mind.[2]

This premise both proved and presupposed a bodily unity between the pregnant woman and the unborn. Nicolas Malebranche had asserted just such a unity in 1688 and was regularly quoted by eighteenth-century proponents of the doctrine of maternal imagination. So intimate is the "union" of mother and child, writes Malebranche, that mothers are surely "capable of imprinting in their unborn children all the same sensations by which they themselves are affected, and all the same passions by which they are agitated."[3] Malebranche is also the source of two anecdotes that frequently surface in writings on the subject: the story of a young Parisian who, at birth, had fractured bones as if broken on the wheel because his mother had attended an execution during her pregnancy, and that of a devout woman who would address her prayers to a portrait of St. Pius and subsequently gave birth to a child in his image.[4]

These vicissitudes of the unborn could be explained by the fundamental assumption that it formed a corporeal unit with its mother. *Ex negativo*, they also demonstrated the regularity of the course of things. This is because the mother-child unit was itself embedded in a principle of premodern natural philosophy that explained the continuity between parents and children as a production of similarity. Without the "communication between the brain of the mother and that of her child" that is entailed by their union, Malebranche concludes from his discussion, it would be impossible to understand "why a mare does not give birth to a calf, or a chicken lay an egg containing a partridge or some bird of a new species."[5]

It was this premise of an organic connection between the pregnant woman and the unborn that prompted the first salvos in the eighteenth-century attack on the theory of maternal imagination. In 1727, physician James Blondel denied that there was a bloodstream, shared between mother and child, through which impressions and shocks can pass from one to the other—quite apart from the fact that he did not see how a fetus in the mother's womb, being still just a fetus, could sense in the same way as she can.[6] In the second half of the century, Albrecht von Haller regarded the error of Malebranche's assertions as an open secret, something he merely needed to point out. A "constant harmony" between the body of the mother and that of the fetus had been proffered as a way of explaining the power of the female imagination, "yet it is very easy to show that none of this can take place, and that there is no possible route by which the affections of the maternal soul can pass to the fetus."[7] After all, no nerve runs from mother to child, and even if, as von Haller believed, blood vessels connect the two, this still does not explain the transfer of impressions.[8]

When they framed their argument in terms of nerve activity and the bloodstream, these authors were not simply deploying well-established physiological knowledge on the relationship between the expectant mother and the unborn; the controversy around maternal imagination was itself what drove them to investigate that relationship. Anyone contesting the power of maternal imagination in the eighteenth century was obliged to tackle nerves and blood vessels and enter upon the question of what paths lead where, what runs how and reaches or does not reach whom, where there is continuity and where there is interruption. When the St. Petersburg Academy of Sciences announced a competition on maternal imagination, the submission of the Göttingen-based obstetrician Johann Georg Roederer was an experimental study on vascular connections.[9]

Critique of the doctrine of imagination and the beginning of a physiology of fetal life were therefore closely intertwined, each confirming the other. The question of whether pregnant woman and fetus share emotionally saturated impressions turned into the question of whether the two are really connected by the bloodstream or nerve fluid. And the more the organic continuum of mother and child became a relationship of two organisms, the more fantastic the traditional belief began to appear. Conversely: the more fantastic the belief appeared, the more probable it became that the union of mother and fetus would have to be reconceived as a connection between two individual organisms.

On the other hand, the physiological criticism of maternal imagination extracted a different form of maternal efficacy from all the things that could *not* be demonstrated. It seemed perfectly plausible that the emotional states of the pregnant woman "influence" the embryonic or fetal organism. At first, as Anke Bennholdt-Thomsen and Alfredo Guzzoni remark, that idea was indeterminable,[10] and accordingly it may at first sight seem to be no more than a chance by-product of the debate. Yet that was far from being the case. The notion of influence unfolded alongside the new view of the relation between the pregnant woman and the unborn, as presented in Part II of this book, which offered it a firmer basis—especially as the maternal-fetal relationship became ever more clearly delineated as one of mediation between two organisms and influence was seen ever more clearly as a disruption to development. However indeterminate the concept of influence may appear today, it was the vehicle of a profound shift in cause and effect: from impression to feeling on the part of the mother and from imprint to pathogeny on the part of the child.

Il faut distinguer

The doctrine of maternal imagination was not transformed into a hypothesis of psychical influence in one fell swoop. A first important step can be found within the doctrine itself, when, in the course of the eighteenth century, attention moved from visual impressions coming from outside to the internal products of the woman's imagination.[11] This internalization was compatible with a premise that even the doctrine's opponents were prepared to accept: that, as Blondel argued in 1727, it was probably not only bodily agitations and uterine spasms that injured the child in the womb, but also fright, anger, distress, and anxiety. These could trigger miscarriage, while unrequited longing could eat away at the child's vital force.[12]

This claim was taken up at the turn of the eighteenth to the nineteenth century by writers who wished to bring some clarity into the dispute around maternal imagination. One of them was the highly productive and well-read physician Jean-Baptiste Demangeon, whose writings exemplify the shift inward. In 1807, Demangeon entered the debate on the doctrine of imagination with the treatise *Considérations physiologiques sur le pouvoir de l'imagination maternelle* (Physiological considerations on the power of the maternal imagination). He did so with a certain note of surprise. As he would explain in a later edition, his interest was sparked by a discussion held in 1806 at the Société de médecine de Paris, when he had been taken aback to find that there was no consensus on the "indirect, mediated influence" of the maternal imagination. Instead, certain colleagues had still postulated a "direct influence" in the sense of a "print modeled by the imagination on a particular object." He began to delve into the topic and wrote a study on it that quickly sold five hundred copies.[13]

In that 1807 text, Demangeon first of all explains that he is not concerned simply to repudiate the doctrine of maternal imagination; his task is to build a "different explanation."[14] To that end, the objective of the investigation must be specified—since there is, Demangeon argues, an "immense" gap between making the imagination "the principal cause of natural deformations" and according it no relevance at all. That space of investigation, the role of "affects in the reproduction of living beings," is far from having been conclusively described.[15] A few pages later, Demangeon explains the point by means of a shrewd provocation. The ancients' belief that looking at beautiful statues creates beautiful children, he writes, is not as absurd as the opponents of maternal imagination make out. However, it is not the sight of well-formed marble that causes the imprint in the child; the child's beauty in fact results from the serenity and happiness experienced by the pregnant woman when she looks at beautiful things. The solution to the puzzle, therefore, lies not in the spectacle but in the feeling. Feeling, Demangeon continues, is capable of affecting the quality of the maternal blood, which, in turn, modifies the unborn's nutrition.[16]

Not wishing to throw the baby out with the bathwater, Demangeon objects to the "inconsistency" of those who completely reject the relevance of the pregnant woman's mental state yet in the same breath lament the corruption of infants by wet-nursing. How can an endangerment of the child through nursing be feared but an endangerment during pregnancy denied? The embryo or fetus is far more "intimately" bound to the mother than the nursling is bound to the breast, and it is also weaker and more delicate, thus more "susceptible to troubles."[17] In this latter point, Demangeon considers himself more hard-hitting than the great von Haller, who had attacked the doctrine of maternal imagination by adducing the neurological discontinuity of mother and child (von Haller is probably the target of Demangeon's polemical accusation of inconsistency). He concludes his discussion by noting: "It seems to me beyond doubt that the maternal imagination is capable of disrupting the form or the original conformation because of the disruption it can cause to circulation and nutrition; but I see nothing that would allow us to presume it is capable of imprinting the form of a particular object."[18]

Starting from the deficits of imaginationism, then, Demangeon arrived at a different notion of the efficacy of the mother's mind by interpreting the correlation between seeing a beautiful statue and generating a beautiful child as an indirect one. The effect arose not from looking at an object, but from the feeling that was produced by looking. A different causality came into play: in the theory of maternal impression, feelings were the medium through which the sight of an object unfolded its formative power; now, feelings themselves were the agent or cause of the child's form. The imprinting of an object was replaced by the influence of a feeling.

At the heart of this shift in argumentation stood the physiology of the maternal-fetal relationship. In the eighteenth century, assaults on the doctrine of maternal imagination had cast doubt on an organic unity between mother and child instituted by the bloodstream or nerve fibers; in the early nineteenth century, the relation between a maternal and a fetal organism made it plausible to postulate the influence of feelings. At stake was no longer which vessels and fibers are continuous or discontinuous, but the existence of a relationship between two individual organisms. This is why Demangeon's line of argument could also draw on situations in which the anatomical connection no longer exists, but a physiological connection still does: the infant that has been born and suckles at the breast.

The turn from maternal imagination to emotional influence became canonical in the early nineteenth century in exactly the form articulated by Demangeon—as the imperative to find a new explanation. Antoine Jourdan, author of the relevant entry in

Charles-Louis-Fleury Panckoucke's authoritative medical dictionary, published in 1818, enjoins his readers: "So you do not believe at all, it will be said, in any influence of the imagination to deform the fetus? One must distinguish" — *il faut distinguer.* Belief in the imitative power of the maternal imagination can safely be dismissed, Jourdan writes, but account must be taken of the "very powerful" and "very real" role of other products of the imagination. He spreads before us a whole panorama of scenarios. Severe fright can cause miscarriage, less intense fears can upset the maternal blood and fluids that supply the embryo, and agitation or strong feelings can provoke uterine spasms that impair the development of limbs and organs. The "storms of passion," Jourdan concludes, in principle pose just the same danger as "physical shocks."[19]

Both before and after this codification, all sorts of speculation proliferated in obstetrical and medical writings when it came to the details of the process. One particularly fine-grained account is found in a book of instructions by the physician Jacques Millot, published in 1809. In Millot's view, all passions are expressed in the contraction and relaxation of nerve and muscle fibers. When the stomach of an agitated pregnant woman contracts, this decelerates her circulation and secretions, so that the embryo's nourishment is blocked or at least delayed. The embryo's vital processes now slow down, disrupting its "primordial constitution and organization" to such an extent that development and growth are hampered. Just as in the case of an inappropriate diet, the problem is exacerbated by changes in the composition of the "lymphatic-milky substance" that the placenta transports to the embryo.[20]

In a detailed and wide-ranging discussion that appeared in 1827, Christoph Wilhelm Hufeland also turns to the workings of emotional influence — motivated not least, he notes a little sarcastically, by the conviction that the most dangerous thing is not maternal impression, but the fear that this superstition arouses in women.[21] Hufeland puts forward almost as many possible explanations as there are manifestations of feeling. Alongside the widespread notion that affects are "imparted" to the fetus through nourishment, for example, he also mentions an "outflow and overflow of nervous force beyond its visible conductors."[22]

In the first half of the nineteenth century, the turn to feeling became omnipresent not only in medical writings, but also in physiology. François Magendie's argument may lack terminological clarity, but his intention is clear when he asserts that the "imagination" is capable of "influencing" the fetus because sudden fright, intense sorrow, and immoderate joy can affect placental supply.[23] Johannes Müller marks the turn by adding a modifier; he postulates a "*reasonable* theory of maternal impression." This holds "that every intense, passionate state of the mother can have an equally sudden influence on the organic interaction [*Wechselwirkung*] between mother and child, and as a result also bring about a retardation of development or a halt to formation at particular stages of metamorphosis, but without the image in the mother's mind being able to influence the place where such obstructions arise."[24]

It is no coincidence that Müller precedes this discussion with his portrait of the maternal-fetal relationship as "the most intimate possible juxtaposition of two beings that are in and of themselves quite independent."[25] Theodor Bischoff likewise takes considerable pains with his rhetoric. He sets up two questions: "First: Can the affects of the mother have an influence on the development of the new organism? Second: Can affects of the mother occasioned by a particular object alter the formation of the new organism in such a way that it becomes the same as, or similar to, that object?"

The second implication is erroneous, writes Bischoff, but the first is not, given that "in many cases it has really been true, and still occurs, that a violent fright or excitement of the mother has occasioned a deformation, but without the shape of that deformation corresponding to the object of the fright." Neither shared blood vessels nor connecting nerves are necessary for this to take place: "The mixture of blood and the supply of maternal blood, the behavior of the uterus, dependent on the mother's outlook on life and mood more generally, can and must affect the ovum and the fetus without any such direct connection."[26]

It is no surprise that fetal physiologists also investigated the turn from the direct to the indirect effect of the mother's mental state, since their work on maternal-fetal communication (presented in Chapter 5) had given them a basis to do just that. More remarkable is the insistent frequency with which the authors of physiological writings up to the mid-nineteenth century invoked the doctrine of maternal imagination — even using the same terminology, as in Müller's "reasonable theory of maternal impression" or Magendie's reference to influence by "imagination." This reflects an argumentative reciprocity between the physiology of the maternal-fetal relationship and pathogenesis through maternal feeling. If feelings really can translate into metabolism and contractions in the pregnant woman, thereby affecting the unborn, this would demonstrate two things at once: that the psychical is physiological and that the maternal organism acts upon the fetal organism. At the same time, the constant reiteration of the postulate of emotional influence reveals just how unstable it still was.

Tellingly, attempts were still being made to rehabilitate *Versehen* in the classical sense as a mechanism of imprinting. In a publication of 1820, Everard Home asserted that he had after all found nerves connecting the mother and the fetus, and accordingly refuted the claim that the theory of maternal imagination is incorrect.[27] The ancient idea of immaterial action was also brought back to the table. In the same year as Home's paper, Carl Gustav Carus, fascinated by the observation "that the fruit lives only in and by the mother," returned to von Haller, only to object pointedly that his denial of "the existence of some nervules" was unhelpful.[28] The mother's mental images can, Carus argues, be transferred to the fetus "deeply asleep in the uterus" in the same manner that "one thinks of the relationship between the somnambulist and the magnetizer."[29] This magnetist reincarnation of imaginationism also made an appearance in Burdach's handbook of physiology.[30]

In physiological writings, and physiologically oriented medical writings, on pathogenesis before birth, such endeavors remained marginal. Bischoff, for example, accused those who "bring magnetism etc. into play" of "trying to explain an obscure matter by means of something even more obscure." Nevertheless, even he was unwilling to "absolutely" deny *Versehen* — partly out of courtesy, since "reliable men" such as Carus still reported incidents that supported the theory.[31] More importantly, "mere denial would cut off a source of research questions that is at least still open, one from which I would certainly expect benefit, all the more so because I strongly believe *Versehen* itself to be erroneous."[32]

In short, although direct physiological influence, and especially immaterial influence, by maternal feeling had become quite implausible — indeed, obscurer than obscure — the notion of maternal imagination itself did not seem to have been conclusively refuted. In turn, emotional influence was plausible, but not demonstrated. The postulate was thus doubly unstable, but when that instability became a springboard for research, it also opened a route toward certainty. In this sense, the highly iterative and gradual winnowing of emotional influence out of maternal imagination during the first half of the nineteenth century was also a call to arms for a new research program.

Crucial to that agenda was the "spirit of observation" that, as Bischoff writes, "makes the trials and observations of physicists and chemists so much more reliable than those of the physiologists" and, in turn, makes those of the physiologists "more reliable than those of the pathologists." What he means is experimentation.[33]

One reason for Bischoff's faith in experimentation's potential was the track record of experimental studies in the first half of the nineteenth century. They had supplied ammunition for the hypothesis of emotional influence — concerning the functions of the placenta, of course, but also the origin of deformities. Teratologists had performed countless experiments on hen's eggs to conclude that congenital anomalies arise as defects in development, enabling an influence on the fetus to be more precisely defined as a developmental impact.

This becomes obvious in the discussions by Bischoff and Müller, who both responded to the teratological postulate of developmental retardation. It is why Müller spoke of inhibited formation and halted metamorphosis, and why Bischoff wove his highly systematic discussion of *Versehen* so tightly into everything that was known about "the incontrovertible fact, brightly illuminated by the theory of malformations, that these can be explained in large part by an inhibition, arrest at a particular stage of development."[34] And when teratologists investigated that incontrovertible fact, they never lost sight of how the pregnant woman feels.

A Natural Experiment on Mental Pain

Before the early nineteenth century, the explanation of monstrosities had hovered between the foibles of nature and the mysteries of divine will, but the theorem of developmental inhibition enabled teratology to insert the monstrous into the lawfulness of all life.[35] Developmental inhibition means that adhesions between the embryo and its surrounding membranes can delay or halt development at that location, so that a transitory form of the embryo becomes permanent and the child's limbs or other body parts are missing or deformed. Malformation, in this view, is a deviation that proceeds according to laws, just as normal development does.

The anatomist Johann Friedrich Meckel the Younger is regarded as the father of the idea. At the beginning of the century, he spoke of *Bildungshemmung*, inhibited or arrested formation, though still cleaving to the notion of an originally malformed germ. In the 1820s, naturalist Étienne Geoffroy Saint-Hilaire subjected the theorem to experiment by provoking deformities in chicken embryos. He disproved the aspect of predetermination and proposed the term "delayed development" (*retardement de développement*) to describe what he considered to be a fundamentally contingent phenomenon.[36]

It is more than a quirk of history that in the years when he was manipulating hens' eggs, Geoffroy Saint-Hilaire also took an interest in the psychological state of unmarried pregnant women. His engagement with "retarded development," he explains, confronted him with the "popular belief" in the "influence of glances on the development of an embryo," which he very soon turned to use in a theory of feeling.[37]

In his anatomical treatise *Philosophie anatomique* of 1822, Geoffroy Saint-Hilaire paints an alarming portrait of pregnancy outside marriage: "What torments of the spirit, what remorse, and as a result what alterations in all the organic processes in a young girl who has been seduced!" With this picture before his eyes, he began to pore over the statistical reports of the city of Paris for 1817. Of the 23,759 births in the city that year, 9,047 were to unmarried mothers, "more than a quarter of whom, but at least two to three thousand" had given birth for the first time. The majority "had no doubt ceaselessly recalled to mind the lamentable circumstances of their seduction," and thus "remained for the long days of their pregnancy oppressed by the most painful of feelings." In a further step, Geoffroy Saint-Hilaire juxtaposed the size of this group with the vanishingly small number of malformed babies born that year. The implication of his natural experiment, the data for which were supplied by life's vicissitudes, seemed entirely obvious: "A profound sorrow is by no means a predisposing cause of monstrosity."[38]

This statistical jaunt is not Geoffroy Saint-Hilaire's last word on the matter. He returns to it a few pages later, in a discussion of delayed development. There, he explains exactly how we should envision maternal influence and draws a further conclusion from the lack of correlation between single motherhood and deformity. In line with the rules of experimental thinking, Geoffroy Saint-Hilaire alters one variable—bringing different emotions than sorrow into play—in order to propose a new hypothesis. Arguing that not only physical injuries, but emotional states may cause the adhesions that alter the embryo's form, he describes the concatenation of effects provoked by a particular state of mind: a "surprise," a "gloomy event," or a "sudden emotion" can set in motion an "action of the nerves" that leads to disruptions in the maternal circulation; this negatively affects the perfusion of the placenta, whereupon membranes detach themselves from it, enter the amniotic fluid, and there adhere to the embryo.[39]

Geoffroy Saint-Hilaire continued to pursue the topic after the appearance of *Philosophie anatomique.* In a paper published in 1825, he coined the term *influence consécutive* for the process I have just described.[40] And in 1826, his *Considérations générales sur les monstres* (General considerations on monsters) once more had the single mothers of Paris step up to disprove the effect of sorrow and distinguish between different states of mind:

> But if we cannot perceive in what has been said that the matter is governed by feelings [*sentiments moraux*], and if on the contrary we are persuaded that neither perturbation of the mind nor suffering of the soul has any grip on the organism, sending it onto unusual and disordered paths, the same cannot be said of cases where bad news is delivered carelessly and throws a pregnant woman into mental turmoil.[41]

This is followed by three examples of pregnant women who, frightened by a toad, subjected to an attack, or surprised by news of a death, subsequently gave birth to children who lacked parts of the skull, cerebral membrane, scalp, or brain. Narrowing the focus down to mental agitation, he finds a modified relationship of cause and effect: rather than a disruption to circulation, he cites uterine contractions that provoke the loss of amniotic fluid and thus the detachment of membranes.[42]

Geoffroy Saint-Hilaire was primarily concerned with examining the genesis of monstrosity by experimental means, and he would enter the annals of science not only for his investigation of delayed development, but also for his typology of deformities. Discussing sorrowful emotions, perturbation of the mind, and turmoil of the senses among pregnant women, his intervention indicates that in the early nineteenth century, the controversy around imaginationism persisted as a debate on the causality of feeling. It was unthinkable for Geoffroy Saint-Hilaire to set aside the "popular belief" in these things entirely, especially given that his denial of predetermined malformation left the question of causes uncomfortably open.[43]

Although his natural experiment on mental pain remained inconclusive, it was still significant in two respects. First, it formulated for teratology the question of how developmental anomalies are caused. Second, it added a new methodological dimension — the evaluation of demographic data — to a dispute that more usually relied on anecdote. At least hypothetically, this yielded a distinction between efficacious and nonefficacious maternal states of mind that was susceptible to further research.

In his own teratological writings, Étienne Geoffroy Saint-Hilaire's son Isidore refined his father's distinctions into a taxonomy of emotional causes. In a publication of 1836, he too notes that deformity can result not only from physical incidents, but also from psychological ones: "Clearly, there are anomalies that are caused by purely mechanical troubles: but there are others that have their first origin in a mental trouble; it is no longer permissible to contest this today, and if present-day science has to admit to a certain impotence, that is only . . . when it comes to determining the proportion of influence attributable to each of these two types of causes."[44]

Adding further detail to this sketch, Isidore Geoffroy Saint-Hilaire proposes a "distinction" whose absence in existing scholarship he deplores. The "mental affections" (*affections morales*), regardless of whether they are "sad or cheerful, of pity or disgust, of desire or fear," should be divided into three groups. "Some are violent and abrupt; others weak and only momentary; others are moderate or weak, but persistent. Each of these three types has its own mode and its specific degree of effect" — namely, "incontestable and even vigorous" in the first case, "slight" in the third, and "nonexistent or at least dubious" in the second.[45] Isidore Geoffroy Saint-Hilaire's typology was much repeated by other writers and became a kind of formula: deformities may be caused by intense and sudden impressions, to a much lesser degree by weak but persistent ones, and not at all by weak and transient ones.[46]

Distinguishing not by the type of feeling but by its mode enabled Isidore Geoffroy Saint-Hilaire to disqualify the doctrine of maternal imagination in its old form once and for all as an "absurd" and "dangerous" prejudice. A faint or only temporary influence has no impact, he asserts, and it is clearly incorrect to claim that "an object seen, feared, or longed for" can "inscribe itself, so to say, on the body of the child."[47] On the other hand, the rare "modern physiologists" who reduce the influence of the mind "to almost nothing" are equally wrong.[48]

Isidore and Étienne Geoffroy Saint-Hilaire followed the imperative to distinguish by differentiating even further. It was important to separate not only efficacious feeling from nonefficacious sight, but also efficacious from nonefficacious feelings; Isidore added the dimension of the feeling's different qualities. In both men's work, the specification was teratological in intent, since both were interested in the causes of deformity. At least in theory, however, distinctions between emotional causes also implied distinctions between different effects. That inevitably raised the question of whether feelings that do not cause deformities are also devoid of other effects.

Étienne Geoffroy Saint-Hilaire's discussion of the noncorrelation between unmarried pregnancies and deformities in his *Considérations générales sur les monstres* of 1826 culminates in just that question. In a short but important note, he moves from his conclusion that sorrow does not generate deformity to a new conjecture: it is quite possible, he contends, that the mother's mental suffering affects the child "in a generalized manner, on all its organs *pro rata* and not separately, on one single organic part, as we see in the case of monsters."[49] It seems that once attention shifted from the cause of deformity to the efficacy of feelings, other effects came into view. They were manifested not in particular damaged or malformed body parts, but in the character of the child.

Certainly, that exceeded the remit of the emerging theory of malformation. It did, though, mean that teratological thinking on the causality of feelings could join forces at a very early stage with a general pathology of emotional influence, the topic pursued at the time by the physicians and physiologists I have discussed. They could contribute insights into the psychophysiology of maternal feeling and the communicative relation between the maternal and the fetal organism, while teratology added to the shift from impression to feeling another shift of emphasis: from imprint in the child to anomaly in development. All these debates were framed by the turn to feeling in the doctrine of maternal imagination, which thus became a bridge to connect fetal pathology with the theory of developmental anomaly. Half a century after Étienne and Isidore Geoffroy Saint-Hilaire, it was at this interface that Charles Féré would realize the experimental synthesis of teratology and prenatal pathology I presented in Chapter 6.

Significant Cases: The Children of the Year of Terror

It was not by chance that Féré intensified his study of developmental anomaly when the *enfants du Siège* so unexpectedly presented him with his own natural experiment. Like Étienne Geoffroy Saint-Hilaire's experiment, Féré's was flanked by statistical data. That was no small advantage in the 1880s. The use of numbers in both clinical practice and research on heredity and degeneration had constantly expanded over the course of the century, and numbers had come to bestow persuasive power on any argument,[50] something no less true for the investigation of prenatal danger. A widely read Swiss study warning of parental alcoholism, for example, correlated statistical data on birth defects with the rhythm of the agricultural year to find that low numbers of "defective" children affected were born nine months after the labor-intensive summer, high numbers nine months after the alcohol-intensive winter and spring.[51]

This intellectual environment helps explain why Henri Legrand du Saulle so firmly emphasizes the systematic and comprehensive nature of the surveys underlying his lecture on the children of the year of terror. His figures, he stresses, are reliable because they were drawn from his medical activities at the Paris police prefecture. There, delinquents were systemically registered, and a kind of psychiatric triage was performed: "You are aware how well the cases in Paris are monitored. In five cases of every six, all those who have lost their senses in Paris pass through the infirmary of the prefecture."[52] As evidence for prenatal endangerment, the *enfants du Siège* were, so to speak, statistically controlled.

Within that framework, Féré's case—the bourgeois lawyer's daughter whose abnormalities he attributed to her mother's trauma at the moment of conception—could become a particularly powerful illustration of prenatal danger. As an anecdote, it fit seamlessly into the litany of episodic incidents that had flourished in nineteenth-century explanations of the influence of feeling and earlier disputes over the doctrine of maternal imagination but which carried no evidential weight in scientific medicine.[53] In this particular story, however, precise knowledge of the circumstances meant that more variables were controlled than for the other children affected. The moment of the mother's trauma was known, as were the family's medical records (which showed no hereditary illnesses), and the influence of malnutrition and alcohol could be ruled out.[54]

Thanks to the isolated, quasi-experimental nature of this episode, far more amenable to scientific use, Féré was able to propose again something that had been postulated in the first half of the nineteenth century in hypotheses on the influence of feeling. He now called it "psychical influence," but was clearly indebted to these earlier studies. Not only did he evoke the older doctrine of impression, but he also alluded to Isidore Geoffroy Saint-Hilaire's formula by distinguishing between the influence of "acute or chronic feelings."[55]

As for the manifestations of the effect caused by that influence, the *enfants du Siège* offered exemplary support for a particular contemporary reading. What Étienne Geoffroy Saint-Hilaire had described as a "generalized" effect on the whole being, Féré describes as "functional stigma." Explaining this concept in an 1896 paper, he gives examples including delays in learning to walk and in language development, motor anomalies, morbid emotivity—all items that, alongside their bodily afflictions, characterized the children of the year of terror.[56] The notion of functional stigma encapsulated a change in psychiatry during the second half of the century, when anomalies that were not deformities began to be thought of not as damage to organs, but as malfunctions of the organism.[57] At the same time, the term indicates how strongly late nineteenth-century perceptions of such anomalies were shaped by the concept of degeneration. As "stigmata," they did not stand for themselves, but encoded another, much larger process.

In many different respects, therefore, the phenomenon of the *enfants du Siège* was a significant exhibit, bristling with signals of methodological validity, in the late nineteenth-century inquiry into the dangers posed by maternal emotion. These children also found their way into the literature of the human sciences, where they served as additional, empirical evidence that the study of the child must commence before birth.

That is particularly striking in work in child psychiatry and pediatrics. The psychiatrist Paul Moreau de Tours, for example, devoted quite some detail to the *enfants du Siège* in his 1888 study *La folie chez les enfants* (Childhood madness), translated into German in 1889: "There was a very marked pathological influence of the events of the siege of Paris and the Commune upon the development of a large number of children conceived during that time."[58]

A few years later, Moreau's portrayal reappeared in another book translated rapidly into German and English, *L'évolution intellectuelle et morale de l'enfant* (The intellectual and moral development of the child) by the philosopher and educationist Gabriel Compayré. Compayré's comments on the *enfants du Siège* lend weight to his call to consider a third aspect, alongside illness and upbringing, when seeking the cause of madness in children: "Most often we must go farther back than birth—to the period of gestation, to the emotions of the mother during pregnancy."[59] The Parisian case was not unique in this respect; in the town of Landrecies in northeastern France, sixteen of the ninety-two children born during a siege died at birth, while "thirty-five languished a few months, [and] ten were idiots."[60]

Such appeals to consider the time before birth often made explicit reference to the doctrine of maternal imagination. An example is the pediatrician Adolphe Combe. Even the ancients, writes Combe, knew that "the state of the parents' feelings" during conception exerts "an unfavorable influence on the nervous system of the child," but "it was left to the modern era to carry out statistical research on the question"—which, Combe argues, is precisely how Legrand du Saulle reached his conclusions about the *enfants du Siège*.[61]

Taking a very different tack, Féré's Scottish admirer John William Ballantyne, pioneer of prenatal pathology, closes a long and knowledgeable historical chapter on the idea of maternal impression by citing the children of the year of terror. Their "bodily and mental stigmata," Ballantyne notes, are due at least partially to "the continued and severe terror and constant anxiety incident to a time of siege in modern warfare." Only in this sense does he accept the "old doctrine of maternal impressions," which he considers "the one grain of truth in an immense mass of fiction and accidental coincidence." In concrete terms, Ballantyne concludes that no, the impressions received by a pregnant woman do not produce defects in her unborn child that resemble the cause of the impression, but yes, the mental state of the mother can influence the development of the unborn.[62]

It is very clear that at the turn of the nineteenth to the twentieth century, the doctrine of maternal imagination acted as a vehicle for the hypothesis of psychical influence. This is underlined by the fact that the word "imagination" appeared again and again, even if—as its users generally either specified immediately or else found perfectly obvious—this referred to emotion, and even if it was not imprinting but influence that was postulated.[63] On the one hand, invoking the doctrine of imagination endowed the hypothesis of psychical influence with the gravitas of tradition. On the other, the formulation of that hypothesis, tailored to contemporary ideas, cleansed the tradition of "fiction" and reoriented the issue of the maternal mind entirely toward the category of the prenatal.

What was still missing was an empirically grounded explanation of the actual mechanism. How exactly does the mother's feeling turn into a disruption of development? That could not be ascertained through statistical surveys; nor were experiments on chicks much use. Of the ideas that proliferated in the early nineteenth century, uterine contraction and, especially, nutritive disturbance had emerged as promising candidates. In fact, Féré supplied experimental proof of the pathogenic effect of altered nutrition, but the psychophysiological connection between maternal feeling and the modification of nutrition remained speculative, as Féré was well aware. Immediately after he heard Legrand du Saulle's lecture on the *enfants du Siège*, and before he began to intensify his experiments with chick embryos, he decided to weigh up what he had at hand so far.

CHAPTER EIGHT

If the Fetus Senses

Experimenting with Fetal Movement

As well as working on degeneration, Charles Féré also researched perception and motion during the 1880s. His point of departure was that "nothing happens in the mind [*esprit*] that is not translated outwardly by movements, modifications of the circulation and consequently of the secretions, etc., and by a general modification of the organic functions."[1] This psychophysiological complex had become the object of experiments in the last third of the nineteenth century.[2] Féré's involvement in such research stemmed from a shared conviction that the mental is both a reality of its own and a function of physiological and cerebral processes.[3]

Two years after he had discussed the children of the year of terror through the case of the lawyer's daughter, Féré published an article extending his psychophysiological deliberations to fetal movement, calling it a "contribution to a psychology of the fetus."[4] It was no coincidence that this 1886 paper appeared in the *Revue philosophique de la France et de l'étranger* (Philosophical review for France and abroad), the mouthpiece of those battling to scientize psychology and promoting its experimental orientation. The *Revue philosophique* was inaugurated by Théodule Ribot in 1876, and after the Société de psychologie physiologique was founded in 1886 with Jean-Martin Charcot as its president and Féré as its first secretary, the journal regularly featured contributions from Société circles. One of those was Féré's paper on fetal movement.[5]

Féré's exercise in fetal psychology rests on the assumption that sensation can be indirectly measured by means of the muscle contractions provoked in an organism by every sensory excitation.[6] He introduces fetal movement into the debate as a mediated version of that logic. It has long been known, Féré remarks, that agitation—"physical or mental shocks, violent emotions"—in the expectant mother provokes movements in the fetus.[7] For example, a pregnant hysteria patient told him that when she entered the Salpêtrière's photographic laboratory with its red lighting, she felt her child move immediately, and it was an easy matter for him to verify that link. All the apparently "active movements" of the fetus, writes Féré, should in reality be regarded as "reflex movements" responding to maternal excitation.[8] If anything, the fetus is "a more sensitive reagent" than the mother, since it responds to excitement that she herself may perceive hardly or not at all. In the case of "psychical excitations," this is true even of "mental representations," such as a "banal dream."[9] The fetus seems to react more strongly to excitations due to its "weakness." It becomes "a kind of multiplier of maternal reactions," and its responses are nuanced: it can "testify" to the varied excitations of the mother through "motions of defense that vary in intensity."[10] The connection is mediated "rather simply" through the uterus: the uterus contracts when the mother is excited, and the fetus responds with a movement. It is thus that the fetus "feels" maternal excitation and "reacts" to it.[11]

Féré's studies on fetal movement underscore the psychophysiological principle that every act of volition and process of consciousness is preceded by a combination of sense impression and reflex movement, which form the basis of all mental events.[12] If, Féré then infers, motion is provoked in the fetus even by those sensations that never enter the mother's consciousness, the fetus may be seen as a kind of measurement device, capable of gauging sensorimotor processes in the maternal organism, and fetal movement itself is a test case that confirms the basic principle. To be sure, this argument works only if spontaneous motor activity in the fetus is regarded as *reactive* movement. In this respect, Féré's psychological sketch also intervenes in a debate, raging throughout the nineteenth century, on how to interpret fetal movement.

The first studies of fetal movement had been carried out in obstetrics as early as the eighteenth century,[13] and it was an obstetrician, Jean-Marie Jacquemier, that Féré cited to back up his position. Jacquemier's work on fetal movement had gone beyond the usual diagnostic interest of obstetrics in signs of fetal vitality to consider, as Féré put it, the "conditions" of fetal movement.[14] In his obstetrical manual of 1846, Jacquemier argues that fetal movements are not automatic, but express sensation and an awareness of needs and of well-being or suffering.[15] The fetus reacts with restlessness to processes in its own interior and to those in the maternal organism, such as when the pregnant woman drinks something cold or when her soul is moved by "keen and sudden feelings." Scattered through Jacquemier's manual are numerous more or less spontaneous observations and experiments on this point—the obstetrician describes, for example, how he registered fetal movement patterns in aborted fetuses, or had pregnant women hold their breath in order to record a subsequent movement of the fetus.[16]

Jacquemier himself cited observations made by his fellow obstetrician Paul Dubois, counting him among the few exceptions (along with Pierre-Jean-Georges Cabanis) who did not reject out of hand the notion that the fetus shows "sensations, instinct, determinations that are, to a certain degree, willed [*volontaire*]," especially toward the end of pregnancy.[17] This inference from movement to "willing" or volition was controversial and remained so for the whole of the nineteenth century. Referring to Jacquemier's most audacious experiment on the matter, his colleague Adolphe Pinard, writing decades later in 1878, praised the "most ingenious" way his predecessor had dissected pregnant rabbits, pinched the paws of their fetuses, and provoked signs of pain[18]—but Pinard was not convinced that these were "phenomena of intelligence and consciousness," even to a limited extent.[19] He was not inclined to attribute anything other than "material" causes and saw the motions provoked by Jacquemier as pure reflexes.[20]

This disagreement on how to interpret fetal movement was a continuation of the early nineteenth-century controversy over fetal sensibility discussed in Chapter 4.[21] Since its beginnings, the physiology of fetal life had tied observations of fetal movement to speculations on the presence or absence of fetal consciousness and will. Partly as a consequence of Jacquemier's vivisectional experiments, by this point — the last third of the nineteenth century — the notion that the fetus senses and that its sensation translates into movement was no longer contentious, and it had begun to interest neurologists, such as the much-read Jules Bernard Luys, as well.[22] What did remain dubious was whether sensation could enter the fetal consciousness and whether the fetus could move voluntarily.

The members of Féré's circle agreed on one point: consciousness is not what determines whether something is relevant to psychology. Contemporaries regarded this as rather a novel insight. Ribot, who co-founded the Société de psychologie physiologique in 1886 and published Féré's paper on fetal psychology the same year, had recapitulated the finding in 1873, describing it as a result of a historical shift. His account was closely based on the psychophysiology of Cabanis. The nervous system having been discovered in the seventeenth century, explained Ribot, the nineteenth century had liberated the mental from its reduction to consciousness and expanded it by adding the phenomenon of the unconscious. The "antithesis of two substances" had been replaced by an "antithesis of two groups of phenomena": at issue was no longer body and soul, but "life" and "self" (*moi*).[23] Ribot describes "life" as a phenomenon conditioned by time and space, "self" as an internal one. Life is unconscious, the ego is conscious, and both are of interest to psychology.

Just a few pages later, this description leads Ribot to the unborn: "The first forms of unconscious life must be sought for in the fœtal life — a subject full of obscurity, and very little studied from the psychological point of view." He finds this priority corroborated in the "trampling" movements of the fetus during the last phase of gestation, which he takes to indicate that "though the external senses are in the fœtus in a state of torpor, and though in the constant temperature of the amniotic fluid the general sensibility of the fœtus is almost null, still its brain has already exercised perception and will."[24]

In this neurologically formulated intimation that the unconscious begins in the unborn, Ribot brings together almost verbatim the claims of two authors who were writing at the start of the century and who, as we saw in Chapter 4, themselves disagreed. Xavier Bichat is the source of the statement (to which Ribot adds the modifying "almost") that the sensibility of the fetus is still "null," Cabanis originated the claim that the fetus's brain "has already perceived and willed."[25] Whereas for Bichat, birth drew a sharp dividing line between different modes of life, Cabanis saw a continuum of volition traversing birth. Ribot, in turn, neutralizes their dissent by interrogating the presence not of sensation, but of consciousness. It is then the specific *manner* of perception and willing — namely, an unconscious one — that defines the trampling fetus.

This shift in interest from the issue of sensibility to the category distinction between conscious and unconscious explains why Pinard could concur with Jacquemier's experiments on sensibility despite being unable to accept his extrapolation from fetal movement to consciousness. It was also the reason why Féré could cite the work of Jacquemier despite, like Pinard, unequivocally defining spontaneous fetal movement as mere reflex. When he discussed Jacquemier, Féré was interested not in the question of consciousness, but in fetal reactivity and how to demonstrate it by experimenting with different conditions.

In this sense, Féré's approach was highly compatible with the contemporaneous research agenda that William T. Preyer outlines in his *Specielle Physiologie des Embryo* of 1883. Preyer infers provisional findings and future tasks from the work on fetal movement carried out in fetal physiology and obstetrics. The fetus, argues Preyer, has at its disposal a growing repertoire of motion types, ranging from irritative, to reflexive, to instinctive movement. Instinctive movement can occur without stimulus, since instinct is inherited, whereas it is only after birth that the child can perform "imagined movements," which may be regarded as "imitations" and "actions."[26]

Preyer prefaces his summary, the most sophisticated in print at the time, with an appeal for the "most thorough investigation" concerning the causes of the fetus's movements. It should be perfectly possible to detect those causes, by "precisely comparing the frequency, strength, speed, change of location, and orientation of the abdominal wall's elevations with the physiological and pathological states of the mother."[27] To this end, procedures are borrowed from obstetrical diagnostics. "Laying one's hand on the bare skin of the abdomen" and "auscultation using the stethoscope . . . or the ear," for example, enable the experienced practitioner "to perceive the noises produced by fetal movements: a peculiar rustling."[28]

Féré did not proceed anywhere near as systematically as Preyer recommended, and he had little more to offer in the way of detecting causes than a verification of what the pregnant patients at the Salpêtrière had told him about the effect of red light. Nevertheless, his sensorimotor digression on fetal movement did follow Preyer's lead in that he, too, was interested in the fetus's reaction to the mother's states and aspired to demonstrate it both theoretically and experimentally. This aspiration involved the maternal-fetal relationship, so fundamental to the physiology of fetal life, in a psychological iteration: as the connection of two subjects of sensation. It recalled the ancient theory of maternal imagination, which also depended on joint experience shared by the pregnant woman and the unborn. Indeed, writing in 1846, Jacquemier expressly left open the possibility of maternal impression: it might not yet be possible to explain communication between the maternal and the fetal nervous system, he remarked, but that is not to say such communication does not exist.[29]

Féré's own study on fetal movement does not proceed directly to a vision of psychological interrelatedness. In a sudden twist, he concludes by announcing his teratological and pathological experiments on the influence of toxic substances upon various developmental anomalies. The pregnant woman's "repeated or violent emotions," he writes, cause "profound disruptions" in the fetus's nutrition and its nervous system. Even so, "the crude facts on the influence of the mother's *mental* state on the *somatic* state of the fetus may set us on the right path to explaining the influence of the mother's imagination on the development of the product of conception."[30] His sketch thus gave succor to those for whom the cue-word "imagination"—by which Féré meant feeling[31]—evoked a maternal-fetal dialogue of sensation. When Julius Preuss, a knowledgeable historiographer of imaginationism, argued in 1892 that maternal influence could be tested experimentally by examining how the fetus "behaves" when its mother is "mentally agitated," Féré was the writer he chose to cite.[32]

Psychogenesis on Page One

At the start of his celebrated career, the psychiatrist Pierre Janet read Féré's comments on fetal movement with some perplexity. Janet found the paper too superficial to be convincing, but his skepticism was pervaded by a note of curiosity.[33] In fact, not only did Féré's interest in the fetus's motor activity have its roots in a time-honored concern within fetal physiology and obstetrics, but his label "psychology of the fetus," audacious though it may have seemed, was likewise no innovation. Writing in the *Revue philosophique*, Bernard Perez, a philosopher interested in educational theory, had discussed fetal movement from just that perspective in 1882, four years before Féré's paper appeared.[34] Perez's article repeats almost entirely the first chapter of a monograph he published the same year, *Psychologie de l'enfant* (Child psychology), which later became a classic of French developmental psychology. This was the second, revised edition of a study that had appeared in 1878 as *Les trois premières années de l'enfant* (translated in 1885 with the title *The First Three Years of Childhood* but reflecting the revised edition).[35] The first edition had included a number of comments on fetal sensibility, but the second presented a whole separate chapter on the subject, the simultaneous publication of which in the journal *Revue philosophique* underlined just how much weight Perez now placed on extending the scope of child psychology to the time before birth. The question he raised was: "Can the fœtus be regarded as belonging to psychology?"[36]

Posed at the start of a chapter that was not present at *Psychologie de l'enfant*'s first printing but would open every subsequent edition, Perez's question was at once rhetorical and in need of explanation. From the physiological point of view, Perez argues, its answer is obvious inasmuch as all the fully formed organs carry out their functions and the embryo is already organically capable of receiving stimuli. On the same basis, "philosophers of the experimental school" have no doubt that the fetus falls under the remit of their own research. The "movements of the intra-uterine stage of existence," only "apparently automatic," are in fact "evident manifestations of sensibility of some sort," and this is bound to interest psychologists since they "accord as much importance to the unconscious as to the conscious life of the mind."[37]

Clearly, the postulate of a fetal psychology constituted the same provocation at the nineteenth century's end as the postulate of a mental life before birth had done at its start. However, whereas the beginnings of a physiology of fetal life, as discussed in Chapter 4, had made the combination of sensibility and consciousness the criterion for deciding whether or not mental activity was already present in the unborn, now the category of the unconscious was added to the mix.[38] It was Perez who formulated this shift with the greatest clarity. His affirmative answer to his own question arose from a conviction that the organic and the psychological begin to interlock before sensation becomes conscious. In this way, Perez embarked on the path that Féré would abandon four years later when he moved from psychophysiological to teratological and pathological explorations of the unborn.

Having begun his book on developmental psychology by noting that, and why, the fetus belongs to psychology, Perez now needed to explain in more detail what exactly was so important about its motor activity from the psychological perspective. "For the observer," he remarks, "the movements of the embryo are doubly interesting: because they express the embryo's mental states, and because of the influence they may have on future dispositions."[39] It is not clear when sensibility commences, and the anatomy of the fetal brain, along with experiments on the fetus's ability to touch, hear, and taste, reveal only "a very limited functional power of the psychical faculties."[40] Yet even if the fetus's psychology were to amount to nothing more than purely "mechanical" movements and "dull excitations of the sensibility," its great receptivity would make up for that. The influence of impressions during the fetal period must be enormous, given that "the deviation of the tenth part of a millimeter, sustained by a nerve fibre during pregnancy, exercises a considerable influence over the . . . general constitution of the mind, as well as on its special dispositions."[41] Even rudimentary sensations such as hunger, pain, or well-being surely "cannot be entirely devoid of importance for future development."[42]

Perez thus adds another pillar to the conviction that fetal movement indicates the prebirth beginnings of sensibility: the idea of the neurological trace, which he conceptualizes within the logic of development. Crucial here is not only whether—and what—the fetus senses, but also that sensations, even the slightest and most fleeting, inscribe themselves into its body and thus contribute to the child's future development in the shape of a predisposition.

In his *Revue philosophique* paper, Perez outlines this notion of the trace in a rather schematic form that is quite characteristic of the period. Citing the neurologist Jules Bernard Luys, he calls it the "conservation" of memories or unconscious ideas.[43] True to his own pedagogical objectives, Perez derives from his account an admonition that, in all the editions of *Psychologie de l'enfant* published after 1886, also references Féré's sketch of a fetal psychology. The fact that every agitation of the mother is "transmitted" to the embryo, Perez tells his readers, imposes a special duty upon caregivers and especially parents: "Even before the child is born, they have a responsibility for its soul [*âme*]."[44]

Perez pursued this idea to different degrees in the various editions of *Psychologie de l'enfant*. Common to them all is that the neurological trace supplied Perez with an argument to vindicate the importance of unconscious sensations. Once and for all, fetal life became more than a psychophysiological "prelude" à la Cabanis. Instead, it was the starting point of the infant mind. Perez inserted the unborn into the process of the child's becoming just as he inserted a chapter on the subject into his volume on child psychology—in both cases, as a beginning. Or, in his résumé: "It is the first page of a book which the psychologist should neither despair of being able, nor be in too great haste to decipher."[45]

In 1887, five years after the second edition of *Psychologie de l'enfant* appeared with its new antenatal chapter, Perez published another contribution on the subject, which he now named "psychological embryology" or "uterine psychology," in the *Revue philosophique*.[46] The paper was a review of two publications by Preyer, recently translated into French, that I introduced in Part II of this book: *Die Seele des Kindes* (1882, translated into English as *The Mind of the Child* in the same year) and *Specielle Physiologie des Embryo* (The special physiology of the embryo, 1883). Perez had found the "first page" in the book of the infant's mental life without Preyer's help, but Preyer now became his most important intellectual companion. In his *Psychologie de l'enfant*, Perez integrated Preyer's research into the section on life before birth, which expanded from edition to edition. He had good reason to cite Preyer, who was not only the first to systematize the physiology of fetal life and link it to the psychology of the child by positing a continuity of sensation across the rift of birth, but also had fundamental things to say about the precise nature of that bridge.

As early as 1880, Preyer had published a paper on "psychogenesis" that contained the agenda for the two books that would follow. Preyer's claims were anything but modest. He promised that by tackling the genesis of the mind, he could overcome the deficits of two unsatisfactory ways of contemplating the human being: on the one hand, the old tradition of body-soul "hypotheses" that were inspiringly imaginative but pure speculation; on the other, a "modern scientific empiricism" that, "blurring the last remaining distinctions between man and animal," gave the animal mind the recognition it had always deserved but wrongly treated the human mind as merely one more stage in the course of natural history.[47]

Essentially, the conflict here was a contest between natural science and the humanities for the right to explain "man," which logically entailed the question of whether psychology belongs to physiology or philosophy. In Preyer's program, that is a false duality. He reasons that either way, one must begin with scientific empiricism, and physiology is "the foundation stone of any empirical teaching on the psyche." However, Preyer's physiology is the study of psycho*genesis*, and thus an essentially historical field of knowledge. The mind, he argues, does not take shape simply as a physiological function, and neither can it be conclusively explained phylogenetically, by the history of the species; instead, it has its own developmental history in each individual, and cannot be understood without that history. And "anyone who wishes to listen in on the genesis of the human mind must, above all, subject the *child's mind* to methodical investigation."[48]

For Preyer, this program of developmental history underlies the psycho-logic of the unborn. In his article on psychogenesis, the point is still rather hazy, but prominently placed references to newborns indicate that the process of psychological becoming begins as soon as the fetus is organically able to exercise its senses, a sensory activity that strives toward the emergence of the child's will, intellect, and emotion.[49] "Above all," Preyer insists in the opening pages of *The Mind of the Child*, "we must be clear on this point, that the fundamental activities of mind, which are manifested only after birth, do not originate after birth."[50] The same enthusiasm that accentuated the infant psyche in the psychogenesis article is here brought to the task of connecting the child's mind with the time before birth.

From the continuity of psychogenesis across birth, Preyer inferred nothing less than a continuity between the sciences of life and the sciences of man. At the end of his *Specielle Physiologie des*

Embryo, looking ahead to the volume on the child's mind, he writes: "Together, the two works intend to illuminate the origins of the human being's vital processes by demonstrating their accordance with animal functions; to indicate the applicability of physiological methods to emerging life; and to prove the great fruitfulness of this kind of genetic investigation for physiology, morphology, pathology, pedagogy, and psychology—in short, for the science of the human being."[51]

Preyer's two books met with differing receptions. *The Mind of the Child* became a catalyst for subsequent developmental studies, a lasting bestseller published in six languages and 180 editions.[52] *Specielle Physiologie des Embryo* was no less momentous for its own field, but it did not attract the same breadth of attention. This divergence somewhat obscures the substantive unity upon which Preyer insisted when, in the volume on developmental psychology, he announced the follow-up volume on fetal physiology as an integral component of the work.[53] "Taken together," as he puts it, the two works examine the psychogenesis that allowed Preyer to make psychology and physiology mutually referential, just as he connected child, fetus, and embryo through the principle of development. He approached the genesis of anatomical form as simultaneously the genesis of physiological function, and, in turn, approached the genesis of physiological function as the genesis of the mind. Perez was quite justified in unceremoniously subsuming both studies into a terminological unit when he titled his review "L'âme de l'embryon et l'âme de l'enfant" (The soul of the embryo and the soul of the child).

A few years later, the educationist Gabriel Compayré proposed a somewhat long-winded but nonetheless striking portrait of the substantive unity between fetal physiology and child psychology. In his 1893 volume *L'évolution intellectuelle et morale de l'enfant* (translated in 1902–1903 as *The Intellectual and Moral Development of the Child*), he notes that "there is no more interesting moment to study in the life of the child than that in which he comes into the world."[54] The plethora of individual questions as to what the child can or cannot yet do must not, Compayré argues, conceal the wider picture, which is important to consider

> for two reasons, drawn, one from the child's future, the other from his past, for he has a past even now; from his future, since it would be impossible to follow the subsequent evolution of his faculties with precision, if we did not begin by forming a clear idea of his starting point; from his past, since only an exact appreciation of the natural gifts that he possesses at his birth can assist us in clearing up, within the range of the possible, the obscure history of the nine months of gestation.[55]

Compayré has little patience with a psychology that claims sovereignty over unconscious matters deeply buried in physiology — "psychology proper has almost nothing to gather from this obscure period when life is being prepared for, not only because it is difficult to find out what is going on, but because, from the mental point of view, nothing, or almost nothing, is going on." But because "consciousness permits of many degrees" and "resembles day succeeding night, after all the shades of dawn and early morning have passed away," he sees "no difficulty in admitting *a priori* that the first stakes of this slow evolution are set before birth."[56]

In other words, even with his low opinion of uterine psychology, Compayré could join Perez in opening the first page of the human psyche's book before birth, since he found the notion of development as a continuum of events plausible and believed that birth could therefore not be an absolute caesura. In Compayré's description, being born forms the transition between before-birth and after-birth in the same way that a moment in time forms the transition between past and future.

But Compayré's deliberations on this matter were no theoretical diversion. He was impressed by the physiologists who had managed to make the labor ward into a laboratory and who extracted from the moment of birth an object that embodied the psychophysiological continuum of development like no other. Compayré praises researchers, such as Preyer (whom he names specifically),

> who lose no time, who do not wait even till the child has had five minutes of existence before taking him to the window to see what effect the light of day exercises upon his eyes; who do still better than this, even, since they anticipate the complete birth, and profit by the fact that the child's head appears first, to experiment upon the force of his instinct of suction by putting the end of his finger in his mouth.[57]

In terms of fetal physiology, research of this kind, discussed in Chapter 5, was an attempt to approach the fetus concealed in the mother's body; in terms of child psychology, it was a technical observation post. Preyer coined a term to describe the object of such practices, temporally most proximate to birth: the "just-born" (*das Ebengeborene*).[58]

The Just-Born as an Object of Research

Isolated observations and experiments on newborn infants appear frequently in writings on antenatal life. Probably the first systematic treatment, much cited in the years that followed, was an 1859 treatise by the eminent physician, neurologist, and writer Adolf Kussmaul, one of the many researchers influenced by Johannes Müller.[59]

Kussmaul's study, *Untersuchungen über das Seelenleben des neugeborenen Menschen* (Studies on the mental life of the newborn human being), was intended as a revision of the fetal physiology proposed by Bichat, who had described fetal life as vegetative life. Kussmaul found it "quite erroneous" to assert that "the life of the unborn child is akin to the life of the plant."[60] By this point in the century, the idea that the fetus leads a purely plantlike life, and specifically that it is incapable of sensation, had been contradicted on many sides. Kussmaul's achievement was to offer more than speculation or unsystematic observations from vivisectional and obstetrical practice, findings that he quite accurately described as being "full of contradictions."[61]

In response, Kussmaul designed a specific methodology for studying "newbornness."[62] He began from the assumption that if fetal life were like the life of the plant, it would show itself as such at the moment of birth. The fact that it does not could be demonstrated satisfactorily only by means of systematic experiments that, first, permit comparison above the individual level and, second, make use of a whole battery of tests. Kussmaul therefore visited a maternity hospital and subjected more than twenty newborns to experiments in which he fed them with quinine and sugar, tickled the surface of their tongues with a glass rod, held a candle flame before their eyes, sounded a bell near their ears, and had them smell acetic acid and ammonia. His research objects proved rather talented in terms of sensory physiology. They were able to distinguish between sweet and bitter, feel stimuli to the skin, and sense warm, cold, and pain (if not yet intensely); they could hear and see, and experienced hunger and thirst.

On the basis of these experiments, Kussmaul evaluates the "mental life" of the newborn as follows. If the human being possesses such a well-formed "sensorium" at the moment of birth, as well as a number of complex "motor apparatuses," that means it must have made use of them "already in the womb," assembled "some experiences," and acquired "skills."[63] This applies in special measure to the sense of touch and taste and the feeling of hunger and thirst, which—again, "already in the womb"—supply the fetus with the sensations and mental images that give rise to consciousness and action. To be sure, such action is initially "only instinctive," consciousness no more distinct than the state of reverie in an adult.[64] Yet what else but "experience" in the amniotic fluid can explain why the thirsty newborn uses its hands and mouth to search for a "something external" from which it expects satisfaction?[65]

Kussmaul drew his psychophysiological conclusions in just the same way as others did before and after him. What makes his study significant and explains its lasting influence is the systematic way in which he proceeded to establish the newborn as a research object, an object by means of which scientists could investigate what already exists in the unborn and what inaugurates the mental life of the born. It was this methodology of newbornness that, around twenty years later, Preyer would reinforce programmatically by embedding it at the interface of child psychology and fetal physiology. Another few years later, Silvio Canestrini, a psychiatrist working in the Austrian city of Graz, took Kussmaul's approach farther and devoted more effort to it than all his predecessors had done. As it turned out, Canestrini's hard work was somewhat disproportionate to the ease with which his conclusions were almost immediately dismissed. This episode is worth detailing, because it sheds light on the great import of a research practice that had the newborn embody the transition from physiology to psychology.

In his 1913 study *Über das Sinnesleben des Neugeborenen* (On the sensory life of the newborn), Canestrini presents his endeavors as an "experimental psychophysiology of the newborn and developing nervous system."[66] A dedicated reader of especially French research, he situates his own studies carefully within existing scholarship. He dismisses the notion of a plantlike life, along with all types of belief in prebirth impressions, as anachronistic remnants of Malebranchian folklore—before birth, he argues, there is no thinking, no acting, and no willing, only reflex and instinct.[67] On the other hand, he does not want to reduce the fetal organism to a pure "apparatus," since an apparatus is not designed to progress toward "perfection."[68] Canestrini declares the object of his experimentation to be processual in just that way—it is the "gradualness with which the infant brain begins to operate" by means of reactions to sensory stimuli—and best explained through an analogy from embryology: "just as the ovum contains the developmental principle of the future body, so, later on, the newborn's brain contains the mental principle in a germinal state, and out of this arises the perfected organ of the human mind."[69]

Canestrini describes his experimental setup thus: "I now conceived the plan of utilizing the gateway to the brain that nature has created (the anterior fontanel) to study psychophysiological changes of the infant's brain movements in response to experimental stimuli."[70] This was by no means an easy matter. It was facilitated by the "happy coincidence" that the psychiatric clinic and the maternity ward were both located on the same floor of the hospital in

Graz, meaning that Canestrini "had newborns at my disposal always and at different times of the day, both before and after feeding," but methodologically the babies proved "extraordinarily difficult material." Despite trying out all sorts of substances — wax, paste, glazier's putty — Canestrini was unable to find a suitable "adhesive mass" for an airtight seal to "keep the metal cap attached to the infant's soft scalp."[71] He solved the problem with a pneumatic spring (Figure 5). This allowed him to measure at the fontanel how the pulse in the brain changed when the newborn was exposed to electric light or the sound of pistol shots, bells, or its mother's whispering, when bitter liquids were administered or cold coins were pressed on its forehead.

Canestrini performed these experiments over the course of three years, on seventy babies aged one to fourteen days, asleep and awake, in seven hundred individual tests that yielded 200 meters of graphs (Figure 6). The curves seemed to confirm his hypothesis that the newborn initially possesses only instinct and reflex, but that these enable the infant brain to gradually begin functioning. Perhaps even more important to him was the methodological finding that the complicated human mind can be explored more effectively "if we use physiological methods to scrutinize psychical life from the cradle onward."[72]

Canestrini's guiding hypothesis meant that he began his experiments after birth, but there was no theoretical difficulty in applying them to a moment *before* birth, if in a significantly simpler version. This was exactly what Albrecht Peiper, a pediatrician interested in anthropology and neurology, set out to do just a few years later. Canestrini's work had piqued Peiper's opposition and convinced him to turn to pregnant women's bodies. He subjected the future mothers to acoustic stimuli and recorded the movements of their fetuses in response, using visual observation of the abdominal wall, questions posed to the women, a kymograph, and above all manual palpation, which he believed to be "most sure."[73] The remarkable result was described in a short article of 1925 on the "child's sensations before its birth": when the stimulus was repeated several times, the intensity of the fetal reaction declined, which Peiper interprets as proof of a "certain, extremely simple memory retention" in the fetus. In view of this, he remarks in a sideswipe at Canestrini, it seems unwarranted to argue that birth is such a dramatic turning point, "since the unborn child, too, is exposed to a constant flow of stimuli."[74]

There is something emblematic about Canestrini's vain attempts to fix metal caps on newborns' heads, for the business of fixing a distinction between fetus and infant in the object of the newborn was equally precarious. After all, from the perspective of psychophysiology and developmental theory, the very reason why the just-born was of interest at all was that it encompassed both—it was an object in which the fetus could be seen to cease and the infant to begin, and one that therefore had no existence of its own, was never present, was always only a pointer toward its past and future being. Not for nothing do the descriptions of experiments on newborn babies swarm with terms that evoke what Reinhart Koselleck calls "historical time": terms that do not simply denote the passing of time, but describe a continuing process of emergence in which the past tips into the future at every instant.[75] The newborn "already" senses, but "as yet" has no consciousness; it is "no longer" sheer body, but "not yet" quite mind. This is what made it so productive as a research object. Studying the newborn enabled scientists to probe how development constantly makes something subsequent emerge from something antecedent. But because research practice required it to embody development, the newborn could not shed light on what exactly ceases and begins at the moment of birth. Its constitution in historical time undermined the very possibility of that question—for it was not only still fetus and already infant, but also no longer fetus and not yet infant. An object so liminal must show birth to be, simultaneously, a caesura and its opposite.

This became a problem whenever findings drawn from the newborn were adduced to support conclusions on why fetal physiology and child psychology did or did not belong together and on the basis of what criteria. The result was a constant back-and-forth of justifications making one case or the other. In 1920s Vienna, it would spark a particularly heated controversy.

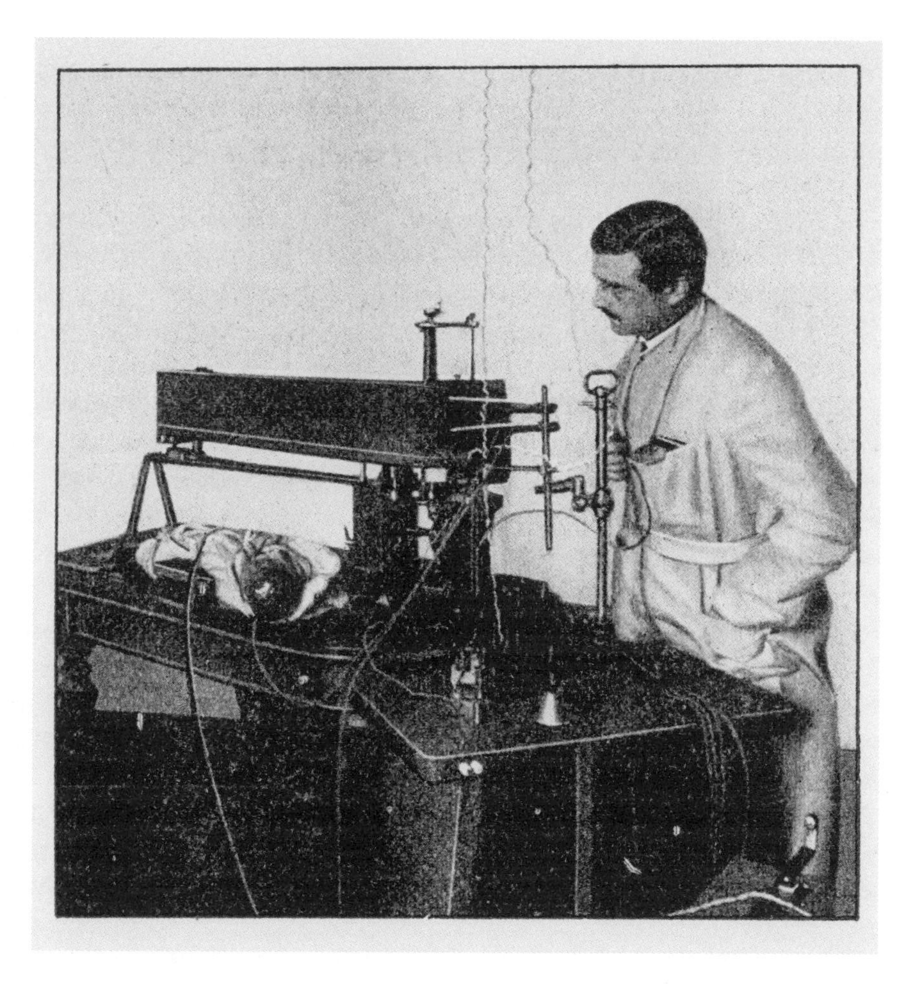

Figure 5. Silvio Canestrini experimenting on a newborn, 1913.

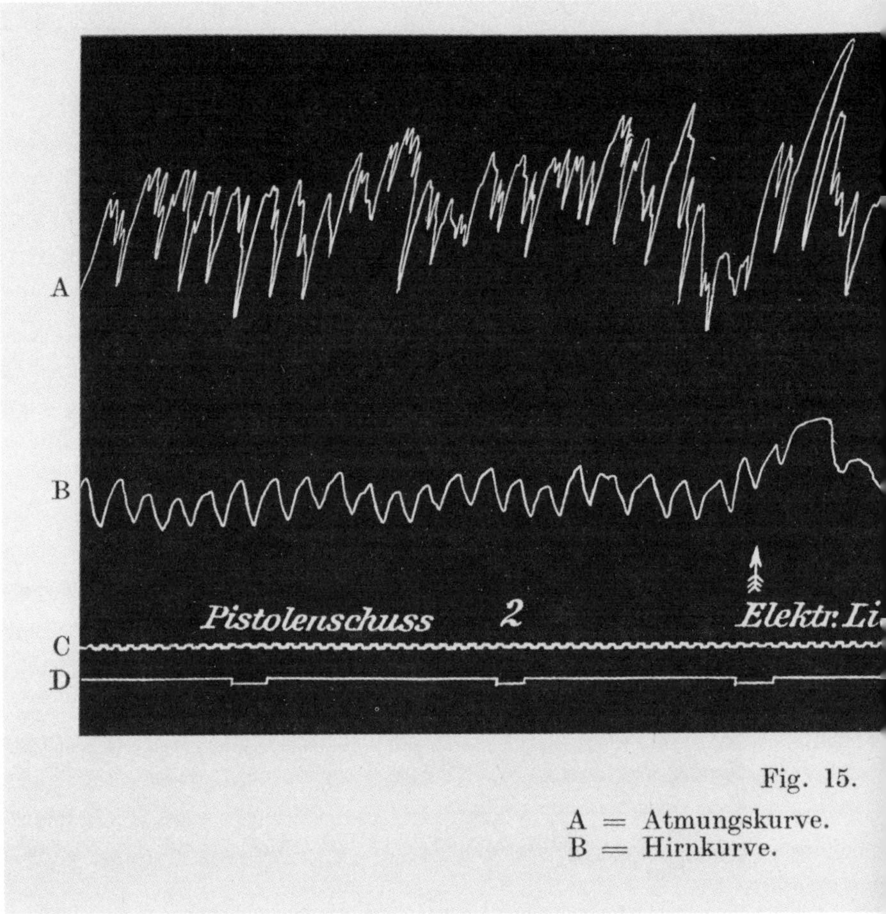

A

B

Pistolenschuss 2 Elektr: Li

C

D

Fig. 15.

A = Atmungskurve.
B = Hirnkurve.

Figure 6. Record of one of Silvio Canestrini's experiments on newborns, 1913.

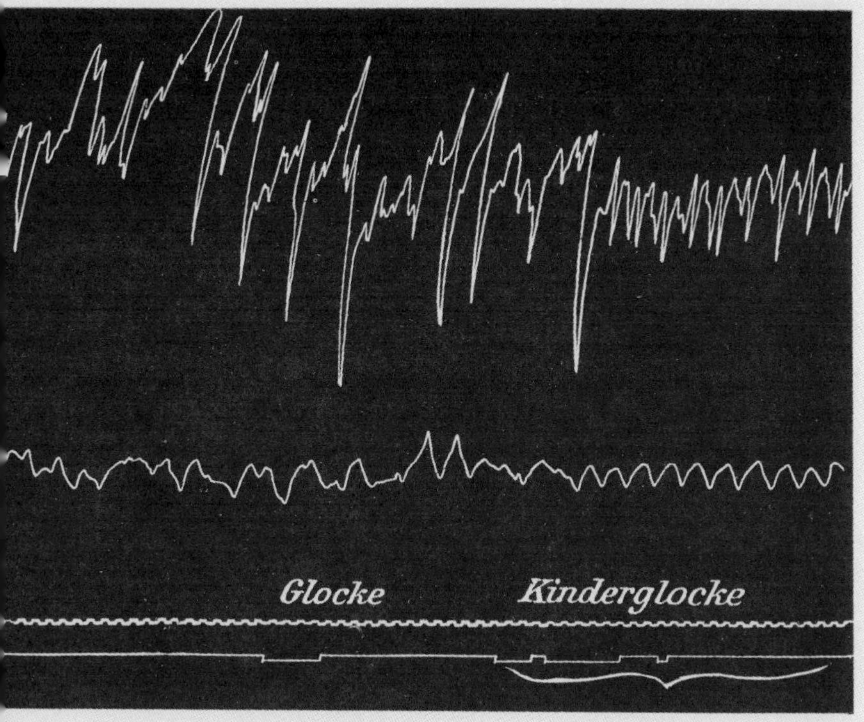

Glocke Kinderglocke

Originalkurve.)

C = Zeiteinteilung auf ½″.
D = Versuchsdauer.

The Anxiety Affect and the Midwife's Sense

When Sigmund Freud attended Charcot's Salpêtrière lectures and his famous soirées in winter 1885/86, it is quite possible that he encountered the debate on the *enfants du Siège*, given that Legrand du Saulle and Féré were also habitual guests and had published on the subject only a year earlier. It is very probable, at least, that Freud knew Féré personally. Féré was the director of neuropathology at the Salpêtrière during the winter semester when Freud, in the first few weeks of his stay in Paris, was examining childhood brain injuries there and noted the importance of the point in development at which a lesion occurred. And it is certain that Freud later read the work of Preyer and Perez, whom he cited in his three treatises on the theory of sexuality in 1905.[76] Writing in 1925 about the "mental make-up of a new-born baby," about which "unfortunately far too little is known,"[77] he did not directly mention these obvious authorities, but in the intervening period, he himself had dedicated great attention to the topic.

In July 1915, Freud told his close confidant Sándor Ferenczi of a "phylogenetic fantasy,"[78] the idea that the cause of neuroses in a patient might make sense as a reiteration of epochs in the history of humanity. In this model, the phase of anxiety hysteria corresponds to the beginning of the Ice Age, that of conversion hysteria corresponds to the speechlessness upon realizing that reproduction will have to be restricted, and so on.[79] This all assumed that experiences must be passed on from generation to generation, something that Freud did indeed describe elsewhere, as the residue of ego experiences in the form of impressions in a heritable "id."[80] It was along these lines of heredity theory that he wrote to Ferenczi about "transference neuroses" in a letter of 1915, and the two men planned to write an article on Lamarck and psychoanalysis.[81]

Though Freud would return to this idea later, in *Moses and Monotheism*, ultimately he found the recapitulation of phylogenetic history in the course of mental illness too speculative to serve as the basis for organizing a theory of neuroses. The Lamarck article never came to fruition and neither was the manuscript on transference neuroses completed. What was already finished was Ferenczi's paper "Stages in the Development of the Sense of Reality," written in 1913. Responding to Freud's letter, Ferenczi refers to the theme as "my *ontogenetic* fantasies," and was happy that these had "so quickly received a phylogenetic sister" in Freud's transference neuroses.[82]

The crux of Ferenczi's ontogenetic thought experiment is the question of when the description of ego development should commence. To answer this, Ferenczi first needs to reiterate the principle that a mature ego arises by gradually forming the capacity to distinguish between wish and reality, thus overcoming the infantile feeling of omnipotence in which pure pleasure is expressed. But what makes the child audacious enough to believe that its wishes are reality and that it can reach either the lamp or the moon simply by stretching out its hand?[83] Ferenczi tells us that Freud regarded this expectation as fictitious from the outset; although the mother's care gives the infant something close to an experience of omnipotence, it remains only "almost realised." Ferenczi is not satisfied: "I might add that there is a stage in human development that realises this ideal of a being subservient only to pleasure, and that does so not only in imagination and approximately, but in actual fact and completely. I mean the period of human life passed in the womb."[84] Where protection, warmth, and nourishment satisfy all needs, the pleasure principle is realized not "almost," but perfectly: "If, therefore, the human being possesses a mental life while in the womb, although only an unconscious one—and it would be foolish to believe that the mind begins to function only at the moment of birth—he must get from his existence the impression that he is in fact omnipotent."[85] Everything begins in the womb, before birth imposes the unwelcome demands of cold or delayed nourishment that confront the child's wishes with its reality.[86]

In these important deliberations, Ferenczi integrates the psychophysiologically grounded postulate of a prebirth unconscious into the psychoanalytical theory of ego development. He interprets the resulting continuum across birth as paralleling the continuity of psychical contents that arises from heredity: "With the same right by which we assume the transference of memory traces of the race's history on to the individual, indeed with more justification than this, we may assert that the traces of intrauterine psychical processes do not remain without influence on the shaping of the psychical material produced after birth." Ferenczi calls this the "continuity of the mental processes," and sees it confirmed in "the behaviour of the child immediately after birth."[87] The baby makes it unmistakably clear that "he is far from pleased at the rude disturbance of the wishless tranquillity he had enjoyed in the womb." This is spontaneously understood by nurses, who "instinctively" seek to simulate for the infant "a situation that resembles as closely as possible the one he has just left" by means of soft blankets, gentle rocking movements, and rhythmical lullabies.[88]

Freud and Ferenczi's conversation about their respective fantasies grew from this conjunction of two continuities: the continuum of phylogenetic experience disposed by heredity, and the continuum of psychical function disposed by development. Embryologists had already had much to say about the connection between phylogenesis and ontogenesis. Ernst Haeckel, writing in 1866, coined the two terms and honed the idea into his famous (or infamous) biogenetic principle that embryonic development recapitulates the evolutionary stages of the species — the inspiration for Freud's later phylogenetic fantasy.[89] And Preyer, who knew Haeckel in Jena, adopted it for his concept of psychogenesis, in which he described the experiential development of the infant psyche as emerging from the phylogenetically determined development of the embryonic body.[90]

Preyer's reluctance to settle upon determination or contingency when defining development is less important here than the fact that his conflation of phylogenesis and psychogenesis amounted to a double rejoinder to the postulate of the tabula rasa. Above and beyond the issue of whether and how phylogenesis organizes psychogenesis, his approach implies a two-pronged assertion about the just-born mind: in the same way that fetal sensibility connects development before birth with that after birth, Preyer argues in *The Mind of the*

Child, "the tablet is already written upon before birth" with the "traces of the imprint of countless sensuous impressions of long-gone generations."[91] Around 1900, this ramification into development and heredity helped to systematize child psychology, but it also helped to constitute a continuity across birth.[92] When Freud twinned his phylogenetic fantasy with Ferenczi's ontogenetic one, therefore, he also endorsed his friend's continuum of the psyche.

More than that: Freud himself had set out a psychophysiological argument that would be cited by all those who subsequently looked at prenatal existence through a psychoanalytical lens. In the second edition of *The Interpretation of Dreams*, published in 1909, we read: "The act of birth is the first experience of anxiety, and thus the source and prototype of the affect of anxiety."[93] Seven years later, Freud elaborated this point for his lecture on anxiety. Every affect, he explained, is a composite of "motor innervations" and "feelings of pleasure and unpleasure" and repeats a primal experience. Furthermore,

> we believe that in the case of the affect of anxiety we know what the early impression is which it repeats. We believe that it is in the *act of birth* that there comes about the combination of unpleasurable feelings, impulses of discharge and bodily sensations which has become the prototype of the effects of a mortal danger and has ever since been repeated by us as the state of anxiety.[94]

It is not without reason, Freud explains, that a person feels their throat constricting when they are afraid. In every subsequent anxiety, they feel the frightening sense of restriction that was caused by the shortage of oxygen during birth. This applies even for cesarean births, because the repetition of the anxiety affect is inscribed so deeply into the organism through intergenerational transmission that it no longer needs an individual cause.[95]

Anyone who had read Preyer would have engaged with prenatal existence without this prompt. The Austrian teacher and psychoanalyst Hermine Hug-Hellmuth, for example, wrote a study on child analysis in 1913 that took Preyer's fetal physiology to explain the beginnings of childhood sexuality.[96] But within the psychoanalytical community, it was Freud's discussion of the experience of birth as a first anxiety affect that authorized attention to fetal life. In his "Stages in the Development of the Sense of Reality," which appeared before Freud's lecture on anxiety, Ferenczi already mentioned a comment of Freud's on the subject, which Ernest Jones would later describe as having been received with "disbelief and much derision."[97]

As Freud himself noted in his lecture, however, this had not been his own idea, and he was not the one who earned derision for it. In fact, he had chanced to hear it at a tavern where a hospital doctor entertained his companions with an anecdote about a midwifery examination. Asked to explain why the baby relieves itself during birth, a candidate answered that this is because it is frightened. The examiners laughed heartily and failed her. Freud, in contrast, kept his ears open (though his mouth closed). "Silently I took her side and began to suspect that this poor woman from the humbler classes had laid an unerring finger on an important correlation."[98]

Interestingly, Dorothy Garley also quotes this story from Freud's lecture at length in a psychoanalytical paper published in 1924, "Über den Schock des Geborenwerdens" (On the shock of being born).[99] Herself a midwife and a pediatric nurse, Garley successfully brought together the practical sense of the obstetrical nurse and the theoretical sense of the fetal physiologist. It is well known, she notes, that the "intelligence center" in premature babies "is already sufficiently developed for it to be able to absorb mental impressions," even if the children do not yet possess a "conscious capacity for memory."[100] For Garley, it is obvious: the violent movements during birth, which she describes in great detail, indicate that the infant genuinely feels a danger of suffocation. This interpretation leads her to describe birth as an experience of shock. She wonders "whether all neurotic reactions — except in cases of innate pathological predisposition — do not reconstruct pleasurable sensations from intrauterine existence or experiences during birth."[101]

This was more than could be extracted from Freud's work. Certainly, he had garnered a more far-reaching idea from the events he studied when seeking the origin of the anxiety affect in 1916: "We shall also recognize it as highly relevant that this first state of anxiety arose out of separation from the mother."[102] But Freud never went as far as to locate ego development or neuroses in the time before birth, as Ferenczi and Garley did. Ten years later, he would reject both claims vehemently.

By then, psychoanalytical studies on birth had accumulated. In the same year as Garley's paper, the *Internationale Zeitschrift für Psychoanalyse* (International journal of psychoanalysis) had published two short papers by Jones, one a discussion of the anxiety affect and the other a speculative note on the unconscious memory of being exposed to cold upon birth.[103] A year after that, two books followed: Gustav Hans Graber's study on birth as the origin of ambivalence drew on the work of Hug-Hellmuth and Ferenczi; and Ferenczi's own *Theory of Genitality* spun the thread of his ontogenetic fantasy farther to interpret orgasm as a renewed enjoyment of "the bliss of intrauterine existence."[104] Most significantly, a book was published in 1924 that brought the discussion to a head and caused turmoil in the psychoanalytical community, *Das Trauma der Geburt* by Otto Rank (translated in 1929 as *The Trauma of Birth*). After some hesitation, Freud pronounced his judgment on Rank's work: "I think you're opening the psychological account too early."[105]

Birth Trauma as a Scene of Conflict

The philosopher Otto Rank belonged to the inner circle of psychoanalysis. After sponsoring his studies, Freud entrusted him with several editorial posts, employed him as secretary of the Wednesday Psychological Society, and made him a member of the "secret committee."[106] Rank firmly anchored his thinking in Freud's notion of dyspnea at birth as the originary moment of the anxiety affect, and in *The Trauma of Birth* he expresses his thanks accordingly.[107] In fact, though, Freud's discussion was little more to Rank than a springboard to a much larger undertaking. He made Freud's intuition about separation from the mother into the cornerstone for a whole new view of psychoanalysis:

> In attempting to reconstruct for the first time from analytic experiences the to all appearances purely physical birth trauma with its prodigious psychical consequences for the whole development of mankind, we are led to recognize in the birth trauma the ultimate biological basis of the psychical. In this way we gain a fundamental insight into the nucleus of the Unconscious on which Freud has constructed [the first psychology] which may claim to be comprehensive and scientific.[108]

Rank's central concern was to reinterpret physiological birth anxiety as a psychological separation trauma. The "primal anxiety-affect at birth," he writes, "is from the very beginning not merely an expression of the new-born child's physiological injuries (dyspnœa — constriction — anxiety), but in consequence of the change from a highly pleasurable situation to an extremely painful one, immediately acquires a 'psychical' quality of feeling."[109]

Pleasure and unpleasure may be physiological, then, but the transition from one to the other is not. Rather, the fact that the transition is "*experienced*" gives rise to "the first psychical act," as the newborn wishes to reinstate the lost situation of pleasure immediately, and the impossibility of fulfilling that wish gives rise to nothing less than "the primal repression."[110] This psychical act makes the uterine situation of pleasure a prebirth union with the mother, and birth an experience of separation. Analytically, there is no longer any need for the heredity theory by means of which Freud had tried to bridge the times before and after birth: "We can provisionally renounce the assumption of an inherited psychical content, for that which is primarily psychical, the real *Unconscious*, proves to be the *embryonal* state *existing unchanged in the adult Ego*. By this embryonal I mean that which Psychoanalysis has recently described in a meta-psychological term as the idea of the sexually neuter 'It.'"[111]

Rank undertook two fundamental shifts. First, unlike Ferenczi, he located the reality of embryonal time not in the physiological satisfaction of needs, but in the relationship with the mother. Second, unlike both Freud and Ferenczi in their phylogenetic and ontogenetic tandem, he instituted continuity across birth not through an inherited id plus psychophysiology, but through the psychical experience of separation. In effect, he was chipping away at the foundations of the edifice built by Freud, since the centrality of birth trauma placed a pre-Oedipal event at the heart of the theory of the unconscious and declared it the event that psychoanalysis must, above all, aim to overcome.[112] Not only did birth trauma form the real content of the adult ego's intrauterine fantasies. In a much more sweeping gesture, Rank imagined the whole of humanity's cultural production as a compensation for the loss of a secure relationship at birth.

Rank dedicated his book to Freud and presented him with a manuscript in 1923. His mentor initially took a positive view, telling Rank in a letter of December 1923: "I enjoy enormously your admirable productivity."[113] In comments on the Oedipus complex published in 1924, he even considered "Otto Rank's interesting study" capable of calling his own ideas into question.[114] Soon, though, ambivalence clouded Freud's fatherly pride in Rank's work. In February 1924, he told Ferenczi of his unease:

> At first, his birth trauma made me distrustful. In the first joy of discovery he seemed ready to see the primal motif of neurosis in birth, so that I told him, jokingly: With a finding like that, someone else would make himself independent. Later he moderated himself, his trauma impressed me greatly, I resolved to look out for it in the analyses, and now, after four weeks of fresh work, I have become quite skeptical again.[115]

Barely two months later, once again in a letter to Ferenczi, Freud's ambivalence had given way to annoyance. He deplored the "crude birth trauma" that dispensed with all phylogenetic logic and threatened to undermine the "ingenious etiological construction" of psychoanalysis. Freud had now realized what "everybody senses": that his disciple had set to work "to put the trauma etiologically in the place of the Oedipus complex."[116] Freud's peers had long since agreed on treating Rank as a dissident, and Freud himself now joined them, in July 1924 accusing Rank of an "elimination of the father in your theory."[117] When Rank's study appeared in print, Freud embarked on a riposte in the form of what, in another letter to Ferenczi, he called a "modified conception of the problem of anxiety."[118]

Freud's critique of Rank appeared in the 1926 essay "Inhibitions, Symptoms and Anxiety." As promised, this paper presents a more precise definition of the notion that had taken on a momentum so disastrous for Freud. Most importantly, Freud here denies that the fetus has the capacity to objectify, which is crucial for psychical processes. He continues to insist that the origin of every subsequent anxiety can be found in the feelings of unpleasure at birth, but this unpleasurable affect has no "psychical content" because the fetus does not realize the objective danger to its life: "It can only be aware of some vast disturbance in the economy of its narcissistic libido." This is what grounds the affect, and the child may retain "tactile and general sensations relating to the process of birth," but not definite sensory impressions, "in particular of a visual kind." It can recollect neither "its happy intra-uterine existence" nor "the traumatic disturbance which interrupted that existence."[119]

Freud marshals similar arguments respecting separation from the mother. That separation takes place, but again it has no psychical content, because in "intra-uterine life," the mother is not a psychical object but merely "the child's biological situation as a foetus."[120] Freud concedes that "there is much more continuity between intra-uterine life and earliest infancy than the impressive caesura of the act of birth would have us believe." In his view, though, what carries that continuity across birth is not the child's psychical activity, but the relationship of physiological nourishment between mother and child: "Just as the mother originally satisfied all the needs of the foetus through the apparatus of her own body, so now, after its birth, she continues to do so, though partly by other means."[121] Rank's attempts to identify the impression left "by the event of birth" in "the earliest phobias of children" must therefore be deemed resolute but unsuccessful.[122]

Freud had asserted his authority. His comments, directed first and foremost at Rank's hubris, aimed to salvage the etiological architecture of a psychoanalytical doctrine of neuroses that refused to exclude the father and could therefore tolerate no other starting point for psychical conflicts than the Oedipus complex. Until his dispute with Rank, Freud had accepted the beginnings of a psychological reinterpretation of the ontogenetic continuum, the notion pioneered by Ferenczi's psychical continuity as a complement to the inherited id. He also remained prepared to accept a psychoanalytical theorization of the birth process, as long as it was limited to a disruption of narcissistic libido in which uterine experience was only a retrospective fantasy. Where Freud drew the line was Rank's postulate that this experience was psychically real and that, as such, it could leave etiological traces. By doing so, Freud sharply and fundamentally demarcated the physiological, before birth, from the psychological, after birth. As he put it, the "biological situation" of a fetus was not the psychical object relation with the mother. In a twofold mission to rescue the father, the author-father of psychoanalysis closed again the psychological account that Rank had opened with birth.

In the history of psychoanalysis, Freud's quarrel with Rank forms part of a series of splits, schisms, and dissidences that were always also conceptual clarifications reinforcing the orthodoxy.[123] In the history of the unborn, however, this particular outcome is a symptom. It indicates the impossibility of conclusively deciding whether or not the unborn belongs to psychology. Like other disciplines, psychoanalysis fell into the vortex that arose when development was conceptualized psychophysiologically as a continuum across the threshold of birth, so that the beginning of the mind could no longer be determined.

When he sliced through this continuum, Freud joined the numerous ranks of those who had tried to pin down the transition from physiology into psychology by fixing a criterion that could distinguish the unborn from the born. He could draw this boundary all the more sharply because his notion of the inherited id acknowledged a continuity that was relevant for all later psychical matters but remained biologically constituted. There was also a disciplinary dimension to his determination to shut down the question of when the psyche begins, for Freud needed to be the one to decree where analytical work must start. Yet such decrees had no chance of capturing the object "development," which makes every criterion distinguishing between phases into one that simultaneously ties them together.

Out of Fear

The symptomatic significance of Freud's dispute with Rank was made evident by the Viennese psychoanalyst and educationist Siegfried Bernfeld, someone who had no father to exclude or defend but was profoundly concerned with an object that he regarded as embodying development. In 1925, before Freud's judgment on Rank appeared in print, Bernfeld published *Psychologie des Säuglings* (translated as *The Psychology of the Infant*, 1929), in which he discussed birth trauma. He ascribed authorship of the idea to Freud's work on the anxiety affect and had read Ferenczi's "Stages in the Development of the Sense of Reality," Garley's "Über den Schock des Geborenwerdens," and Preyer's *The Mind of the Child*. Just before his manuscript went to press, he also read Rank's *The Trauma of Birth* and Ferenczi's *Theory of Genitality*. Bernfeld broadly agreed with both authors and granted them the claim to originality that he had squandered by delaying his own publication.[124]

In his chapter on birth trauma, Bernfeld defines his project as a "developmental psychology which traces psychic manifestations back to their first origins." In that endeavor, the methodologically vital "boundary" is birth. Above all, this means that one must not set to work any later: "The first phenomena of the life of the new-born are psychic," and even if these are only "preliminary stages of the mental life," they are nevertheless "such that one cannot say when they merge into the actual mental life because they are so similar to it." Of course, this immediately raises the question of whether one should not start even earlier: "Perhaps some time later we may, with our knowledge of the mental, penetrate beyond birth into the fœtal being."[125] To be sure, even after such an advance, birth "will retain its methodological and systematic significance" because it brings to a close the specific conditions of "fœtal life."[126] But it is equally true that the "fœtal mental manifestations" themselves must be counted as part of a process that subverts the caesura by conjoining things that are adjacent in time: "The mental preliminary stage of the first day embryo will be further removed from post-natal mental manifestations than that of the last day embryo."[127]

This logic of development makes it impossible for Bernfeld to dismiss Rank's neurosis-oriented interpretation of birth trauma, even though he accepts Freud's view that the child does not subjectively experience the objective separation from the mother as such.[128] In the same way — because development connects things that are adjacent in time, even if birth interrupts them — the question of consciousness, too, defies resolution. For a psychoanalytically oriented developmental psychologist like Bernfeld, unconscious material is enough to make the infant interesting.[129] Even so, he delves into the problem of consciousness at the beginning of his book, tellingly in a chapter dedicated to the newborn child. There, he indirectly illustrates the undecidability of the conflict between Rank and Freud. He also describes more explicitly than anyone before him what, for more than a century, had been at stake in the question of when mental life begins.

Bernfeld's discussion of consciousness proceeds from two observations. First, the infant expresses consciousness within its first two or three months; second, the neurological foundations for this are present at birth. As for the period between, "one cannot specify the moment of the psychical development in the first few weeks of life in which an essential change occurs, which gives one the right, the opportunity, to speak of it as the hour of the birth of consciousness." From this Bernfeld derives his justification for "assuming its [i.e., consciousness's] existence in some degree at the beginning."[130]

This assumption is well-founded, but it remains an assumption. Anyone who lays claim to certainty must be motivated by a "nonscientific attitude." This is true of those who hold "the view which assumes on principle the absence of consciousness in the new-born";

they are "biassed" and deny such consciousness "because its admission has unwished-for consequences . . . *e.g.*, there is no reason against employing the same considerations for the fœtus of more mature stages of development, and for certain groups of animals." Conversely, it is true of those who assert that consciousness in the newborn is "indubitable": they speak out of a "fear of denying the psychical in the new-born." Those, in contrast, who keep an open mind, engage with science, and observe with care will receive the impression "that the child in the first days makes upon us—namely, that it is indeed more animal than human, but that it is by no means a machine."[131]

The observer of the newborn as envisaged by Bernfeld sees a being that is not quite what it is, but is not something else, even if there were a criterion named "consciousness" that could distinguish one from the other. Bernfeld's portrait of this situation has far-reaching consequences. Anyone who denies a consciousness to that being declares it an animal, and anyone who does not, declares the animal a human. The fears and desires of those who persist in fixed attitudes and shy away from observation are perfectly understandable, because the scientific impression of the newborn child threatens to violate a double taboo: that the human at the moment of birth might be nothing more than an animal, and that the animal might also be a human. This was the ontological maelstrom into which knowledge of development was bound to tumble because it excluded every beginning of psychical life. The uncertainty as to what is already, or is not yet, present at birth was more than a matter of not being able to watch the beginnings of consciousness and not yet knowing anything much about the psychical life of the fetus. It also entailed a question that was impossible to resolve conclusively: What *is* an organism that *will be* a subject?

If Bernfeld contributed nothing to the outcome of the dispute between Freud and Rank, that may have been partly because he turned the spotlight on its undecidability. After Freud laid down the law in 1926, Rank emigrated to Paris and in 1935 traveled on to the

United States, where he settled permanently and continued to work on the prenatal from his dissident exile.[132] And when Graber founded the International Study Group for Prenatal Psychology in 1971, Rank's *The Trauma of Birth* (along with occasional mentions of Preyer) was the work that supplied the foundations of a distinct psychological perspective on the unborn. In the agenda of this study group, the barrier of birth had fallen almost entirely. Pregnancy was defined as a "period of active and continuous dialogue between the prenatal child, the mother, and her psychosocial environment."[133] The fetal situation here was an interpsychical interaction, the most interesting aspect being the dialogical shaping of the psyche of the "prenatal child" through the expectant mother's emotion.

When the study group eventually expanded its remit to become the International Society for Prenatal and Perinatal Psychology and Medicine in 1986, it yet again accentuated the psychosomatic interface where Féré had situated his hypothesis of psychical influence more than a hundred years before.

CHAPTER NINE

What the Environment Does

Relational Chemistry

In his observations on fetal movement, Charles Féré had invoked an unborn that reacts to the pregnant woman and her experience. However, he did not go further than positing a sensorimotor connection, leaving others to draw conclusions for developmental psychology. What interested Féré was the pathology of psychical influence in the form he had observed in the lawyer's daughter. It was this that induced him to move his work on the unborn into teratological experimentation, as discussed in Chapter 6.[1]

Féré's two enterprises – a psychophysiology of fetal movement and an experimental developmental pathology – cannot be explained by his preoccupation with the *enfants du Siège* alone; they both belonged to larger undertakings. The first project was part of his work on perception and motion, the second grew from his interest in degeneration. At the same time, both contributed to a single, more specific task of demonstrating psychical influence (Féré said as much in 1903, when he titled a short paper "On the Reactions of the Fetus to the Mother's Emotions").[2] Between them, though, there is a missing link. If Féré's observations of fetal movement proved maternal emotion exerts effects, that proof remained limited to sensorimotor phenomena; and if teratological experiments proved toxins and microbes damage development, that only revealed an influence of substances.

There had long been conjectures that the mother's emotion momentously affects the nutrition of the fetus through the conduit of her nervous system, but, Féré lamented in 1894, this was exactly the point that still lacked experimental proof.[3] In the decades to follow, such evidence remained elusive. Instead, a new substance was found.

In 1904, a medical weekly published a note by the Viennese gynecologist Josef Halban, who, when studying the reproductive organs of newborn babies, observed what he calls "pregnancy reactions": the mammary glands, uterus, and prostate glands appeared enlarged compared to their later size, and in isolated cases he found milk secretion and menstruation-like changes to the endometrium. He concludes from this peculiar phenomenon that "the chemical substances which circulate in the mother's blood during pregnancy also pass into the circulation of the fetus, and elicit in the fetal organism changes analogous to those in the body of the mother." This, Halban adds, is probably the result of "internal secretion" by the placenta. The placental secretions "pass into both organisms, that of the mother and that of the child, and a little while after the birth of the child and the detachment of the placenta, they disappear again from both."[4]

Physiology's long-standing postulate that bodily processes are regulated chemically was what enabled Halban to interpret analogous phenomena in the maternal and fetal organism as a correspondence mediated by particular substances. These would acquire a name one year later, when the messenger chemicals that are secreted by glands, circulate in the bloodstream, and exert effects elsewhere in the body came to be called "hormones."[5]

Halban's work on the endocrine function of the placenta, regarded as his most important achievement, initiated the endocrinological turn that would later make pregnancy a hormonal event.[6] In that process, the connection between the maternal and fetal organisms was also articulated in a novel way. This is exemplified in a study carried out by the Halle-based gynecologist Hugo Sellheim, a creative investigator of all sorts of gynecological questions. Sellheim's paper, which appeared in 1924 with the title "Mother-Child Relationships on the Basis of Internal Secretionary Connections," asked how the correspondence between fetal growth and the enlargement of the pregnant woman's organs comes about.

The work of Halban and others had shown Sellheim that both the fetal organism and the pregnant organism react to the placenta secretionally. From this, Sellheim infers the principle of a link through secretion, which in Halban's "pregnancy reactions" was an indirect one, mediated through an organ that belongs to neither the maternal nor the fetal organism. Sellheim, in contrast, also finds indications of a direct connection. His colleague Werner Lüttge had discovered a "testicular alloreaction" in the mother, a response to the testicular secretions of the male fetus, that proved the transfer of secretions from the fetus to the woman. In turn, chemical traces of this reaction could be found "in the blood of the boy inside the womb." The circle closed: there was not only a maternal and a fetal reaction to the placenta, but also "a reciprocal exchange of organ-specific substances between mother and child." This explained why the same phenomenon appears in both organisms. From the fetus, a "master builder" specializing in growth, substances "in the sense of hormones" find their way into the "mother organism," where they unfold the same "specific effect."[7]

Sellheim's conclusion speaks to both endocrinology and the physiology of pregnancy. He considers it confirmed, first of all, that hormones "really" do exist. From this, he infers something so significant that it requires typographical emphasis: "*Without detriment to the independence of the fetal and the maternal organism, mother and child grow together during pregnancy like the parts of one organism.* They stand in an intimate interrelationship [*Wechselbeziehung*] that is mediated by internal secretionary products."[8]

In nineteenth-century fetal physiology, this concept of interrelationship and interaction described a mediated connection — nutritive supply — that was a function of the placental organ, as I discussed in Chapter 5. Now, in the early twentieth century, the same concept seemed apt to describe a chemical version of the link, not mediated by an organ: a material mechanism of reciprocally reactive secretion and circulation. This shifted the maternal-fetal relationship out of the mode of mediation and influence into a mode of correlation and correspondence. Substances exchanged between the two organisms affected each in the same way.

In Sellheim's comments, this nexus culminates in talk of "one organism," and that is not simply a rhetorical ruse to underline the closeness of the relationship: organic unity arises from the logic of the matter itself. Interrelatedness is realized in a direct mutual engagement of maternal and fetal functions, so that we find precisely the entwinement of functions that defines an organism as an organism. Yet speaking of "one" organism in this way does not only modify the physiological view of the maternal-fetal relationship. It also reactivates the ancient belief in an organic unity between the mother and the coming child.

Sellheim himself alludes to this historical echo. In his conclusions, it resonates more than lexically when he discusses the theory of maternal imagination and *Versehen*. What can now safely be assumed, he writes, "thanks to the successful proof of an exchange of secretions between child and mother and mother and child, and of mutual influence between analogous parts of the mother and child," is "in itself a *very ancient idea*." The same applies to the equally ancient idea that "through an impression on one of her parts, the mother can elicit a malformation at the corresponding part of the child in her womb." Although this "*Versehen* of pregnant women" itself must be banished to the "realm of fables," it has now "gained at least a certain substrate." That is because the falsely posited and never-discovered "'nervous correlation'" can now be replaced by the supposition that there is "a 'chemical correlation along the path of hormonal exchange' between the organs, not only within the fetal organism and not only in the maternal one, but also reciprocally between the two intimately connected organisms."[9]

If the pregnant woman and the unborn correlate with each other through the circulation of hormones in just the same way as the different organs do within one body, then—at least theoretically—the same thing can happen to both of them in the same way. "On this point," Sellheim continues, "the child, as independent as it is within the maternal organism, indeed appears to be connected to the mother just as one part of the body is connected with the whole."[10]

At the beginning of the twentieth century, then, hormones supplied the erstwhile missing link, or at least a "substrate," required for the efficacy of the maternal mind. This interpretation was retrospective, since the search had hitherto been not for a material substance exerting such power, but for a functional relationship between emotion and nutrition. In that sense, hormones did not deliver the missing piece of the puzzle and complete the picture. Neither did Sellheim leapfrog past the physiologically constituted influence of emotion, as his semantic recourse to the "ancient" unity of experience might suggest. Rather, the unity of experience became plausible when the nineteenth-century maternal-fetal relationship took on a new shape: hormones materialize influence, and correlation makes influence into experiential unity.

But if the maternal and the fetal organism are integrated by hormones into an organic whole, then their respective milieus must also become a single, shared environment. In the immediate aftermath of the Second World War, more than half a century after the Parisian year of terror, it was argued once again that wartime events "may irreparably injure a child even before he is born," though this time, fetal life was affected directly by the "war environment."[11]

Fetal Environment as a Developmental Factor

This claim was made in a paper published in a journal of family studies in 1944, "War and the Fetal-Maternal Relationship." The author, physician Lester Warren Sontag, did not know (or at least did not point out) that he was following a thread first spun in 1880s Paris. By the 1940s, the *enfants du Siège* had largely disappeared from the literature, though it is fair to suppose that anglophone physicians, including Sontag, would have encountered them in John William

Ballantyne's manual of prenatal pathology. At any rate, the question had presented itself again, and Sontag, like Féré before him, saw wartime events as a natural experiment in prenatal endangerment. Unlike Féré, however, the relationship between pregnant woman and fetus on which he grounded his argument was one not of physiological mediation, but of chemical correlation.

The conceptual scaffolding of Sontag's article is an explanation of the factors that impact upon fetal development and together form the fetus's "chemical environment." He describes as "exogenous" factors all the substances that keep the embryofetal organism alive—the "nutrition" that passes through the maternal organism when the pregnant woman breathes and eats. "Endogenous" factors are substances produced within the maternal organism, thus also the material correlates of the "stresses" that typify experiences of war. First of all, therefore, the interdisciplinary readership must understand that "emotions" such as anger, fear, or anxiety have "a physiological component," because the stimulated nervous system releases chemical substances such as "acotylcholine" or epinephrine into the bloodstream. During pregnancy, these pass through the placental filter into the fetal circulation, where they "act as stimulants" on the fetus's nervous system.[12]

This process has both physical and psychological consequences. First, there is an "increase in body activity," draining energy from the task of forming fat—for Sontag, that means the significant drop in average birth weight during the First World War was not due to maternal malnutrition alone. Second, extreme nervous stimulation in a mother gives rise to a "neurotic infant": a "hyperactive, irritable, squirming, crying" baby who "generally makes a nuisance of himself," for example by refusing to observe "his four hour feeding."[13] Whether this "pre-natal neurosis" is irreversible remains to be seen, Sontag notes. What *has* already been ascertained is that an "unfavorable fetal environment" results in constitutional defects.[14]

Just as Sellheim hinted, Sontag sees substances such as the neurotransmitters acetylcholine and epinephrine as enabling the maternal and fetal nervous system to correlate in such a way that the fetus participates somatically in the experience of the mother. In this case, the experience is the stress that she perceives as anxiety, and it generates a disposition to neurotic behavior in the fetus. In the larger context of Sontag's research, this connection between oppressive experiences and the prenatal formation of neurosis appears as an exemplary case of prenatal imprinting by an environment that is composed of numerous and heterogeneous factors. Four years before he published his findings on the factor of war, Sontag had described similar processes in a gynecological journal under the title "The Significance of Fetal Environmental Differences." For the fetus, he commented, noise, mechanical pressure, nutrition, oxygen, toxins, and hormones make up an environment that is no less formative than the social and natural environments surrounding infants and children. The fetus receives protection from the pregnant body, but that protection is counteracted by the structural imprint left by the environment on the organism's development.[15]

The environment had been pivotal in Féré's work on psychical influence as well, in the shape of a milieu that imposes potentially disastrous conditions on development. That milieu had interested him as the mode of transmission, a "degenerative" connection between generating mother and generated child. For Sontag, however, the environment was the point of departure. As a bundle of factors impacting on development, it explained traits that are not determined any other way—in practice, ones that are not determined genetically. Between Féré and Sontag, we find a shift from environment as the medium of intergenerational molding to environment as the Other of genetics.

What nonetheless connected the two viewpoints was an enduring conceptualization of the maternal-fetal relationship, articulated in historically specific ways: as a physiological mediation for Féré and as a chemical correlation for Sontag. Methodologically, furthermore, Sontag picked up just where research from the nineteenth and early twentieth century, some of which he knew, left off. He sought statistical correlations by examining birth-weight data in wartime years; he subjected pregnant mothers to sensory stimuli and measured the fetal movements that resulted. Sontag also performed experiments on animals that he could not perform on human beings, provoking audiogenic seizures in pregnant rats to generate developmental pathologies in their young.[16] He certainly viewed all these approaches as individual components of one grand research project that would shed light on the developmental impact of environmental factors of all kinds. Fittingly enough, his paper on war and the fetal-maternal relationship in *Marriage and Family Living* was preceded by an essay on marriage among the Hopi and followed by one on working mothers.

The Natural History of Childhood and the Longitudinal Study

Ambitious research enterprises need generous amounts of money, and such funding fell into Sontag's lap when he took on the directorship of a project that can now be described as the longest of its kind worldwide: the Fels Longitudinal Study, started in 1929 with funding by the soap manufacturer Samuel L. Fels in the rural town of Yellow Springs, Ohio, far from the hothouses of research schools and university hospitals. The impulse came from the president of Antioch College in Yellow Springs, Arthur Morgan, who posed a question that sounded as simple as it was in fact complex: "What makes people different?"[17]

The Yellow Springs investigators hoped to find answers to that question by means of an experimental setup in which "a large group of children" was studied over such a long period that "the whole landscape of childhood" could be opened to the "biomedical" gaze. Concretely, this meant the study would begin "before the child is born and continue to maturity."[18] The physician Sontag, who continually updated his training in psychiatry and pediatrics and attracted staff from many different disciplines, adhered to the framework set out by his sponsor, recruiting his human research objects before they were born by persuading expectant parents to take part. The program accepted ten to twenty new "participants" every year, and at the time of my writing, its data cover several generations of the same families.

The Fels Longitudinal Study is one of a group of such studies that began in the late 1920s and early 1930s at the intersection of pediatrics, developmental psychology, and behavioral science. In 1971, Sontag looked back at the founding phase in a retrospective essay that also set out the project's conceptual stall. Longitudinal studies, Sontag wrote there, had had to fight for recognition, since they were often misunderstood as theoretically naive and empirically overambitious descriptions of the course of life "from womb to tomb." Yet that was never their intent. The "longitudinal study" meant primarily the "longitudinal method": the "employment of the same measure or measures of statistically, and sometimes causally, related characteristics on two or more occasions on the same individual, the measures separated by a sufficient period of time so that changes . . . may have occurred."[19]

Another scientist to stress the distinction between longitudinal study and womb-to-tomb chronicle was Jerome Kagan, a Harvard psychologist who had previously worked with Sontag. His 1964 survey of longitudinal research in psychology provides a detailed derivation of the methodology that closely follows the logic of the object. At issue, writes Kagan, are the "historical forces" first discovered by embryologists, whose "natural history approach to the child" was subsequently adopted by psychologists. In that perspective, cognitive and emotional capabilities, just like the form of the body, emerge not through simple growth, but by means of "lawful transformations." Kagan gives the same embryological theorem a behaviorist spin: early reactions lead to later behavior that does not resemble them "phenotypically," yet is generated by them. As such, development is a "cryptograph," and its decoder is the longitudinal method. Looking at two chronologically separated and unlike phenomena, the method can identify the developmental connection that enables one to emerge from the other.[20]

By studying development among large groups of children, the longitudinal projects brought to the fore variations in development's course. Of interest here was not the mere discovery of variety, but the identification of factors that cause development to vary—that, as the sponsors of the Fels Institute put it, explain why people are different. As well as heredity, this included the environment, encompassing everything that placed conditions on development, whether nutrition or neighborhood, war, marriage culture, or family organization. Once bundled in the research setting of the large-scale longitudinal study, the specific lives of the test subjects accumulated to form one huge natural experiment on development.

It was probably on the basis of this experimental profile that, in his retrospect of 1971, Sontag locates the historical roots of longitudinal research not in the child observations of the nineteenth century, but in the work of an unnamed "French scientist" who, "two hundred years ago," perforated the shells of hens' eggs to generate abnormalities.[21] The vagueness of Sontag's chronology makes it impossible to know for sure whether he means Étienne Geoffroy Saint-Hilaire, who was working in the 1820s, or Camille Dareste, who continued Geoffroy Saint-Hilaire's work in the 1860s, or even Féré, who extended teratology into pathology during the 1890s. All three would fit Sontag's description; all three would be incorrectly dated. Either way, Sontag was pointing to the beginnings of an experimental "natural history of the child," which did indeed go back to nineteenth-century teratology. When teratologists produced abnormalities by altering the conditions of development, they established in principle that environmental influences can cause divergences in the course of development; Dareste specifically related that to the idea of variation. The longitudinal projects expanded the teratological model of physical and psychosocial development and reversed its narrow focus on pathology by distinguishing between functional and dysfunctional developmental conditions. This also explains the close collaboration of medicine and psychology that Sontag pursued with such dedication in Yellow Springs.

Projects of this kind carried into the twentieth century the fascination with fetal life, now articulated in the concept of prenatality, and secured a systematic place for the unborn in the human sciences.

A research disposition that arose as the heir to a notion of development drawn from the study of embryonal formation was bound to begin its natural history of the child where embryonal formation commences: before birth. Tellingly, the editorial note preceding Sontag's review of the Fels study's foundation bore the title "On Organizing for Continuity."[22] The twentieth century's transdisciplinary longitudinal projects sustained the interest in developmental continuity that had previously tied the prenatal to the postnatal and ontogenesis to psychogenesis. Now, this had become a biopsychosocial development that began before birth.

Not Such a Vulgar Error

We do not know whether Sontag had actually read Geoffroy Saint-Hilaire, Dareste, or Féré, and indeed it seems unlikely. But the longitudinal methodology of the twentieth century and the experimental teratology of the nineteenth are so closely intertwined in their logic that we do not need to reconstruct Sontag's exact authorities in order to see why they hit upon the same ideas — which they did to a remarkable extent. For before Sontag began to think about maternal stress in wartime, he had stumbled upon the situation of unmarried mothers, the very case that had spurred Geoffroy Saint-Hilaire's research ambitions. It was just an anecdote, the story of a woman who discovered during her pregnancy that the baby's father was already married and, in great distress, gave birth to a child with nutritional problems. Nevertheless, Sontag found the sketch so "fascinating" that he suggested further studies should be made on the subject.[23]

And they were. From the mid-1950s on, rivers of questionnaires flowed through maternity wards, data from school registers and statistical offices piled up on the desks of epidemiologists, and observers swarmed out through neonatal units, where they took notes on wails, bowel movements, and battles with the breast. The research logic of all these enterprises exactly resembled the natural experiment in developmental variation that was modeled in the longitudinal projects: they related the "now" of childhood anomaly to the "then" of the prenatal environment and extracted a developmental connection. Before long, this undertaking would generate the historical echo with which we are by now familiar. In the very new notion of maternal stress, concerned with the organism's exertion to keep its balance in adverse circumstances,[24] we hear the resonance of the very old concept of maternal impression, concerned with emotional excess.

The first in a rapid succession of studies that referred either to Sontag or to each other was published in 1955. A Baltimore team comprising a professor of nursing, an epidemiologist, and a psychiatrist—Martha E. Rogers, Abraham M. Lilienfeld, and Benjamin Pasamanick—had collected the "prenatal and birth histories" of 2,053 schoolchildren.[25] In 1,151 cases, they found correlations between "behavior disorders of childhood" and "intrauterine events," supplying them with an argument against the reduction of abnormality to "genetic causation."[26] Rogers and her colleagues cited Sontag, of course, but also an earlier study on the etiology of infantile cerebral palsy that Lilienfeld had carried out with Elizabeth Parkhurst. Lilienfeld and Parkhurst had scoured the literature for indications of a "continuum of reproductive casualty" across birth that might explain the prenatal or neonatal causes of neurological damage.[27] It is not clear whether that search led them to Féré, perhaps through Ballantyne's manual, but the idea of the casualty or accident points to Féré, as does the notion of congenital "stigmata" that Lilienfeld and Parkhurst use to describe the mental and physical anomalies in need of explanation (in contrast, their references to the "biosocial" environment as an explanatory factor are clearly a product of their own era).[28]

Just one year later, a study was published by the Australian hospital pediatrician Elizabeth K. Turner, who had noticed an unnamed "syndrome" among newborns who were restless, oversensitive, and irritable, prone to vomiting and diarrhea, who cried excessively and slept badly. The nurses had alerted her to the fact that this was especially common in babies destined for adoption. Turner herself was asked again and again by mothers from "normal families" whether emotional stress they had experienced during pregnancy might be causing their babies' restless behavior. After some initial skepticism, Turner took that question seriously and had one hundred newborns systematically observed, while the mothers supplied information about the history of their pregnancy in questionnaires and interviews. She found little in the way of conclusive evidence but determined that the three babies in the sample who were born to unmarried mothers behaved "differently from the normal." Turner had also come across studies by the Fels Institute explaining that the "maternal fœtal circulation" creates a "common endocrine pool." This point, along with her own observations, lent at least "some justifiable basis" to the proposition that "prenatal emotional stress might affect the reactivity of the fœtal nervous system and alter the whole pattern of postnatal behaviour."[29]

Another year on, in 1957, the psychologist Denis Herbert Stott published a paper in *The Lancet* that drew on the work of Rogers, Lilienfeld, and Pasamanick. Stott had interviewed the parents of 102 "educationally subnormal" children and "mental defectives" about their memories of pregnancy and discovered cases of quarreling and conflict, illness and death in the family, shocks, and accidents. The study did not proceed entirely smoothly, for the parents tended—in line with the "contemporary folklore"—to favor hereditary explanations, against which Stott now wished to range "psychosomatic influences."[30]

In 1960, Antonio J. Ferreira, another psychologist, contributed to the well-established citation carousel with the results of questions administered to 268 pregnant women and the observation of 235 newborn babies. He found a clear correlation between "deviant behavior in the newborn" and "a 'negative' maternal attitude in evidence *prior* to delivery," which he interpreted as further proof of the "prenatal influences" that Sontag and his colleagues had already rescued from the exile of "folklore and magic."[31]

A year later, a 1961 article by the psychologists Anthony Davids, Spencer de Vault, and Max Talmadge described how, inspired by

Sontag and Stott, they subjected pregnant women to batteries of tests, scored their feelings, and measured their attitude to pregnancy, then compared this information with subsequent events to find a connection between "emotional disturbance during pregnancy" and birth complications or physical disabilities in childhood.[32]

Inspired by this last study, Jane Oppenheimer, an embryologist and historian of science, suggested in 1968 that the theory of maternal impression might actually not be such a "vulgar error" as was previously thought.[33] Oppenheimer's cautious rehabilitation of a long-discredited motif should not be co-opted too hastily to the idea of an unbroken historical continuity, or even of a discontinuous history of constant recurrence. Stress as a prenatal environmental factor in the mid-twentieth century both resembled and differed from the maternal impression of the seventeenth and eighteenth centuries. On the one hand, both were concerned with how the woman's experience shapes the child; on the other, over the course of two centuries and via the nineteenth-century notion of psychical influence, the shaping force of outer or inner images became developmental anomaly, encrypting the relation between cause and effect in such a way that the cause (maternal stress) and the effect (childhood anomalies) no longer look alike.

In the scientific imagination, the historical difference between these two articulations of a formative force of the maternal mind can be calibrated by assessing the amount of positive knowledge that each provides. What separates and connects the two is then a long path from error to truth, from speculation to empiricism, from idea to fact. That is how the sciences themselves see it, finding in the historicity of their object a kernel of truth shot through with error and waiting to be known. Not for nothing are references to folklore, old wives' tales—and, indeed, to mothers—just as common in these texts as references to erudition, scientific verification, empirical proof, and theoretical plausibility. But something other than a road from error to truth comes into view when, instead of a history of "progressive refinement" and "continuously increasing rationality," we seek a history of "successive rules of use" and "fields of constitution and validity," to cite Michel Foucault. Then, again following Foucault, the issue becomes not the accrual of more, and more correct, knowledge, but the distinction between theoretical milieus within which concepts take shape.[34] This Foucauldian perspective on the historicity of concepts may also be applied to the historicity of objects, and the conditions of their emergence can be extended from theoretical to practical milieus. For the purposes of the present study, this is to say that Féré's "psychical influence" was not historically singular, but it was historically specific.

In this third section of my book, I have tried to come closer to that specificity by drawing an arc from the maternal imagination of the seventeenth and eighteenth centuries, by way of the psychical influence of the nineteenth century, to the prenatal stress of the mid-twentieth century. Passing through three articulations of the efficacy of the maternal mind, that arc mingles continuity and discontinuity. The continuity arises not from a certain thing remaining the same, but from actors tying their object to historically antecedent terms and concepts (particularly the notion of impression or imagination) while, in their practices and descriptions, they transform the object and thereby generate discontinuity.[35] The formative power of the mother's mind is a historically constant "thing" only in the sense of what Gianna Pomata calls a "polythetic category." It courses through time like—to continue Pomata's figure of thought—a single rope composed of different strands at different moments.[36]

The three articulations can be described as conjunctions, perhaps like thicker points on that rope. The efficacy of the mother's mind enters the stage whenever it is called upon, in an intensifying form—almost every passage on the matter is introduced with an "even"—to represent a particular conception of pregnancy and the unborn. There are three ways of looking at these different conjunctions. The first is the relationship between the pregnant mother and the fetus. So intimate is their union that the child safe in the womb even experiences, with all its senses, the same as the mother—thus the seventeenth-century theorists of the maternal imagination. So close is the relationship between the maternal and the fetal organism that its development is even influenced by her feelings, say the physicians and physiologists of the nineteenth century. So deeply does the prenatal environment mark postnatal behavior that it can even transform maternal stress into childhood neurosis; that is the observation of psychologists and epidemiologists in the twentieth century.

Second, a connection between the body and its Other is always constituted. The child even suffers its mother's pain, so closely are their two souls bound together in one body; the child even experiences her traumas, so physical is the psyche; it is even chemically thrown off balance by her stress, so indistinguishably biosocial are environmental events.

And third, each articulation contributes to explaining the particular features of the child. The child's form even mirrors just what the mother saw; its stigmata are the outcome of events even in pregnancy; its neuroses have their roots even in the prenatal environment.

This last dimension — the explanation of the child's characteristics — links each articulation of the maternal mind with its era's understanding of the rules followed by Nature when she has a genitor give rise to something that comes from it yet differs from it. For the seventeenth century, the maternal imagination creates similarity and dissimilarity just as every other circumstance surrounding the procreative act does; for the nineteenth century, psychical influence creates a generational connection just as "true heredity" does; for the twentieth century, maternal stress is one of the multifarious modifications of genetic disposition by environmental factors.

In the concluding part of the book, I return once again to the link between the efficacy of the maternal mind and the natural laws of generation. It is a link that inserted fetal life, already enclosed in the individual time of development, into the social time of history.

PART FOUR

Politics of the Unborn

CHAPTER TEN

Thresholds in Time

Eventful Children

Not everyone was pleased with Henri Legrand du Saulle's and Charles Féré's portrayal of the *enfants du Siège*, who were also children of the Commune. In 1885, just after Legrand du Saulle's lecture appeared in print, a certain Dr. Dupouy gave vent to his indignation in the professional journal he edited, *Le médecin.* As a military and welfare physician, Dupouy knew the population of Paris just as well as Legrand du Saulle, and asked: "How can the children of such men, the sons of these brave people, be degenerates?"[1] He had no sympathy at all with Legrand du Saulle and others when they spoke of Parisians who, despite their "swagger," were "in a state of mental trauma."[2]

Above all, Dupouy objected to generalizations. During the Prussian siege of winter 1870/71, he pointed out, alcoholic excess had still been rare; it only became widespread under the rebellious Paris Commune in May 1871. These later alcoholics were the parents of the "degenerate children" whom Legrand du Saulle had seen in the prefecture, Désiré-Magloire Bourneville had treated in the psychiatric wards of the Salpêtrière, and Dupouy himself had cared for in the city's welfare institutions. But sons, Dupouy noted, had also been born to the men who had proved their worth in the siege. These boys were now thirteen years old, and he knew after questioning their teachers that they constituted an especially "fervent, vigorous, intelligent, and patriotic generation." Reading their school compositions had convinced him of the profundity of these youngsters' ideas, "the energy of their expression, the virility of their thoughts." They were, wrote Dupouy, the children of "extempore soldiers" and of mothers who "did not tremble" before the noise of the Prussian shells, the clamor of the bugles, the rolling of the drums. No differently than the children of 1792 and 1815, the lads were "drops of a fiery blood that flowed for their country and for freedom; they were born of war, for war, for struggle, for revenge."[3]

This rival image, too, made its way into the literature. The child psychiatrist Paul Moreau de Tours drew liberally on Dupouy's short text when, a few years later, he picked up the controversy in his treatise on childhood insanity, a work that would appear in thirteen French editions and in a German translation. Moreau de Tours found the hypothesis of psychical influence both plausible and enlightening, and quoted at length the case of the lawyer's daughter published by Féré (though he erroneously attributed it to Legrand du Saulle). But, warned Moreau de Tours, it would be wrong to think that all the children conceived or born in the relevant six months of 1870/71 were in the same position. Those whom Legrand du Saulle had encountered at the police prefecture were the offspring of the "unfortunates" who, "under the flag of the Commune," had sought in "absinthism" the arousal "required for the role they were made to play." Of course there had been such inebriated and "idle workers," whose children were now "puny, neuropathic, imbalanced," and "really" marked with the "indelible stigmata of the Siege." However, Legrand du Saulle's statistics had failed to include the offspring of that "great majority" of the Parisian population who had shown themselves "worthy of admiration."[4]

When Dupouy and Moreau de Tours distinguished drunken Communards and idle workers from steadfast mothers and brave impromptu soldiers, they were partly addressing the problem of differential diagnosis that Legrand du Saulle himself had raised—how to determine the relative importance of alcoholism, malnutrition, and trauma.[5] More than that, though, their skepticism was a form of outrage, and their objection part of a politics of remembrance. They countered the vision of a traumatized Parisian population with one showing fearless Parisian fathers and mothers, and thereby queried whether a few "really" stigmatized children actually represented what had happened. In this way, the same *enfants du Siège* who provoked experiments in laboratories and observations in clinics became a political vehicle to draw past events into the present.

In the young and unstable Third Republic, that was a highly charged undertaking. The humiliation of defeat by Prussia continued to fester, and the Paris Commune remained vividly present in the confrontation between authoritarian, hierarchical models of society and egalitarian ones. In this explosive situation, collective memory was a way of securing an illusion of national unity and the continuity of a national character. The mortification of defeat was turned into talk of revenge, and the Paris Commune was stigmatized, neutralizing the persistent and divisive memories of the Commune's violent suppression while also marginalizing socialist positions.[6] Setting patriotic mothers and spontaneous soldiers apart from alcoholic Communards could play precisely that political role. It cordoned off workers manipulated by revolution, who conceived "degenerate" children, as a pathological deviation from the continuity of a heroic nation that endured, undaunted, in a "patriotic" generation whose members Dupouy specified as *sons*.

But Dupouy's objection was more than simply political interest confronting a fundamentally apolitical scientific object. The politics of memory is about how the past—and, indeed, *which* past—is significant for the present. Memory was one of the conditions that made children fascinating in 1884 because they had been conceived in 1870/71. Scientific preoccupation with the children of the year of terror coincided with national self-affirmation through politicized memory, in one and the same question: Which present arose from which past?

Legrand du Saulle and Féré were writing under these conditions as well. Legrand du Saulle embedded his discussion in a detailed chronology of the events and descriptions of everyday life in wartime; Féré's two-page story of the lawyer's daughter dramatically staged a pregnancy accident as the invasion of guardsmen into a family home.[7] Yet the narrative atmosphere they created was very different from that evoked by Dupouy and Moreau de Tours. They wrote not of workers and soldiers, but of a unified experience of siege and rebellion that made those born in 1870/71 the children of an event and formed a pathogenic historical connection of national significance.

This was the generalization that provoked Dupouy's protest. As a widespread fear, it lent sociopolitical urgency to the investigation of developmental anomalies that had been pursued since the early nineteenth century and would culminate near the end of the century in a medically informed, state-administered system of prenatal care. Conversely, the existing stock of knowledge about developmental anomalies guided attention to events of the past whenever childhood problems were noted in the present. It is this nexus that I discuss in the rest of this chapter, showing that the construction of fetal life in nineteenth-century science made pregnancy an organic threshold of social time.

Describing the Suffering of All

When French psychiatry began to stake its own political claims in the first half of the nineteenth century,[8] they were grounded to a significant extent in the idea of revolutionary madness. In that schema, the violence that occurred between 1789 and 1848 was a pathological phenomenon and was responsible for the growing number of the "insane," or *alienés*, as they were called by psychiatrists on all sides.[9]

With the next Paris insurrection, in 1871, this pathologization of political and social unrest was immediately reactivated. The postulate of revolutionary madness now appeared in a heightened form—partly because psychophysiological theories on the manipulability of "the masses" were now in circulation; partly because the constant succession of uprisings seemed to confirm the heritability of just about everything.[10] Insurgency was seen as something both epidemic and endemic, a process that infects people of the present and is transmitted to those of the future. From that perspective, the Commune seemed to be the latest manifestation of a pathology that had long since lodged in the national body. The very next year, physician and teetotaler Jean-Vincent Laborde presented a study on the "morbid psychology" of the events that had just taken place. The Parisian revolt had not, he insisted, arisen out of political and social circumstances. No, the population had fallen prey to a "collective madness" that had flourished in the soil of an inherited "organic predisposition." The events of Paris, and the inhabitants of Paris, should therefore be the concern "of the physician and mental specialist [*moraliste*] rather than that of the historian."[11]

Legrand du Saulle saw things rather differently, though he was equally quick to integrate what was happening around him into his psychiatric research. In spring 1871, he was occupied in finishing a treatise on paranoia, and the circumstances of the day prompted him to add a chapter on the "mental state of the residents of Paris during the events of 1870–1871." His first step in the added chapter is to dispel the myth that "political events" increase the frequency of madness in a population.[12] Certainly, he writes, they affect "predisposed individuals," but that is just as true of any other dramatic and unexpected event, and "great social upheavals" are much too "temporary" to exert "disastrous action upon the intellectual faculties of the nation."[13] In fact, Legrand du Saulle continues, the number of mental patients in Paris was considerably lower than usual in summer 1871. Even given the reduction in the total population due to extremely high death rates and the departure of many foreigners and miscreants, this proves "once more" that even the most serious political events "in no way produce an increase in the number of the insane." For Legrand du Saulle, the Paris Commune was the work not of a mad population, but of "a few fanatics."[14]

Legrand du Saulle's attack on the hypothesis that madness increased in the wake of the revolutionary events may be interpreted as an attack on the idea of revolutionary madness in general.[15] On the link between endemic and epidemic processes, he shares the assumption that catastrophic events may activate a predisposition for mental illness, but believes that political occurrences are too ephemeral to play that role. Above all, he shifts the perspective by focusing on a different collective dimension of the 1870/71 events. In Legrand du Saulle's view, what connected everyone in the Parisian year of terror was not madness, but suffering.

Framed in an insistent present tense, the new chapter in Legrand du Saulle's 1871 study of paranoia again and again describes a "profoundly intimidated population."[16] To be sure, this intimidation has affected some more painfully than others. Those "with an impressionable imagination, weak intelligence, hypochondriacal worries, melancholic tendencies, or inherited threats to the brain" are likely to be "overwhelmed by the terror" that others can resist. They reach the municipal psychiatric ward with "bowed bodies" and "in an attitude of the most pitiful pain, weeping, moaning."[17] As to the impact of the "sudden extraordinary event," however, these unfortunates differ from other Parisians only in the intensity of their reaction, for in besieged and insurgent Paris, "everyone suffers."[18]

Legrand du Saulle reiterated this interpretation thirteen years later in his lecture on the *enfants du Siège*, and Féré took the same view. His minutely detailed description of the conception of the lawyer's daughter was designed not to denounce the incursion of crazed Communards into a bourgeois idyll, but to emphasize that not a single Parisian was safe from trauma. The tic-tormented, bed-wetting, sleepwalking, melancholy girl was evidence that events had burst in upon revolutionary laborer and loyal bourgeois alike — in the shape of an "anxiety and constant fear" that only some drowned in alcohol, but that afflicted everyone. This was what Féré defined, in line with Legrand du Saulle, as the "net influence on the mental state of the Parisian population": the transformation of their "conditions of existence" by siege and rebellion.[19]

When they used a vision of the past to institute a unity in the present, then, Legrand du Saulle and Féré, too, were intervening in the politics of memory. But unlike Dupouy and Moreau de Tours, whose supratemporal "national character" depended on splitting off political dissidence as something pathological, Legrand du Saulle and Féré found unity in the fact that the disastrous change of conditions had been suffered by everyone together.

It would fall short to attribute that distinction to different political preferences. Féré was a staunch republican who also read Karl Marx and Pierre-Joseph Proudhon,[20] but he too believed that the leaders of the century of uprisings were no more than history's will-o'-the-wisps, and he did not omit to mention that they were undergoing psychiatric treatment or at least had hereditary defects.[21] As for Legrand du Saulle, he maintains in his 1871 treatise on paranoia that politics does not exist for the physician, who must never set foot in the "inflamed arenas" where people "begin with speeches and end up with infamies."[22] That this alludes to the Paris Commune becomes clear in his supplementary chapter on the events of 1870/71, which details the execution of the judge Louis-Bernard Bonjean with such deference as to constitute a memorial to a loyal statesman.[23]

When Legrand du Saulle and Féré set about building a different relationship with the past than that proposed by Dupouy and Moreau de Tours, their aim was clearly not to rehabilitate the Paris Commune, either politically or in terms of national history. The important point for them was that the past events of 1870/71 had left an impression in the present of 1884, not only as damage to the progeny of workers drunk on rebellion, but as a correlation suffered by all the nation's offspring alike.

In 1871, Legrand du Saulle had already voiced a fear that the disaster might perpetuate itself into the future in just this way. His concern arose from more than just the impression of the events around him. In the book on paranoia, he also brings historical experience and physiological knowledge to bear, writing: "Revolutions are capable of causing terror, and terror can not only alter the intellectual condition of present generations, but also continue, through heredity, to weigh heavily on the mental disposition of future generations."[24]

Here, heredity does not mean the transmission of political madness—Legrand du Saulle refutes the existence of such insanity in the same book. Instead, it means a sequence of events that turn the temporary circumstances of conception into the permanent characteristics of the resulting child: "We know, in fact, that a child conceived when one of his progenitors is surrounded by those particular conditions is at far more risk than all others of irritability, instability, melancholy, imbecility, epilepsy." Such effects are still rare, Legrand du Saulle remarks, and it is not yet known whether the "catastrophes experienced by the population" will significantly increase the future incidence of mental illness.[25] Yet that is precisely the correlation he would later observe in the psychiatric clinic of the Paris police prefecture. From the frequency with which the children born in 1870 and 1871 presented there, Legrand du Saulle concludes in his 1884 Salpêtrière lecture that the dramatic events caused a disastrous alteration to the "innermost conditions of fetal life" in Paris.[26]

Talk of the *enfants du Siège*, in short, was not about a diagnosis applicable to individual children, nor about one statistical group among others. At stake was a particular thing, "generation," that was always invoked when these children were mentioned. They formed a generation by being not only the children of their parents, but also the momentary realization of a supratemporal unity—in this case, the nation. That is why some writers wanted the children of 1870/71 to underwrite an unbroken line of heroic national community, whereas others saw in them the disastrous irruption of historical events into that very continuity.

Generation/Degeneration

The appearance of the label "generation" in all the writings on the *enfants du Siège* had more historical preconditions than might be obvious at first. As Ohad Parnes, Ulrike Vedder, and Stefan Willer have shown, it was only in the nineteenth century that the term acquired the meanings it conveys so emphatically in studies of the children of 1870/71. There, it locates individuals or living beings within a twofold temporal context: a diachronic one, as the successors of predecessors and themselves the predecessors of their own successors; and a synchronic one that connects them with others in the present. "Generation" combines these two dimensions into a unit whose members do not so much share a purely formal contemporaneity as occupy the same position in a sequence.[27]

The beginnings of this concept of generation can be found in late Enlightenment political theory. The Enlightenment replaced royal genealogy with the succession of egalitarian units, each of which brings together all the citizens of one period as the carriers of a political sovereignty that they inherited from their forebears and will pass on to their descendants.[28] The legal notion of inheritance was admirably suited to describing this transmission-like sequence. In the nineteenth-century life sciences, inheritance became heredity, a concept that made it possible to think of species continuity as a succession of temporary realizations of species characteristics.[29] The generational configuration produced by transmission also appealed to social and historical philosophies that were confronted by upheavals and alarming ruptures. They used it to constitute the continuity of society or the nation as a historical process in which each generation is conditioned by the situation bequeathed to it by the one before and each conditions the next generation through its own legacy.[30]

The kernel of the modern concept of generation is the idea that both property and properties are transmitted through succession, a principle that joins opposites in both the social and the biological realm. On the synchronic level, it assures the membership of individuals and living beings in their nation or species, however much they differ from one another. On the diachronic level, it assures the continuity of the nation or species, however much this may change — for each political generation alters its legacy, biological generations are subject to variation, and sociohistorical generations are differently conditioned.

In his 1871 speculations on the inheritance of fear, Legrand du Saulle invokes this connection across time, as instituted by generational succession. It is impossible to say whether he contemplated a direct transmission of the content of experiences, in this case the experience of fright. Some of his contemporaries postulated exactly that, arguing that an "organic memory" transmits psychical contents from one generation to the next — Théodule Ribot, for example, writing in the 1870s about the recurrent revolutionary events of the nineteenth century, argued that instincts and habits are the expressions of inherited, unconsciously remembered, ancestral experiences.[31] At any rate, by the time Legrand du Saulle returned to his 1871 premonitions in 1884, he had observed a different process occurring among the children he studied, one that led him to abandon the term "heredity" and speak instead of "influence." In these children, horror had not passed from one generation to the next; instead, the suffering caused to the parent generation by horrific events had influenced the development of fetal life in such a way that multifarious anomalies had arisen in the child generation.

Nevertheless, as I argued in Chapter 6, the shared element of these two processes or modes had not disappeared.[32] No differently than inheritance, influence constituted generational relations both synchronically and diachronically: shared conditions of existence in the preceding generation leave their mark on characteristics in the successor generation and are then shared by all that generation's members. In this way, Féré could describe the pathologies of the *enfants du Siège* as something generational even as he distinguished them from hereditary disorders: "Today we even recognize several illnesses . . . that afflict several children of the same generation without something similar, analogous, being present in the maternal or paternal line."[33] In other words, the *enfants du Siège* resemble each other and not their forebears because they are shaped by the shared experiences of their parents' generation.

The generational connection in this case is a pathogenic one, and that is the very definition of "degeneration." Dupouy was among many to use this terminology, which helped him to pick out the offspring of drunken Communards from an otherwise fine generation.

It was as a "degenerative" phenomenon that the children of 1870/71 attracted Féré's attention in the first place. Legrand du Saulle sounded the same note, closing his lecture with the hope that peace in Europe — along with the amelioration of working-class living conditions — could create a basis for "the race" to "move back toward its original type."[34] Although there was disagreement on whether the *enfants du Siège* were representative of a whole generation, their physical form and behavior was generally held to reveal morphological and, especially, functional "stigmata" that signaled the presence in individuals of a supraindividual process named "degeneration."[35]

Like generation, degeneration had numerous historical preconditions. It emerged over the course of more than a hundred years, as the eighteenth century's milieu-based models of human variation coalesced with nineteenth-century concepts of generational transmission to form a distinct process with its own logic. The verb *to degenerate* as used in this context gave rise to the noun *degeneration* — *dégénérescence* in French, *Entartung* in German. A civilizational evil became a pathological decay, and an event affecting particular groups became a tendency inherent in the whole of a nation or species.[36]

The version of the degeneration model that would come to prevail in French psychiatry was set out in Bénédict Augustin Morel's 1857 study of "physical, intellectual, and moral degeneration," *Traité des dégénérescences*. There, Morel defines degeneration as the process by which physical and psychological anomalies accumulate across generations of a genealogical line, up to the point of complete sterility and thus the extinction of the lineage. Starting from a neurological weakness in one individual, a pathological constitution and disposition pass from one generation to the next, manifesting differently each time but gradually worsening overall. An increasingly dense web of different anomalies and pathologies settles over a lineage — the "unhealthy varieties" that Morel mentions in his title and regards as a pathological form of generative events.[37]

This specter could be objectified by the novel procedures of statistics and by tracing family histories of hereditary illness, but its mechanism — degeneration — was a physiological one, just like the mechanisms that Geoffroy Saint-Hilaire, Dareste, and later Féré had explored in the teratological lab by producing developmental anomalies in their experimental animals.[38] Morel's account of degeneration is inspired by toxic conditions: swampy terrain, alcohol and narcotic substances, epidemic diseases and nutritional imbalances or deficiencies, dirty factories and unsanitary housing, sickly constitutions, infections and ill-health among pregnant women, unsatisfactory care for infants. What unifies these multifarious factors is that they are all environmental influences upon the events of generational change. They also include heredity, which Morel calls "hereditary influences."[39] His model is organized around the question of how the environment, as a condition, produces enduring variations. Well into the nineteenth century, that approach served to explain not only species variation as such but, in Morel's terms, the "pathological transformation of the species" by an "unhealthy deviation from the normal type of humanity."[40]

The degeneration scenario had already acquired alarmist overtones in the first half of the nineteenth century, and after the events of 1870/71, it became a standard trope in anxious debates on the state of the nation. Degeneration was now a common denominator for all sorts of perceptions of decline, and it sharpened them into nationalist antagonisms.[41] The industrialized cities and their slums took on ever more unnerving prominence as the malign *milieu social* that Morel had identified as a crucial cause of degeneration.[42] Additionally, the civilizational critique intrinsic to the old notion of degeneracy was now reactivated as cultural critique, the etiology of a modern society fallen prey to degeneration.

When Legrand du Saulle, Morel's one-time assistant, sought the causes of the "extremely worrying" increase in mental illness in 1871, he cited a whole collection of cultural phenomena — not ephemeral events such as particular political revolts, but long-term conditions: "lax upbringing, perverted teaching, the absence of all belief, the lack of all moral sense in literature, the cult of egoism, the desire for material enjoyment, the thirst for money, unbridled speculation, incessant anxiety resulting from a strained commercial situation, playing the stock market, sudden changes to the state of persons or fortunes, worries about the future, the constant progress of alcoholism, and the shameful refinements of debauchery" — altogether a "slow, continued, and progressive deterioration of education, habits, and public mores."[43] In the Third Republic, this spectrum of factors was joined by the challenge of coping yet again with social division and the national humiliation of defeat by Prussia. Moreover, those turbulent six months seemed to have given an objective foundation to Morel's vision of a nation changing pathologically in the direction of ultimate sterility: not only had stillbirths risen in the period, but the birth rate had stagnated ever since.[44]

Ab utero

In French medicine, with its physiological orientation, the concept of degeneration remained embedded in the theory of milieu until the early twentieth century. Degeneration was blamed not on a proliferation of "wrong" people, but first and foremost on a "wrong" culture, creating the conditions under which epidemic diseases (specifically, syphilis, alcoholism, and tuberculosis) constantly threatened to tip into endemic morbidity by damaging germ, embryo, and fetus.[45] The last third of the nineteenth century, however, saw a certain convergence between heredity-oriented and physiological notions of degeneration, resulting in what Féré considered a hopeless confusion that cried out for clarification. In an 1894 paper published in the *Revue des deux mondes* (Review of the two worlds), a journal addressing a general French and American readership, he attempted to disentangle these theories of degeneration while also countering what he saw as their mistaken popularization.

Fundamentally, Féré's article criticizes all those who have interpreted the pathological transformation described by Morel as an atavistic relapse into earlier evolutionary stages.[46] In Féré's eyes, that interpretation confuses heredity with degeneration. Just as he did in *La famille névropathique*, discussed in Chapter 6, Féré here distinguishes two main forms of pathogenic connection between the generations. Such connection is heredity when progenitors transmit natural or acquired diseases to their offspring; it is degeneration when an infection of the mother passes over to the unborn child or when the parents' inebriation, the mother's violent feelings or chronic distress, or nutritional deficiencies in the womb produce anomalies and pathologies in the child. There is also "degeneration via heredity," in which not the disease itself, but a predisposition to disease, is transmitted from parents to children.[47] Unlike heredity, degeneration always involves a dissimilarity in the next generation, so that the characteristic features of the "race" are lost. "Degenerated" children resemble neither their parents and siblings nor an "ancestral type."[48]

Féré underpins his distinction between degeneration and heredity with several pages describing the developmental pathologies that he has studied experimentally. He asserts that the "processes of embryogenesis" are the same as the "processes of nutrition," and that "influences of the milieu" are "modifications of nutrition." At the end of the paper, Féré mentions Darwin, but not with reference to evolutionary theories. It was Darwin, he writes, who found the "reproductive function" to be "the most delicate" of all the organic functions.[49] Reacting to so many factors, it makes its product the product of circumstances.

In this text, as so often, the children of the year of terror supply a persuasive illustration of the point. Féré calls on them to exemplify an originary degenerative process, in which neither heredity nor hereditary degeneration plays a part. In *Sensation et mouvement* of 1887, he frames that idea terminologically by describing the children as "congenitally degenerate (*ab utero*)."[50]

Everything that anchored evolutionary notions of degeneration in the last third of the nineteenth century, and ultimately pushed them toward eugenicist aggression, can be found in Féré's writing as well: atavism, inherited disease, the defective germ.[51] If Féré advocated a physiological model of degeneration, that was because he wanted to highlight environmental influences rather than narrowing the definition of decline to the framework of heredity theory. His interest in possible ways of remedying the situation makes that evident. For Féré was by no means fatalistic about degeneration, and the postulate of selection as propounded by eugenicists was not his resource of choice. In this, too, he remained true to Morel, for whom both hygiene and a certain measure of *métissage* promised to deliver regeneration. Féré argued that although pathogenic influences insert morbidity into the generational sequence, that process can be interrupted by improving nutrition.[52] Accordingly, he derived an expanded notion of "prophylaxis" from his teratological experimentation, and also envisaged mixing healthy with diseased families.[53]

This in no way detracts from the hostility inherent in his talk of children who are degenerate *ab utero*. Elsewhere, Féré declares "the degenerate" to be "unproductive" or even "destructive" from "the social point of view."[54] And in his paper in the *Revue des deux mondes*, directed to a colonialist French and a racially segregated American public, he presents an analogy between the distinctions in society caused by degeneration and the consequences of "introducing foreigners of too-divergent races."[55] Even in Morel's tradition, the concept of degeneration was a discourse of order, dominated by psychiatry, that somatized everything wayward, intractable, and disorderly and, as Foucault argues, slotted it into a transgenerational process of "metasomatization."[56]

Féré's distinction between degeneration and heredity is nonetheless crucial if we are to understand the implications of regarding generalized emotional distress as a risk factor that threatens the nation. The milieu-oriented notion of degeneration emphasized the aspects of event and collectivity—not by accident, a best-selling American book on degeneration later presented the *enfants du Siège* in the company of children born after the Great Chicago Fire of 1871, the stock market crash of 1873, and the financial crisis of 1875–80.[57] Morel himself, in the wake of the 1848 revolutions, had ascribed a "degenerative" effect to the shared experience of political unrest, which he considered substantially more significant than revolutionary madness as such.[58] Legrand du Saulle returned to Morel's hypothesis after 1870/71, and in the years that followed, he kept an eye out for evidence. When the children of the Paris siege seemed to supply just such proof, Féré set off on his quest to find the mechanisms by moving his experimental focus from teratology to prenatal pathology.

This case shows paradigmatically that in the nineteenth century, historical events were negotiated in terms of the life sciences; equally importantly, biological events were constituted as a historical process. Evolutionary theory in the second half of the century had detached species history from the older theory of stages and opened it to contingency—in other words, to a nondirectional historicity.[59] At the same time, the development of the individual organism was approached as a process conditioned by contingent circumstances that were sometimes collectively shared. The reproduction of the collective became a doubly open-ended process, and one in which specific events played a part in determining how the past transitions into the future.

This explains why, in 1884, Legrand du Saulle replaced the horror of which he had spoken in 1871 with a term more suited to describing a temporal connection between event and repercussion: "mental trauma" (*traumatisme moral*).[60] The temporal dimension of the concept of trauma was what made it relevant for clinical psychiatry in the last third of the century. The term could describe a shock-like experience that is inflicted on the psyche just like a physical wound and whose injurious effects persist over time—not merely as a sedimentation of things past, but as a complex composed of event and subsequent effect, since it is only through the continuing impact of memory that trauma unfolds.[61] Trauma in this sense is a process in the individual mind, but under Jean-Martin Charcot's direction, the concept was reinforced by the collective reverberations of national defeat and uprising.[62] They simultaneously prompted theories on the inheritance of trauma and attracted attention to pathogenic development.[63]

Seen from this perspective, the fascination that the children of 1870/71 exerted at the Salpêtrière was not due solely to a psychiatric interest in childhood disorders. It also reflected an interest in the mode of subsequence that was at work when anomalies in one generation were interpreted as an effect of experiences in the generation before. The more highly charged the events of the year of terror were for the present day, the more obvious it appeared that a temporary event can be experienced in a manner that does not simply leave a trace but sets off a whole cascade of effects. The anomaly of the children is not the suffering of the parents, just as the trauma is not the shock, but both transcribe something temporary into the future. In both, the event remains present by means of an effect that no longer contains it.[64]

When physicians debated the enduring presence of past besiegement and rebellion in a nation that drew its identity from generative continuity, they were touching on a fundamental question: How can an event connect the past, present, and future? French society in the Third Republic was preoccupied with the fear that the events of 1870/71 might have rent apart that continuity, making the past destructive, the present precarious, and the future uncertain—a fear that gave the generational significance of the *enfants du Siège* its remarkable momentum. The diagnosis of a "congenital degeneration" caused by events evoked the image of a nation threatened *ab utero*. That is why Dr. Dupouy, a skeptic on the matter of these children, concluded his call to rehabilitate the 1870/71 generation by conjuring up national continuity across the abyss of defeat and unrest—using gendered expressions and a future tense: "They will be men."[65]

Embodying Trauma

The children of the Parisian year of terror indicated most immediately that a mother's suffering was connected to her child's anomaly, yet at the same time they cast doubt on the exact extent of that connection. Their high profile reveals the degree to which, in the course of the nineteenth century, the individual had become part of a whole, whether that whole was "degeneration" or "present-day society." This distinguishes the case of the lawyer's daughter, explained by Féré as exemplifying psychical influence, from the many incidents of maternal impression or *Versehen* that had previously populated the scholarly literature. The distinction brings me back to the case with which I began this book, because it reveals the process by which fetal life became a way of thinking about societal time.

Like the many pre-nineteenth-century anecdotes about the power of maternal imagination, in the account of the lawyer's daughter an individual episode draws its explanatory power from being part of a series. However, the two kinds of series differ in terms of their procedure, ascribing different epistemological functions to the individual episode. Every reported incident of maternal *Versehen* is one more example added to a cumulative series of incidents, whereas the significance of Féré's story depends on the fact that the girl's case isolates just one factor—psychical influence—within a statistical sample in a quasi-experimental form. It exemplifies the close mutual dependence of experiment and large numbers in the search for causal relationships. But the statistical series of the *enfants du Siège* is also significant in itself, because it produces a double collective: the synchronic collective of a generation of children similar to each other but dissimilar to their parents; and the diachronic collective of a nation whose generational sequence has been corrupted by deviation.

One might say that the statistical series of the children made Paris a "menagerie" of generations, like the collection that Camille Dareste envisioned when he extended his teratological laboratory experiments into the question of species variation.[66] Whereas the cases of *Versehen* had testified to a process that could happen, or fail to happen, at any time or place, the story of the lawyer's daughter was about a process that was both widespread and extended across time. It was not the isolated occurrence of something possible in nature, but the individual exemplar of a collective biological process.

Étienne Geoffroy Saint-Hilaire had linked bioscientific investigation with statistical rationality in the study of the maternal mind when, in the early 1820s, he scoured the demographic data for a correlation between developmental anomalies and single motherhood. As I argued in Chapter 7, what interested him was causality.[67] But as well as sketching a configuration of risk, correlating frequencies also constituted populations: a maternal population defined by its circumstances and a fetal population defined as the product of those circumstances. Foucault has shown that the concept of populations goes back to the eighteenth century, when both human beings and physiological events began to be counted—how frequently or seldom births took place; how many living, dead, or deformed beings were born; which illnesses spread among the living and how; when deaths occurred and why. Regularities and peculiarities emerged in the rates of birth, disease, and mortality, so that the totality of people living in a territory appeared not "as a collection of subjects of right, nor as a set of hands making up the workforce," but as a "population" within which the vital processes of a species are accomplished.[68] This idea was the precondition for Geoffroy Saint-Hilaire's venture into the statistical records. By extracting a natural experiment from them, he circumvented the problem that human beings are not chicks that can be consumed for research. He also ordered a physiological, developmental process within the life of a population, by interlocking the question of how childhood anomalies are caused with the question of how to ascribe causes to particular *groups* and of how those groups should be related to the population as a whole.

At the time, pregnancy had already begun to be colonized by what Foucault calls "biopolitics," which promoted hygiene, sanitation, and public health campaigns on a grand scale in the course of the nineteenth century and would later integrate obstetrics and pediatrics into the "demographic maneuvers" of the state.[69] The institutional and epistemological foundations for this change arose as two epochal processes came to fruition. First, pregnancy moved from the hands of midwives into those of medical authorities, physicians who saw themselves as the agents of state interests; second, statistical rationality was introduced into obstetrics, where the identification of fetal normality and perinatal risk factors began to compete with the traditional orientation on individual cases.[70]

Much earlier, in the eighteenth century, there had been appeals to mothers like the advice that Jacques A. Millot, former obstetrician to Marie-Antoinette, formulated in his 1801 self-help book on the art of procreation, *L'art d'améliorer et perfectionner les générations humaines.* The widely read manual secured Millot's livelihood after the Revolution. In his preface addressing the "ladies of France," he tells them that "the good of the country" rests in their loins because man acquires his physical and mental characteristics not only through education, but especially "in" the mother, where he is subject to the influence of her climate, her diet, and her lifestyle.[71] Such vital tasks cannot safely be left to the pregnant woman alone. Millot exhorts expectant mothers to "be guided by knowledgeable men" who are committed to the goal of "perfecting the human being."[72] For Millot, this objective can unite the parents' joy in their child with the patriotic interest in "improving and perfecting" future human generations, as the title of his book proclaims.

Millot's handbook continued an aesthetics of procreation, dating back to antiquity, that offered guidance on making beautiful and talented sons and daughters. From this tradition, he extracted a bundle of venerable hygienic measures. Whether the choice of spouse, the constitution of the semen, the degree of arousal during conception, the stories that a pregnant woman listens to and the foods she eats, or the composition of the mother's milk, there was much that could be influenced.[73] Millot disposed these generative actions within a conceptual framework informed by the "science of man" and physiological ideas of the organism: the notion of the womb as the human being's first milieu. In the eighteenth century, this idea was invigorated by the Enlightenment's aim to unify human perfection and patriotic duty.[74] The whole of the nation was at stake, since, in contemporary perceptions, unfavorable conditions — insalubrious terrain, toxic urban fumes, careless wet-nursing — were ubiquitous and afflicted the population with a veritable epidemic of death, sickness, and enfeeblement.

More than fifty years after Millot's publication, obstetricians and gynecologists similarly articulated the duty of each individual pregnancy toward a larger whole. Their most theoretically ambitious representative, Adolphe Pinard, echoed Millot's formulations, writing of "loins" and "improvement," but his program of *puériculture intrauterine*, the intrauterine care of children, pursued very different goals. In 1865, the pediatrician Charles A. Caron had coined the term *puériculture* to stake out childcare as a particular domain within the field of public health.[75] Thirty years later, Pinard expanded its terrain by adding the intrauterine environment. He presented his agenda in an 1895 lecture at the Academy of Medicine, and a few years later would put it into political and governmental practice as president of the state depopulation commission.[76]

Pinard had made a name for himself in the 1870s with a statistical investigation of the links between infant mortality and the fetus's presentation at birth. His conclusion was that abdominal palpation, for which he provided precise instructions, should form part of preventive care in pregnancy.[77] This study was firmly rooted in practical obstetrics, but simultaneously Pinard was writing a highly erudite entry on the fetus for Amédée Dechambre's authoritative medical dictionary, incorporating everything then known about the physiology of fetal life and the pathology and teratology of fetal development.[78] Against that backdrop, he initiated a study to underpin empirically his call for a state-led intrauterine puericulture. The birth weights of two groups of newborns were compared, five hundred whose mothers had worked, standing or sitting, until the birth and five hundred whose mothers had been able to rest before the birth in specially designed hospital bed. The comparison showed a clear correlation between overexertion in the mother and feebleness in the child, which bolstered Pinard's call for pregnant women in the third trimester to be supported by a state pension that would free them from all kinds of paid employment.[79]

As he moved from one study to the next, Pinard did not merely expand his scope, but altered his object. By drawing an arc from infant mortality to congenital morbidity, and from the conditions under which individual mothers gave birth to the living conditions of whole populations, he constituted pregnancy in a way that continued down a path prepared over the nineteenth century by the work on fetal physiology and fetal pathology he knew so well. Pregnancy in that guise is a series of developmental events, susceptible to pathogenic influences, that is simultaneously a generative connection. Pinard calls these events "uterine heredity," in contradistinction to germinal heredity.[80] An epidemic milieu that damages numerous individual acts of generation across the nation became an endemic degeneration that disrupts the continuity of a nationally defined collective. Though Millot's notion of improvement in the womb still resonates in Pinard's intrauterine puericulture, at the end of the nineteenth century talk of the womb was no longer concerned with a place. Instead, it described a connection that transmuted a present society's conditions into its future members' organic properties.

The era of medically and scientifically orchestrated prenatal care with state involvement had now begun. The constitution of the unborn came to a preliminary conclusion in what I have described as the making of fetal life. In the eighteenth century, the womb had become a place in which—"early on," as Albrecht von Haller put it—an organism lives because it is nourished by a milieu. In the course of the nineteenth century, this topological nexus, embodied by the pregnant woman, became a temporal nexus embodied by fetal life. Fetal life as an epistemic thing is not only genuinely temporal on its own account, in that it accomplishes development; it also connects different times. In the open-ended development of the organism, the mother's present becomes the child's future. Social time occurs in this combination of biographical times because both depend on conditions that, as environment, define populations and, as influence, connect generations. Accordingly, the adjective through which this concept of the unborn would come to prevail in the early twentieth century was no longer a spatial one, *intrauterine*, but a temporal one, *prenatal*. It is no coincidence that John William Ballantyne, who launched that term in his manual of prenatal pathology and hygiene, identified kindred spirits not only in Féré, but also in Pinard.[81]

In the course of the nineteenth century, then, the unborn became a threshold in time in a dual sense. As an organism in the process of developing, it was no longer just germ and not yet quite child; as the embodiment of a generational connection, it held in abeyance the future of a society that drew its unity from an imagined national character and its supratemporal identity from the continuation of that character. Nothing evoked all this more vividly than the children of 1870/71. Their disposition and behavior were taken by their mothers to be the legacy of a disaster, while on the national stage, they became the object of disputes in which the past was negotiated and fears for the future brought to bear. No influence on fetal life, moreover, made all this more impressively concrete than the afflicted mind of the pregnant woman. That injury was read differently over time. For Millot it had been the excessive emotion of the negligent, for Geoffroy Saint-Hilaire it was the malaise of a population, and it now became a generation's trauma.

CHAPTER ELEVEN

Of Human Born

It is February 27, 1821, and the young physiologist Johannes Müller is "opening" pregnant animals. In the company of his friends, and perhaps at his kitchen table (that would not have been unusual for an early experimental physiologist), he cuts open the abdomen of sheep and rabbits, drains off the amniotic fluid, picks out the fetuses, lays them under glass receptacles, administers and removes oxygen, records their movements, yawning, and fear, and lets them die on the table.[1] The scene brings everything together: the experimenter, the fetal organism, and a milieu that conditions its life. Fetal life is made here on Müller's experimental bench, when the incision into the mother animal's body displaces the excavated fetus into a glass container. Knife, vessel, and fetus form the intersection of object and instrument where, in Hans-Jörg Rheinberger's account, we find the epistemic thing—the material entity or process to which the effort of inquiry is addressed.[2]

Müller's physiological experiments on fetal oxygen requirements were intended to complement his anatomical study of embryonic development. Anatomical embryology might have discovered how a germ gradually takes on human form, he believed, but it would remain incomplete until scientists knew how that form came alive by gradually acquiring the capacity to process nutritive substances, form blood and make it circulate, set its limbs in motion, keep itself warm, perceive things external to itself, and feel. And indeed, it was only physiological investigations such as Müller's that completed the momentous epistemological transformation by which the child inside its mother's body became a biological object. What was once a unit comprising mother and child now became a relationship between a fetal organism growing toward life and a maternal organism that is its milieu. This milieu produces both the substances that the fetus needs to live and the disruptions that cause its anomalies and pathologies.

In the case of human beings, however, a biologically objectified unborn continued to elude the scientist's grasp. Between the physiologist and the fetal organism stood the doubly impassible barrier of the pregnant woman, who is part of fetal life as its milieu and inaccessible as a personal body. Every study of fetal life in the human being is forced to do without its own "*material*," complains William T. Preyer in 1883, devoting the first pages of his book on fetal physiology to his yearning for visual access to the womb — though even that would not be enough for the physiologist, since functions appear only through experimentation. Preyer's hunger for transparency shows what was missing: the chance to observe the living fetus "in its natural surroundings" while it "still continues to develop in the uterus."[3] This deficit gave rise to a research practice of approximation, drawing closer to the living fetus through substitute objects (animals, premature babies, newborns) and substitute practices (listening to and palpating the mother's belly).

Returning to this and other aspects of my study so far, and to passages from sources I have already discussed, my aim is not to assess the failure or success of such approximations. In this concluding chapter, I wish to add an analysis of the relationship between the scientists of fetal life and the unborn itself. That relationship is not self-evidently one between a subject of knowledge and an object of knowledge. It is a relationship in which the scientist's efforts are bound to fail because something impassible conceals something unavailable.

This unavailability becomes obvious in an event, the moment when the pregnant woman's body relinquishes the object. Those who are quick enough may catch one last scrap of fetal life in the "just-born."[4] They stimulate the newborn's skin, tongue, ears, and eyes within more or less systematic experimental setups in order, as Gabriel Compayré puts it, to discover "the natural gifts that he possesses at his birth," which can "assist us in clearing up, within the range of the possible, the obscure history of the nine months of gestation."[5] Yet when those gifts are studied, the just-born turns out to be already-child and still-fetus in equal measure. It is as ephemeral as the womb is inaccessible, and dissolves into the continuum that makes birth an epiphenomenon. That is development: an uninterrupted process turning one life form into another — a "shining current," says Preyer, "that becomes ever broader, flows ever faster, and into whose clear waters I peer without ever finding the riverbed, even when no errant wave is wrinkling the surface."[6]

In that case, who can know what kind of being arises from the womb? It is "more animal than human, but . . . by no means a machine," noted Siegfried Bernfeld in 1925.[7] The physiologist hoped for an object that would reveal what had happened in the time of obscurity, but it proved impossible even to say what the resulting creature *is*. More than a hundred years of research failed to yield a definition. On the contrary, the more fetal life was studied as a genesis of functions, the less possible it was to determine when the defining property of the human being begins — when the human organism already lives but does not yet sense, when it senses but does not yet feel, when it feels but does not yet have consciousness, when it has consciousness but not yet a will. The physiologist was left with his own affects. Preyer felt dizzy at the attempt "to pin down developmental processes," while it was "fear" that prevented Bernfeld from "denying the psychical in the new-born."[8]

One might object that none of this should have worried the sciences. Their interest was in development, not in the beings that develop. But the various attempts to split off the unborn (as something physiological) from the born (as something psychological), and the converse insistence on a psychogenetic unity between the unborn and the born, may just as well be interpreted as different reactions to the fact that the object resists knowledge. Bernfeld sees both responses as arising "when one has arrived at the end of factual knowledge" and finds oneself surrounded by speculation and "non-scientific attitude."[9] Perhaps the border is only provisional, he suggests. Perhaps, though, it is not.

A creature that seems to be something other more than it is itself raises the question of what is not known, but it also confronts the beholder with a particular kind of non-knowledge, a knowledge that is discounted.[10] As Foucault explains, such invalid speech is talk that scientific discourse does not even consider erroneous, because there is no way of judging whether it is true or not. Instead, it is simply "null and void."[11] In the study of life before birth, then, we find a lack of knowledge as to what the fetus can or cannot yet do; and we find unqualifiable talk as to what the unborn is. What draws the line between these two kinds of deficit is not an epistemological gap, but a power gap. The border between not knowing and non-knowledge takes shape when certain talk is banished from the domain of knowledge.

That talk is, first and foremost, mothers' talk. James Sully, a self-declared authority on the science of babies, warned in 1881 that the mother is an "inaccurate observer" whose accounts are not to be trusted.[12] Here, once more, the developmental sciences come up against the mother: she encloses the object yet again, this time not with her body, but with words. Before birth, the opacity of the mother's body withdrew the unborn from the scientific grasp; after birth, the unreliability of her words blocks potential knowledge of the born. This logic devalues maternal speech on the grounds that a mother's statements must result from a specific relationship to a specific object.

Sully's admonishment forms part of a ceaseless stream of complaints about that relationship. They began just where the objectification of the unborn was initiated, in eighteenth-century anatomy. At that time, the Italian anatomist Giovanni Battista Morgagni lamented that the "tenderness of mothers" made it impossible for the pathologist to perform autopsies on stillborn babies.[13] In 1828, Charles Billard announced that he had fulfilled "Morgagni's wish" and autopsied the corpses of newborns, but this had been possible only for babies unprotected by maternal affection: foundlings.[14] Those who wanted to lay their hands on the living object were no better off. Preyer's aspirations to the systematic observation of large numbers of newborns were sabotaged by the "often insurmountable difficulty in having the infant separated from its mother or nurse."[15] At the beginning of the twentieth century, Silvio Canestrini was able to overcome that impediment, but only because the authoritative arrangement of bodies in the hospital allowed it — which is why he does not thank mothers for handing him their children, but the board of the maternity clinic for "granting me the infant material."[16]

The reason why mothers and nurses deprive the scientists of the object is that their relationship to the child springs from a gaze that does not objectify. "Her way of looking at babies," writes Sully, makes the mother "a formidable obstacle" to research. The hindrance is even "more invincible" in the case of the nurse, who, determined to shield her "protégé," is capable of "barricading the cradle against his [i.e., the father's] scientific approaches." Even if, as Sully recounts of his own wife, "the mother herself gets in time infected with the scientific ardour of the father," she is apt to cause havoc with the study because "her maternal instincts impel her" always to see in the child her *particular* child.[17] What for the scientific gaze is *a* child among others, a window into abstract childness, for the tender and protective gaze is *this* child. And anyone who, like the mother, does not see the childness of a child because she insists on recognizing *this* child can have nothing to say about children that is capable of being either correct or incorrect.

But it is not only the words of mothers that cannot be true because they flow from a relationship of recognition. What, Preyer asks, could "the great Immanuel Kant" know when he interpreted the cries of a newborn child as a protest against its encounter with the world? "Had he ever, in his long life as a bachelor, seen and heard just-born children?"[18] Kant's words are just as invalid as those of a mother, because they too arise from recognition, albeit on entirely contrary grounds: not because Kant always recognizes a child as *this* child, but because he knows *no* child at all. In the child, he sees only himself. In Preyer's account, Kant thinks the cries of the newborn baby carry a "tone of outrage and furious anger" that indicates "an expression of volition." For Kant, it is "not because something pains him, but because something vexes him, that the just-born human being must cry, and he cries because he wants to move and experiences his incapacity to do so as a fetter, robbing him of freedom!"[19] Only a philosopher who has never seen and heard a newborn infant, Preyer concludes, could infer from the baby's demeanor an expression of its will to freedom, the definition of the human subject as which he sees himself.

"For thousands of years, children have been born and lovingly tended and observed by their mothers, and for thousands of years, the erudite have argued about the child's ensoulment without studying the child themselves," writes Preyer.[20] By conjoining Kant and mothers in a single sentence as the two poles of invalid discourse, he opens up a space of knowledge between them. In the human sciences, which locate truth in controlled experiential knowledge, the words of the deskbound philosopher are too empty of experience, those of the loving mother too full. Between these two figures — one of whom only wants to know and does not experience, while the other only experiences and does not want to know — stands the one who both experiences and knows, and who can therefore be "in the true."[21] He is what Sully calls "the scientific father."[22]

Unlike the childless scientist, the scientific father has access to the object, enabling him to do what can be done "only in the nursery." Almost every day from the moment of birth, Preyer tells us, and "at least three times a day," he studied "the child" that was his son. Unlike the mother in the nursery, he found "almost every day some fact of mental genesis to record" in his diary.[23] He observed how the erratic twitching of hands and legs became the choreography of a mastered body, how periods of sleep shortened and wakefulness extended, how sucking became eating, reflex became will, and cries became babbling that culminated in speech. One child is as good as another for this purpose, for even though each one is different, the "important matter" is the "order in which the steps [of development] are taken," and these steps "are the same in all individuals."[24] The father is scientific inasmuch as he detects childness in his child because he has available *a* child that he can insert into comparisons whenever "several men with a thorough physiological training" independently do the same, or when "fathers who are friends of mine" exchange and mutually verify "observations made upon their own children."[25] Speech can be in the true where, rather than *this* child being wrapped in a relationship of recognition, a relationship of objectification allows *a* child to manifest the sequence of developmental moments that define childness. Women are quite capable of objectifying too, and so they do, encouraged and praised by Preyer—as long as they do not speak as mothers.[26]

But objectification and recognition have the same object, and in that sense they coincide. Taking a more conciliatory tone, Preyer notes: "Mothers will not be cheated of the magic of the child's gaze if fathers also take an interest in the manner by which the infant eye moves, in which resides pure truth."[27] The child as object divides into eye and gaze, but whereas the mother is enchanted by the gaze alone, the scientific father sees both. Just five minutes after the birth, Preyer held his son to the window in the dusk light, monitoring the blinking of his eyes, and yet the "poetry of the infant soul" remained "inviolate."[28]

For the scientist studying functions, the eye is a thing, but he is aware of the instinct of the mother, who sees in the child's eye a response to her own gaze. That is because every moment of observation repeats an event in the history of science. As Sully puts it, only a "harmless conspiracy" among mothers was able, over the course of millennia, to make the "sluggish brain of man" receptive to "the mysterious charm that surrounds the baby."[29] It was through the detour of the maternal gaze that late eighteenth-century science found the puzzle that it then set about deciphering as a sequence of developmental moments. "Before we know it," writes Preyer, "the helpless infant is transformed into a being the same as us." For this proponent of the human sciences, the charm of *this* child is the magic of the human being. And so, just like the mother and like Kant, he is connected to his object by an act of recognition preceding all descriptions: the object is like us; it already *is* the human subject that it will become before we realize it. Preyer concludes: "We wonder at our development and do not understand it."[30]

Truths about the unborn, therefore, are mingled in the countenance of the born. The physiologist detects in the eye's motion knowledge about his object, the mother finds in her child's gaze certainty about something commended to her care, and Kant draws in the sand at the edge of the sea the law that humanity must always already be present in the human being. The developmental scientists undertook much, even everything, to cleanse their truth from the truths of the others. They cut up pregnant animals, pinched the toes of premature babies, glued sensors to newborns' scalps. But setting out to read from an object how the human emerges from a mother's body, they found themselves faced with a being that demands to be recognized as one of their own.

In the long nineteenth century, "development" was probably one of the things that, as Philippe Descola writes, "finally liberated humans from the matrix of objects both animate and inanimate" and slotted them into the laws of nature. Yet the same thing tied the rationality of the human sciences to the "dark muddle of the experience of others."[31] A profound ontological indeterminacy of the unborn installed itself in the sciences themselves. The transformation of that indeterminacy into an epistemological question about the beginnings of subjectivity was bound to remain unfinished. Just as the human being in the process of development became a liminal creature, always on the border between no longer and not yet, the human scientists' talk about the human child wavered on the threshold between knowledge and non-knowledge. Perhaps they could have found their balance in the words of mothers.

> My daughter doesn't yet have a tongue: I'll speak for her. My daughter already has eyes; what does she see? Does she have ears, I don't know.
>
> — Rahel Hutmacher, *Tochter*, 1987

Acknowledgments

My research for this book began many years ago in Princeton. I was working on something quite different when, in a footnote, I stumbled across the children of the French *année terrible*, 1870/71. The Institute for Advanced Study's intellectual generosity and the trust it places in its members meant my detour could become a destination. In the years that followed, I continued this research at various locations: at the German Historical Institute in Paris and the Research Center for Social and Economic History in Zurich, both with the support of the Swiss National Science Foundation; in Berlin at the meetings of the German Research Foundation's "Economies of Reproduction" network; and finally at the history department of the University of Basel. With working conditions that magically mirrored the project's beginnings in Princeton, the Institute for Advanced Study Konstanz gave me the time I needed to create a book out of a project beset by interruptions. My thanks go to everyone in these institutions who believed that something could grow from my fascination with a seemingly minor episode from medical history.

Many friends and colleagues accompanied my research with an open ear and nourished this book with both inspiration and hands-on support. I thank Ina Boesch, Tatjana Buklijaš, Caroline W. Bynum, Barbara Duden, the late Esther Fischer-Homberger, Duana Fullwiley, Christine Loytved, Andrew MacFadyen, Rachel Mader, Marika Moisseeff, Claudia Opitz, Florence Rochefort, Martin Schaffner,

Regina Schulte, Joan W. Scott, and Marianne Sommer. Lea Bühlmann, Jennifer Burri, and Joana Burkart found out-of-the-way sources and information for me and brought order to my bibliography. Katharina Böhmer supervised the making of the German edition with the greatest dedication.

It continues to fill me with gratitude that the editors of Zone Books added my book to their program, and I thank Ramona Naddaff and Meighan Gale for their warm welcome to the Zone community. Thanks, as well, to the anonymous reviewers, who helped me by identifying weak points and sharing their perspectives on the subject matter. Kate Sturge translated the book with outstanding sensitivity; her care and good humor have made our collaboration a most joyous journey. I am grateful to the Börsenverein des Deutschen Buchhandels for the award that funded the translation.

All along the way, Simon Teuscher could be relied upon to make me laugh. Saskia Walentowitz gave me her hospitality in Paris, and later supplied me with all things anthropological. Urs Hafner and Claudia Honegger were both gracious and ruthless in reading and criticizing the manuscript at various stages. Dagmar Herzog's friendship has supported my writing for many years. Andrea Feller, Michael Herzig, and the Arni and Wehrli families are the village that it takes to raise a child. Stefan Wehrli is always there, considerate and generous in all things.

Nora Arni and Julien Arni have taught me the optimism that every mother needs, and I offer this book to them. It is also dedicated to all those who had to struggle to keep their seat when photographic portraits were wrested from them in May 1880 at the Parisian Bicêtre hospital—images that were intended to document anomalies but in fact show children. Those children challenged me to make the decisions that gave *Of Human Born* its present shape.

Notes

PREFACE

1. Translator's note: Webs of terminology relating to the mind do not transfer easily between English, French, and German. Thus, the French adjectives *psychique*, *mental*, and *moral* — and equally the nouns *esprit* and *âme* — map only partially onto the German *psychisch*, *mental*, *geistig*, and *seelisch* or the English *mental*, *psychological*, and *psychical*, the *mind*, *spirit*, *intellect*, and *soul*. I have translated according to the context, adding the source-language term where the semantic boundaries seem particularly unclear.

CHAPTER ONE: PARIS 1870/71

1. Charles Féré, "Les enfants du Siège," *Progrès médical* 12.13 (March 29, 1884), p. 246.

2. Féré, "Les enfants du Siège," p. 246.

3. Ibid., p. 245.

4. On the context of this photograph, see the periodical edited by Désiré-Magloire Bourneville, *Recherches cliniques et thérapeutiques sur l'épilepsie, l'hystérie et l'idiotie: Compte-rendu du Service des enfants idiots, épileptiques et arriérés de Bicêtre* (Paris: Progrès médical, 1881–1908). Bourneville's observations are cited by various authors, including Paul Moreau (de Tours), *La folie chez les enfants* (Paris: J.-B. Baillière, 1888), pp. 37–38. Moreau also mentions observations of his own. On Bourneville's role in child psychiatry, see Gerhardt Nissen, *Kulturgeschichte seelischer Störungen bei Kindern und Jugendlichen* (Stuttgart: Klett-Cotta, 2005), pp. 109–10.

5. Henri Legrand du Saulle, "Influence des événements politiques sur les caractères du délire et anomalies physiques et intellectuelles que l'on observe chez les enfants conçus

pendant le siège de Paris," *Le praticien* 7 (1884), pp. 160–63 and 184–86. I follow the usage in the sources, where the term *enfants du Siège* also encompasses the *enfants de la Commune*. On the historical events surrounding the period from fall 1870 to spring 1871, see Francis Démier, *La France du XIXe siècle, 1814–1914* (Paris: Éditions du Seuil, 2000), pp. 289–300.

6. Legrand du Saulle, "Influence des événements politiques," p. 185.

7. Jean-Martin Charcot, *Leçons du mardi à la Salpêtrière: Policlinique 1888–1889* (Paris: E. Lecrosnier & Babé, 1889), pp. 115 and 116.

8. Legrand du Saulle, "Influence des événements politiques," p. 185.

9. Ibid., p. 184.

10. Ibid. On the food supply in besieged Paris, see Bertrand Taithe, *Defeated Flesh: Welfare, Warfare and the Making of Modern France* (Manchester: Manchester University Press, 1999).

11. Legrand du Saulle, "Influence des événements politiques," pp. 184–85.

12. On these physicians in the context of Charcot, see Julien Bogousslavsky (ed.), *Following Charcot: A Forgotten History of Neurology and Psychiatry* (Basel: Karger, 2011).

13. See, for example, Émile Forgue, "Traumatisme," in *Dictionnaire encyclopédique des sciences médicales*, ed. Amédée Dechambre, 3rd series, vol. 18 (1885), pp. 39–52. Forgue discusses the course of a traumatic lesion in the case of neurological illnesses but does not yet have a concept of psychological trauma.

14. On the rise of the psychological concept of trauma, see Mark S. Micale and Paul Lerner, "Trauma, Psychiatry, and History: A Conceptual and Historiographical Introduction," in *Traumatic Pasts: History, Psychiatry, and Trauma in the Modern Age, 1870–1930*, ed. Mark S. Micale and Paul Lerner (Cambridge, UK: Cambridge University Press, 2001), p. 10; Esther Fischer-Homberger, *Die traumatische Neurose: Vom somatischen zum sozialen Leiden* (Bern: H. Huber, 1975), p. 104.

15. See Mark S. Micale, "Jean-Martin Charcot and *les névroses traumatiques*: From Medicine to Culture in French Trauma Theory of the Late Nineteenth Century," in Micale and Lerner, *Traumatic Pasts*.

16. Fischer-Homberger, *Die traumatische Neurose*, p. 53, and for detail on shock theory, pp. 34–36 and 47–57. For Féré's definition of shock as a concussion of the brain, see Charles Féré, "Contribution à l'histoire du choc moral chez les enfants," *Bulletin de la Société de médecine mentale de Belgique* 74 (1894), pp. 333–40.

17. Féré, "Les enfants du Siège," pp. 245 and 246. On differential diagnostics, see also Legrand du Saulle, "Influence des événements politiques," p. 185; Charles Féré, *La famille*

névropathique: Théorie tératologique de l'hérédité et de la prédisposition morbides et de la dégénérescence (Paris: F. Alcan, [1894] 1898), p. 22.

18. Féré, "Les enfants du Siège," p. 245.

19. On Féré's biography, see Roger Courtin, *Charles Féré (1852–1907), médecin de la Bicêtre, et la "Néo-psychologie"* (Paris: Connaissances et Savoirs, 2007); Fréderic Carbonel, "Le Docteur Féré (1852-1907): Une vie, une œuvre, de la médecine aux sciences sociales," *L'information psychiatrique* 82 (2006), pp. 59–69.

CHAPTER TWO: CHILDREN, FETAL LIFE, AND THE PRENATAL

1. See Erna Lesky, *Die Zeugungs- und Vererbungslehren der Antike und ihr Nachwirken* (Wiesbaden: Franz Steiner, 1950), p. 103.

2. See, among others, Sarah S. Richardson and Hallam Stevens (ed.), *Postgenomics: Perspectives on Biology after the Genome* (Durham, NC: Duke University Press, 2015); Hans-Jörg Rheinberger and Staffan Müller-Wille, *The Gene: From Genetics to Postgenomics*, trans. Adam Bostanci (Chicago: University of Chicago Press, 2017); Eva Jablonka and Marion J. Lamb, *Evolution in Four Dimensions: Genetic, Epigenetic, Behavioral, and Symbolic Variation in the History of Life* (Cambridge, MA: MIT Press, 2006). Discussing the political implications: Maurizio Meloni, *Political Biology: Science and Social Values in Human Heredity from Eugenics to Epigenetics* (New York: Palgrave Macmillan, 2016), and more recently Meloni, *Impressionable Biologies: From the Archaeology of Plasticity to the Sociology of Epigenetics* (New York: Routledge, 2019). On the fuzziness of the nature/nurture duality, see Evelyn Fox Keller, *The Mirage of a Space Between Nature and Nurture* (Durham, NC: Duke University Press, 2010).

3. See the *Journal of Developmental Origins of Health and Disease*, which has been in print since 2010. A historical contextualization is offered by Peter D. Gluckmann, Mark A. Hanson, and Tatjana Buklijaš, "A Conceptual Framework for the Developmental Origins of Health and Disease," *Journal of Developmental Origins of Health and Disease* 1.1 (2010), pp. 6–18. On the history of the field, see also Tatjana Buklijaš, "Food, Growth and Time: Elsie Widdowson's and Robert McCance's Research into Prenatal and Early Postnatal Growth," *Studies in History and Philosophy of Biological and Biomedical Sciences* 47, Part B (2014), pp. 267–77.

4. See, among others, Peter D. Gluckmann, Mark A. Hanson, and Tatjana Buklijaš, "Maternal and Transgenerational Influences on Human Health," in *Transformations of Lamarckism: From Subtle Fluids to Molecular Biology*, ed. Snait Gissis and Eva Jablonka

(Cambridge, MA: MIT Press, 2011); Sarah S. Richardson, "Maternal Bodies in the Postgenomic Order: Gender and the Explanatory Landscape of Epigenetics," in Richardson and Stevens, *Postgenomics*. Sarah S. Richardson has recently published a book-length history and critique of the science of "maternal effect" and "fetal programming" as it fed into epigenetic strands of DOHaD: Richardson, *The Maternal Imprint: The Contested Science of Maternal-Fetal Effects* (Chicago: University of Chicago Press, 2021).

5. Annie Murphy Paul, *Origins: How the Nine Months before Birth Shape the Rest of Our Lives* (New York: Free Press, 2010). Initial insights into the practical significance of epigenetically oriented DOHaD research can be found in Luca Chiapperino, Francesco Panese, and Umberto Simeoni, "L'épigénétique et le concept DOHaD: Vers de nouvelles temporalités de la médecine 'personnalisée'?," *Revue Médicale Suisse* 13 (2017), pp. 334–36. On the contemporary fetus as something under constant threat, see Deborah Lupton, "'Precious Cargo': Foetal Subjects, Risk and Reproductive Citizenship," *Critical Public Health* 22.3 (2012), pp. 329–40.

6. Of the rapidly growing literature, see, for example, Rachel Yehuda et al., "Holocaust Exposure Induced Intergenerational Effects on FKBP5 Methylation," *Biological Psychiatry* 80.5 (2016), pp. 372–80.

7. Henri Legrand du Saulle, "Influence des événements politiques sur les caractères du délire et anomalies physiques et intellectuelles que l'on observe chez les enfants conçus pendant le siège de Paris," *Le praticien* 7 (1884), p. 161.

8. Legrand du Saulle refers to *brûlantes arènes*. Henri Legrand du Saulle, *Le délire des persecutions* (Paris: H. Plon, 1871), p. 504.

9. On the figure of interiority and childhood, see Carolyn Steedman, *Strange Dislocations: Childhood and the Idea of Human Interiority, 1780–1930* (Cambridge, MA: Harvard University Press, 1995), p. 12; on developmental psychology and child observation, Sally Shuttleworth, *The Mind of the Child: Child Development in Literature, Science, and Medicine, 1840–1900* (Oxford: Oxford University Press, 2010); André Turmel, *A Historical Sociology of Childhood: Developmental Thinking, Categorization and Graphic Visualization* (Cambridge, UK: Cambridge University Press, 2008); Dominique Ottavi, *De Darwin à Piaget: Pour une histoire de la psychologie de l'enfant* (Paris: CNRS Éditions, 2001). On the psychiatrization of childhood: Michel Foucault, *Abnormal: Lectures at the Collège de France, 1974–1975*, trans. Graham Burchell (London: Verso, 2003), esp. the lecture of March 19, 1975, pp. 291–322. Generally on the nineteenth century's interest in the child, see, among many others, Catherine Rollet, *Les enfants au XIXe siècle* (Paris: Hachette, 2001).

10. Reinhart Koselleck, *Futures Past: On the Semantics of Historical Time*, trans. Keith Tribe (New York: Columbia University Press, [1979] 2004), esp. pp. 267–88 and 92–104. A paradigmatic formulation of this concept of time is found in Georg Simmel's 1916 essay "The Problem of Historical Time," in *Essays on Interpretation in Social Science*, ed. and trans. Guy Oakes (Totowa, NJ: Rowman and Littlefield, 1980).

11. Staffan Müller-Wille and Hans-Jörg Rheinberger, *A Cultural History of Heredity* (Chicago: University of Chicago Press, 2012), p. xi. See also Staffan Müller-Wille and Hans-Jörg Rheinberger, "Heredity: The Formation of an Epistemic Space," in *Heredity Produced: At the Crossroads of Biology, Politics, and Culture, 1500–1870*, ed. Staffan Müller-Wille and Hans-Jörg Rheinberger (Cambridge, MA: MIT Press, 2007).

12. On the rather slow transition from *generatio* to reproduction, see Nick Hopwood, "The Keywords 'Generation' and 'Reproduction,'" in *Reproduction: Antiquity to the Present Day*, ed. Nick Hopwood, Rebecca Flemming, and Lauren Kassell (Cambridge, UK: Cambridge University Press, 2018); Susanne Lettow, "Generation, Genealogy, and Time: The Concept of Reproduction from *Histoire naturelle* to *Naturphilosophie*," in *Reproduction, Race, and Gender in Philosophy and the Early Life Sciences*, ed. Susanne Lettow (Albany: SUNY Press, 2014); Staffan Müller-Wille, "Reproducing Species," in *The Secrets of Generation: Reproduction in the Long Eighteenth Century*, ed. Raymond Stephanson and Darren N. Wagner (Toronto: University of Toronto Press, 2015); Ludmilla Jordanova, "Interrogating the Concept of Reproduction in the Eighteenth Century," in *Conceiving the New World Order: The Global Politics of Reproduction*, ed. Faye Ginsburg and Rayna Rapp (Berkeley: University of California Press, 1995).

13. Charles Féré, *La famille névropathique: Théorie tératologique de l'hérédité et de la prédisposition morbides et de la dégénérescence* (Paris: F. Alcan, [1894] 1898), p. 18.

14. See, for example, Evelyn Fox Keller, *Refiguring Life: Metaphors of Twentieth-Century Biology* (New York: Columbia University Press, 1995), pp. 4–5. With an emphasis on gender history: Helga Satzinger, *Differenz und Vererbung: Geschlechterordnungen in der Genetik und Hormonforschung 1890–1950* (Cologne: Böhlau, 2009). Discussing the emergence of reproductive sciences in terms of disciplines, Adele E. Clarke speaks of a "three-way split" into genetics, developmental embryology, and reproductive sciences in *Disciplining Reproduction: Modernity, American Life Sciences, and "the Problems of Sex"* (Berkeley: University of California Press, 1998), pp. 66–69.

15. On the currency of this distinction, see J. Andrew Mendelsohn, "Medicine and the Making of Bodily Inequality in Twentieth-Century Europe," in *Heredity and Infection:*

The History of Disease Transmission, ed. Jean-Paul Gaudillière and Ilana Löwy (London: Routledge, 2001), p. 40.

16. Jane Maienschein calls them an "intimately intertwined couple" in "Heredity/Development in the United States, circa 1900," *History and Philosophy of Life Sciences* 9.1 (1987), p. 93.

17. On dealing with the *longue durée* in the history of science, see also Heiko Stoff, "Der aktuelle Gebrauch der *longue durée* in der Wissenschaftsgeschichte," *Berichte zur Wissenschaftsgeschichte* 32.2 (2009), pp. 144–58.

18. Hans-Jörg Rheinberger, *Toward a History of Epistemic Things: Synthesizing Proteins in the Test Tube* (Stanford, CA: Stanford University Press, 1997), p. 28. See also Rheinberger, *An Epistemology of the Concrete: Twentieth-Century Histories of Life*, trans. G. M. Goshgarian (Durham, NC: Duke University Press, 2010), p. 2.

19. Hans-Jörg Rheinberger and Michael Hagner, "Experimentalsysteme," in *Die Experimentalisierung des Lebens: Experimentalsysteme in den biologischen Wissenschaften 1850/1950*, ed. Hans-Jörg Rheinberger and Michael Hagner (Berlin: Akademie Verlag, 1993), p. 9.

20. On the nonnatural object of the history of science, see Georges Canguilhem, "The Object of the History of Sciences" (1966), trans. Mary Tiles, in *Continental Philosophy of Science*, ed. Gary Gutting (Oxford: Blackwell, 2005); Lorraine Daston (ed.), *Biographies of Scientific Objects* (Chicago: University of Chicago Press, 2000).

21. Methodologically, this approach is inspired by an ethnographic precept, that "the things encountered in fieldwork are allowed to dictate the terms of their own analysis." Amiria Henare, Martin Holbraad, and Sari Wastell, "Introduction: Thinking through Things," in *Thinking Through Things: Theorising Artefacts Ethnographically*, ed. Amiria Henare, Martin Holbraad, and Sari Wastell (London: Routledge, 2007), p. 4. This approach aims not to explain things through an external "context of action," but to examine the "context in action" within them. See Steve Woolgar and Javier Lezaun, "The Wrong Bin Bag: A Turn to Ontology in Science and Technology Studies?," *Social Studies of Science* 43.3 (2013), pp. 323–27; Kristin Asdal and Ingunn Moser, "Experiments in Context and Contexting," *Science, Technology & Human Values* 37.4 (2012), pp. 291–306.

22. On the French context, see John E. Lesch, *Science and Medicine in France: The Emergence of Experimental Physiology, 1790–1855* (Cambridge, MA: Harvard University Press, 1984).

23. Ludwik Fleck, *Genesis and Development of a Scientific Fact*, ed. Thaddeus J. Trenn and Robert K. Merton, trans. Fred Bradley and Thaddeus J. Trenn (Chicago: University of Chicago Press, [1935] 1979).

24. On the complex history of embryology, see Nick Hopwood, "Embryology," in *The Cambridge History of Science*, vol. 6: *The Modern Biological and Earth Sciences*, ed. Peter J. Bowler and John V. Pickstone (Cambridge, UK: Cambridge University Press, 2009), pp. 285–315. On the related case of morphology, see Lynn K. Nyhart, *Biology Takes Form: Animal Morphology and the German Universities, 1800–1900* (Chicago: University of Chicago Press, 1995). Also on the history of embryology, see Scott Gilbert (ed.), *A Conceptual History of Modern Embryology* (Baltimore, MD: Johns Hopkins University Press, 1994); T. J. Horder, J. A. Witkowski, and C. C. Wylie (ed.), *A History of Embryology* (Cambridge, UK: Cambridge University Press, 1986); Jane M. Oppenheimer, *Essays in the History of Embryology and Biology* (Cambridge, MA: MIT Press, 1967).

25. Barbara Duden, "Zwischen 'wahrem Wissen' und Prophetie: Konzeptionen des Ungeborenen," in *Geschichte des Ungeborenen: Zur Erfahrungs- und Wissenschaftsgeschichte der Schwangerschaft, 17.–20. Jahrhundert*, ed. Barbara Duden, Jürgen Schlumbohm, and Patrice Veit (Göttingen: Vandenhoeck & Ruprecht, 2002), p. 13; Barbara Duden, Jürgen Schlumbohm, and Patrice Veit, "Vorwort," in Duden, Schlumborn, and Veit, *Geschichte des Ungeborenen*, p. 7.

26. On these parameters of biology, see William Coleman, *Biology in the Nineteenth Century: Problems of Form, Function, and Transformation* (Cambridge, UK: Cambridge University Press, 1979).

27. Lorna Weir, *Pregnancy, Risk and Biopolitics: On the Threshold of the Living Subject* (London: Routledge, 2006), p. 36.

28. As far as I am aware, the only systematic studies address the history of particular disciplines. See Lawrence D. Longo, *The Rise of Fetal and Neonatal Physiology: Basic Science to Clinical Care* (New York: Springer, 2013); Joseph Needham, *History of Embryology*, rev. ed. (Cambridge, UK: Cambridge University Press, 1959). Sara Dubow discusses fetal physiology but concentrates on the twentieth century in *Ourselves Unborn: A History of the Fetus in Modern America* (New York: Oxford University Press, 2011). On "producing development," see Nick Hopwood, "Producing Development: The Anatomy of Human Embryos and the Norms of Wilhelm His," *Bulletin of the History of Medicine* 74.1 (2000), pp. 29–79.

29. Gabriel Compayré, *The Intellectual and Moral Development of the Child*, vol. 1, trans. Mary E. Wilson (New York: Appleton, [1893] 1900), p. 30. The French original appeared in 1893 as *L'évolution intellectuelle et morale de l'enfant*; a German translation was published in 1900 as *Die Entwicklung der Kindesseele*. On the manner in which interest in the child extended into prenatal life, see Rollet, *Les enfants au XIXe siècle*, p. 189.

30. See Michel Foucault, *The Order of Things: An Archaeology of the Human Sciences*, trans. from the French (London: Routledge, [1966] 1974), p. 219. On the contribution of embryological issues to the emergence of the historical time of living beings, see Georges Canguilhem et al., *Du développement à l'évolution au XIXe siècle* (Paris: Presses universitaires de France, 1962).

31. I take a broad view of "the social," based on a social anthropology that defines itself as study of "the variation of social relations." Eduardo Viveiros de Castro, "The Relative Native," *HAU: Journal of Ethnographic Theory* 3.3 (2013), p. 483.

32. See Duden, "Zwischen 'wahrem Wissen' und Prophetie"; also from this perspective, Barbara Orland, "Labor-Reproduktion: Die Identität des Embryo zwischen Natur, Technik und Politik," in *Sexualität als Experiment: Identität, Lust und Reproduktion zwischen Science und Fiction*, ed. Nicolas Pethes and Silke Schicktanz (Frankfurt am Main: Campus, 2008).

33. As overviews: Pierre Charbonnier, Gildas Salmon, and Peter Skafish (ed.), *Comparative Metaphysics: Ontology After Anthropology* (London: Rowman & Littlefield, 2017); Martin Holbraad and Morten Axel Pedersen, *The Ontological Turn: An Anthropological Exposition* (Cambridge, UK: Cambridge University Press, 2017). For groundbreaking studies of present-day reproductive practices, see, among many others, Marilyn Strathern, *Reproducing the Future: Essays on Anthropology, Kinship and the New Reproductive Technologies* (Manchester: Manchester University Press, 1992); Sarah Franklin, "Fetal Fascinations: New Dimensions to the Medical-Scientific Construction of Fetal Personhood," in *Off-Centre: Feminism and Cultural Studies*, ed. Sarah Franklin, Celia Lury, and Jackie Stacey (London: HarperCollins Academic, 1991); Charis Thompson, *Making Parents: The Ontological Choreography of Reproductive Technologies* (Cambridge, MA: MIT Press, 2005); more recently also Sarah Franklin, "*In Vitro Anthropos*: New Conception Models for a Recursive Anthropology?," *Cambridge Anthropology* 31.1 (2013), pp. 3–32.

CHAPTER THREE: THE UNBORN AND THE HUMAN SCIENCES

1. Bernard Perez, "Les facultés de l'enfant à l'époque de la naissance," *Revue philosophique de la France et de l'étranger* 13.1 (1882), p. 135. For a critique of the "precursor" idea, see Georges Canguilhem, "The Object of the History of Sciences" (1966), trans. Mary Tiles, in *Continental Philosophy of Science*, ed. Gary Gutting (Oxford: Blackwell, 2005).

2. On this Aristotelian notion, see Erna Lesky, *Die Zeugungs- und Vererbungslehren der Antike und ihr Nachwirken* (Wiesbaden: Franz Steiner, 1950), pp. 141–46.

3. See Jacques Roger, *The Life Sciences in Eighteenth-Century French Thought*, ed. Keith

R. Benson, trans. Robert Ellrich (Stanford, CA: Stanford University Press, [1963] 1998), pp. 53–62; G. R. Dunstan (ed.), *The Human Embryo: Aristotle and the Arabic and European Traditions* (Exeter: University of Exeter Press, 1990); Luc Brisson, Marie-Hélène Congourdeau, and Jean-Luc Solère (ed.), *L'embryon: Formation et animation; Antiquité grecque et latine, tradition hébraïque, chrétienne et islamique* (Paris: J. Vrin, 2008), especially the chapter by Maaike van der Lugt, "L'animation de l'embryon humain et le statut de l'enfant à naître dans la pensée médiévale." For more detail on the Middle Ages: Maaike van der Lugt, *Le ver, le démon et la vierge: Les théories médievales de la génération extraordinaire* (Paris: Les Belles Lettres, 2004), pp. 43–93. On the multifarious ensoulment theories in legal debate, see Esther Fischer-Homberger, *Medizin vor Gericht: Zur Sozialgeschichte der Gerichtsmedizin* (Darmstadt: Luchterhand, 1988). On seventeenth-century theological and scientific discussions of the moment of animation: Adriano Prosperi, *Infanticide, Secular Justice, and Religious Debate in Early Modern Europe*, trans. Hilary Siddons (Turnhout: Brepols, 2016), pp. 231–319. From an anthropological perspective on the nonsingularity of modern Western notions of transnatal continuity, Enric Porqueres i Gené, "Individu et parenté: Individuation de l'embryon," in *Corps et affects*, ed. Françoise Héritier and Margarita Xanthakou (Paris: O. Jacob, 2004), pp. 146–48.

4. This is Barbara Duden's argument. Duden, "Die 'Geheimnisse' der Schwangeren und das Öffentlichkeitsinteresse der Medizin: Zur sozialen Bedeutung der Kindsregung," in *Frauengeschichte — Geschlechtergeschichte*, ed. Karin Hausen and Heide Wunder (Frankfurt am Main: Campus, 1992), pp. 117 and 124.

5. On the pregnant body as a protective "corps-filtre" and an endangering "corps-conducteur," see Jacques Gélis, *L'arbre et le fruit: La naissance dans l'Occident moderne, XVIe–XIXe siècle* (Paris: Fayard, 1984), p. 118. Marie-France Morel refers to the "transparent body" of the pregnant woman in "Grossesse, foetus et histoire," in *La grossesse, l'enfant virtuel et la parentalité*, ed. Sylvain Missonnier, Bernard Golse, and Michel Soulé (Paris: Presses universitaires de France, 2004), p. 26. On the widespread notion of the fetus as being open to influence, see Saskia Walentowitz, "La vie sociale du foetus: Regards anthropologiques," *Spirales* 36.4 (2005), p. 137.

6. Shirley Roe, *Matter, Life, and Generation: Eighteenth-Century Embryology and the Haller-Wolff Debate* (Cambridge, UK: Cambridge University Press, 1981), p. 152. More generally, this portrayal of the controversy is not intended to suggest a linear development. The core concepts of epigenesis can also be found in Aristotle, and those of preformation in modern genetics. Generally on the emergence of the modern concept of development:

Georges Canguilhem et al., *Du développement à l'évolution au XIXe siècle* (Paris: Presses universitaires de France, 1962). On the preformation versus epigenesis controversy, see also Justin E. H. Smith (ed.), *The Problem of Animal Generation in Early Modern Philosophy* (Cambridge, UK: Cambridge University Press, 2006).

7. Canguilhem et al., *Du développement à l'évolution*, pp. 3–4.

8. Georges Canguilhem, "On the History of the Life Sciences since Darwin," in *Ideology and Rationality in the History of the Life Sciences*, trans. Arthur Goldhammer (Cambridge, MA: MIT Press, 1977 [1988]), p. 108; François Jacob, *The Logic of Life: A History of Heredity*, trans. Betty E. Spillmann (Princeton, NJ: Princeton University Press, [1973] 1993), p. 130.

9. Canguilhem et al., *Du développement à l'évolution*, pp. 11–17. For a very detailed study, see Janina Wellmann, *The Form of Becoming: Embryology and the Epistemology of Rhythm, 1760–1830*, trans. Kate Sturge (New York: Zone Books, 2017).

10. This is what historians of science mean when they refer to the rise of a four-dimensional nature of life and a historical nature of development. See Jacob, *Logic of Life*, p. 123. On the conceptual link between organic and historical time, see Owsei Temkin, "German Concepts of Ontogeny and History around 1800," *Bulletin of the History of Medicine* 24.3 (1950), p. 234.

11. Seminal accounts of this point are Michel Foucault, *The Order of Things: An Archaeology of the Human Sciences* (London: Routledge, [1966] 1974), and Jacob, *Logic of Life*; see also William Coleman, *Biology in the Nineteenth Century: Problems of Form, Function, and Transformation* (Cambridge, UK: Cambridge University Press, 1979), p. 10. On time and knowledge of reproduction, see Bettina Bock von Wülfingen et al. (ed.), "Temporalities of Reproduction: Practices and Concepts from the Eighteenth to the Early Twentieth-First Century," special issue, *History and Philosophy of the Life Sciences* 37.1 (2015); Susanne Lettow, "Generation, Genealogy, and Time: The Concept of Reproduction from *Histoire naturelle* to *Naturphilosophie*," in *Reproduction, Race, and Gender in Philosophy and the Early Life Sciences*, ed. Susanne Lettow (Albany: SUNY Press, 2014).

12. On the beginnings "in enclosure," see Barbara Duden, "Zwischen 'wahrem Wissen' und Prophetie: Konzeptionen des Ungeborenen," in *Geschichte des Ungeborenen: Zur Erfahrungs- und Wissenschaftsgeschichte der Schwangerschaft, 17.–20. Jahrhundert*, ed. Barbara Duden, Jürgen Schlumbohm, and Patrice Veit (Göttingen: Vandenhoeck & Ruprecht, 2002), p. 18. On the time of anticipation, Duden, *Die Gene im Kopf — der Fötus im Bauch: Historisches zum Frauenkörper* (Hannover: Offizin, 2002), pp. 51–65. My description is also

inspired by Walentowitz, who analyzes the spatiotemporal constitution of the unborn among the Tuareg. See, for example, Saskia Walentowitz, "L'enfant qui n'a pas atteint son lieu: Représentations et soins autour des prématurés chez les Touaregs de l'Azawagh (Niger)," *L'Autre: Cliniques, cultures et sociétés* 5.2 (2004), pp. 227–41.

13. On the interlocking of *logos* and *chronos* in the concept of development, see Canguilhem et al., *Du développement à l'évolution*, p. 9.

14. See Fernando Vidal, "La 'science de l'homme': Désirs d'unité et juxtaposition encyclopédiques," in *L'histoire des sciences de l'homme: Trajectoire, enjeux et questions vives*, ed. Claude Blanckaert et al. (Paris: L'Harmattan, 1999), pp. 66 and 70. The following very brief sketch also draws on Sergio Moravia, "The Enlightenment and the Sciences of Man," *History of Science* 18.4 (1980), pp. 247–68; Moravia, "From *homme machine* to *homme sensible*: Changing Eighteenth-Century Models of Man's Image," *Journal of the History of Ideas* 39.1 (1978), pp. 45–60; Wolf Lepenies, "Naturgeschichte und Anthropologie im 18. Jahrhundert," *Historische Zeitschrift* 231.1 (1980), pp. 21–42.

15. Pierre-Jean-Georges Cabanis, *Rapports du physique et du moral de l'homme et lettres sur les causes premières*, 2 vols. (Paris: Crapart, Caille et Ravier, 1802), translated into German in 1804 as *Über die Verbindung des Physischen und Moralischen in dem Menschen* and available in English as *On the Relations between the Physical and Moral Aspects of Man*, 2 vols., ed. George Mora, trans. Margaret Duggan Saidi (Baltimore, MD: Johns Hopkins University Press, 1982). On Cabanis and his reception: Claudia Honegger, *Die Ordnung der Geschlechter: Die Wissenschaften vom Menschen und das Weib, 1750–1850* (Frankfurt am Main: Campus, 1991), pp. 151–64. On Cabanis in the context of the human sciences, see also Elizabeth A. Williams, *The Physical and the Moral: Anthropology, Physiology, and Philosophical Medicine in France, 1750–1850* (Cambridge, UK: Cambridge University Press, 1994).

16. Michael Hagner, *Homo cerebralis: Der Wandel vom Seelenorgan zum Gehirn* (Frankfurt am Main: Suhrkamp, 2008), p. 238.

17. Société des Observateurs de l'homme, quoted in Lepenies, "Naturgeschichte und Anthropologie im 18. Jahrhundert," p. 38. For a counterpart to this in German pedagogy, see Pia Schmid, "Väter und Forscher: Zu Selbstdarstellungen bürgerlicher Männer um 1800 im Medium empirischer Kinderbeobachtungen," *Feministische Studien* 18.2 (2000), pp. 35–48. More detail on the Parisian Société des Observateurs de l'homme and its natural history–based justification of a dualist human nature can be found in Jean-Luc Chappey, *La Société des Observateurs de l'homme (1799–1804): Des anthropologues au temps de Bonaparte* (Paris: Société des études robespierristes, 2002).

18. Jacob, *Logic of Life*, pp. 74–129, quotation p. 83. My comments here and in the rest of this section also draw on Foucault, *Order of Things*.

19. Jacob, *Logic of Life*, p. 143.

20. See Staffan Müller-Wille, "Evolutionstheorien vor Darwin," in *Evolution: Ein interdisziplinäres Handbuch*, ed. Philipp Sarasin and Marianne Sommer (Stuttgart: Metzler, 2010), esp. p. 69.

21. On the significance of contemporary embryology for Darwin, see Canguilhem, "History of the Life Sciences since Darwin," p. 108. On the historical arc between the concept of development of the organism and that of the evolution of species: Canguilhem et al., *Du développement à l'évolution*. The significance of evolution as an undirected process lies not in the term itself, but in its usage: in the past, *evolvere* had also meant the very opposite, the unfolding of something preexistent, with which *development* was contrasted as the genesis of something new. See Dominique Ottavi, *De Darwin à Piaget: Pour une histoire de la psychologie de l'enfant* (Paris: CNRS Éditions, 2001), pp. 27–28. On the turn from directed "development" to undirected "genealogy" in the notion of evolution, see Peter J. Bowler, *Life's Splendid Drama: Evolutionary Biology and the Reconstruction of Life's Ancestry, 1860–1940* (Chicago: University of Chicago Press, 1996).

22. See Foucault, *Order of Things*. A survey of the historiography of the human sciences is given by Claude Blanckaert, "L'histoire générale des sciences de l'homme: Principes et périodisation," in *L'histoire des sciences de l'homme: Trajectoire, enjeux et questions vives*, ed. Claude Blanckaert et al. (Paris: L'Harmattan, 1999); in an updated and expanded form, taking into account the practical turn: Florence Vienne and Christina Brandt, "Einleitung," in *Wissensobjekt Mensch: Humanwissenschaftliche Praktiken im 20. Jahrhundert*, ed. Florence Vienne and Christina Brandt (Berlin: Kulturverlag Kadmos, 2008), p. 11–14. For an overview of questions in the human sciences from the perspective of the history of biology, see Peter J. Bowler, "Biology and Human Nature," in *The Cambridge History of Science*, vol. 6: *The Modern Biological and Earth Sciences*, ed. Peter J. Bowler and John V. Pickstone (Cambridge, UK: Cambridge University Press, 2009), pp. 285–315.

23. Philipp Sarasin, "Der öffentlich sichtbare Körper: Vom Spektakel der Anatomie zu den *curiosités physiologiques*," in *Physiologie und industrielle Gesellschaft: Studien zur Verwissenschaftlichung des Körpers im 19. und 20. Jahrhundert*, ed. Philipp Sarasin and Jakob Tanner (Frankfurt am Main: Suhrkamp 1998), p. 422.

24. Quoted in Canguilhem et al., *Du développement à l'évolution*, p. 45.

25. On the entanglement of phylogenesis and psychogenesis, see ibid., pp. 44–56.

Darwin made use of his child observations in *The Expression of the Emotions in Man and Animals* (London: John Murray, 1872); he also presented his results as "A Biographical Sketch of an Infant," *Mind* 2.7 (1877), pp. 285–94, in response to a publication by Hippolyte Taine on an infant's intellectual development. In the emerging field of developmental psychology, this essay became a key point of reference; see Ottavi, *De Darwin à Piaget*, pp. 103–28; Fernando Vidal, Marino Buscaglia, and J. Jacques Vonèche, "Darwinism and Developmental Psychology," *Journal of the History of the Behavioral Sciences* 19.1 (1983), pp. 81–91.

26. It is not possible to detail here the many variants of this reading, the most spectacular of which is Ernst Haeckel's recapitulation theory. Though crucial to discussions in embryology, recapitulation theory was less so for the physiology of fetal life, or only when physiology was turned to the purposes of developmental psychology. See, especially, psychoanalytical interest in the unborn as discussed in Chapter 8. On recapitulation theory, see Steven Jay Gould, *Ontogeny and Phylogeny* (Cambridge, MA: Belknap Press, 1977).

27. It is here that we find the origin of the polarization into naturalistic monism ("natural culture") and culturalist relativism ("cultural nature"), discussed by Philippe Descola in *Beyond Nature and Culture*, trans. Janet Lloyd (Chicago: University of Chicago Press, 2014). The latter is the option chosen by the philosophical anthropology that, in the twentieth century, would define the human as a "deficient being," made human by its unfinished organic state and insufficient instincts. This was the view propounded by the conservative German sociologist Arnold Gehlen with reference to Louis Bolk's fetalization theory, which in turn drew on the ideas of John Fiske. See Canguilhem et al., *Du développement à l'évolution*, p. 44.

28. See Descola, *Beyond Nature and Culture*, pp. 57–89.

29. Eduardo Viveiros de Castro, "Exchanging Perspectives: The Transformation of Objects into Subjects in Amerindian Ontologies," *Common Knowledge* 10.3 (2004), p. 483.

30. Hans-Jörg Rheinberger and Michael Hagner, "Experimentalsysteme," in *Die Experimentalisierung des Lebens: Experimentalsysteme in den biologischen Wissenschaften 1850/1950*, ed. Hans-Jörg Rheinberger and Michael Hagner (Berlin: Akademie Verlag, 1993), p. 22.

31. Descola speaks of a combination of a "continuity of physicalities" and a "discontinuity of interiorities." Descola, *Beyond Nature and Culture*, p. 172. Generally on this structural analysis of naturalism, see ibid., pp. 172–99, and see pp. 91–127 on ontologies as schemas of practice that are rooted in the identification of continuity and discontinuity with respect to the physicality and interiority of beings.

32. Ibid., pp. 174 and 178. On the program of an ethnography of naturalism that starts

from this assumption but also criticizes Descola's rather monolithic stance, see Lys Alcayna-Stevens and Matei Candea (ed.), "Internal Others: Ethnographies of Naturalism," special issue, *Cambridge Journal of Anthropology* 30.2 (2012). It should be noted, though, that Descola is interested in an ontological "grammar," and describes naturalism as one of four possible ontological types that are quite capable of coexisting. My reference to Descola's point is strategic: it helps me to show what "makes" developmental knowledge. A pioneer in the strategic deployment of historically or locally specific metaphysics is Marilyn Strathern, *The Gender of the Gift: Problems with Women and Problems with Society in Melanesia* (Berkeley: University of California Press, 1988), p. 12.

33. This is the research desideratum proposed by Vienne and Brandt, "Einleitung," p. 19.

34. Van der Lugt, "L'animation de l'embryon humain," p. 233.

35. Luc Brisson, Marie-Hélène Congourdeau, and Jean-Luc Solère, "Préface," in Brisson, Congourdeau, and Solère, *L'embryon*, pp. 13–14.

36. See Jean-Claude Dupont, "Un autre embryon? Quelques relectures classiques de l'embryologie antique," in Brisson, Congourdeau, and Solère, *L'embryon*, p. 269.

37. For the argument that an "aestheticization of the idea of development" compensated for the naturalization of the human being, see Michael Hagner, "Vom Naturalienkabinett zur Embryologie: Wandlungen des Monströsen und die Ordnung des Lebens," in *Der falsche Körper: Beiträge zu einer Geschichte der Monstrositäten*, ed. Michael Hagner (Göttingen: Wallstein, 1995), pp. 93–94. On Soemmerring: Ulrike Enke, "Von der Schönheit der Embryonen: Samuel Thomas Soemmerrings Werk *Icones embryonum humanorum* (1799)," in Duden, Schlumbohm, and Veit, *Geschichte des Ungeborenen*, pp. 224–25. On the idealization of natural truth: Lorraine Daston and Peter Galison, *Objectivity* (New York: Zone Books, 2007), chap. 2. On the embryo's humanity, see also Hagner, *Homo cerebralis*, pp. 194–201.

38. I argue here that the manner in which these questions were posed and addressed is historically specific, but not that the questions themselves were new. Descartes, for example, asked as much when he postulated the fetus's capacity for thought. See Rebecca Wilkin, "Descartes, Individualism, and the Fetal Subject," *Differences* 19.1 (2008), p. 106.

39. The same question arises for the organism itself. See Laura Nuño de la Rosa, "Becoming Organisms: The Organisation of Development and the Development of Organisation," *History and Philosophy of Life Sciences* 32.2–3 (2010), pp. 289–316. *Animatio* too had faced the problem of how to pinpoint the commencement of becoming. See G. R. Dunstan, "Introduction: Text and Context," in *The Human Embryo: Aristotle and the Arabic*

and European Traditions, ed. G. R. Dunstan (Exeter: University of Exeter Press, 1990), p. 7.

40. See Bruno Latour, *We Have Never Been Modern*, trans. Catherine Porter (London: Harvester Wheatsheaf, 1993). Irrespective of the ontological worries around the knowledge of the life and human sciences that I outline here, such knowledge was much used for ontological postulates in the fields of law, theology, and politics. Enric Porqueres i Gené, for example, shows how twentieth-century developmental biology's portrayal of individuation would make the embryo an icon of humanity. Porqueres i Gené, "Personne et parenté," *L'Homme* 210 (2014), p. 36. On the biophilosophical debates, see, among others, Susanne Lettow, *Biophilosophien: Wissenschaft, Technologie und Geschlecht im philosophischen Diskurs der Gegenwart* (Frankfurt am Main: Campus, 2011), esp. pp. 71–86 on the psyche as what enables the human being to be interpellated as a person. On debates around personhood, see Lynn M. Morgan, "Fetal Relationality in Feminist Philosophy: An Anthropological Critique," *Hypatia* 11.3 (1996), p. 50.

41. On the "discovery of the environment" in the *science de l'homme*, see Moravia, "The Enlightenment and the Sciences of Man."

42. Claude Bernard, "Cours de physiologie générale de la Faculté des sciences: Leçon d'ouverture — Exposition de la méthode," *Le moniteur des hôpitaux: Journal des progrès de la médecine et de la chirurgie pratique* 1–2.54 (1854), p. 410.

43. Here, "Western" is intended more in a conceptual than in a geographical or cultural sense. In Marilyn Strathern's work, the term plays a strategic role in an ethnographic comparison that aims to uncover the otherness of Melanesian conceptions of personhood. Strathern, *Kinship, Law and the Unexpected: Relatives Are Always a Surprise* (Cambridge, UK: Cambridge University Press, 2005), pp. 29–30; Strathern, *Gender of the Gift*.

44. Hélène Rouch, "Le placenta comme tiers," *Langages* 21.85 (1987), p. 75. It is not by chance that the placenta was at the focus of twentieth-century attempts in cultural studies to ground the relationality of pregnancy in a "material basis." See JaneMaree Maher, "Visibly Pregnant: Toward a Placental Body," *Feminist Review* 72.1 (2002), p. 96.

45. This adds historical depth to Strathern's observation that for Euro-American societies, which think in terms of entities, birth marks not only the beginning of a new person, but also the end of a relationship that in other cultural contexts continues after birth through care. Strathern, *Kinship, Law and the Unexpected*, p. 169 n. 43; on the physiological mother–child relation, also Strathern, *Reproducing the Future: Essays on Anthropology, Kinship and the New Reproductive Technologies* (Manchester: Manchester University Press, 1992), p. 150. For the pregnant woman, this relation means that she is "divisible," as Ilana

Löwy puts it. Löwy, *Imperfect Pregnancies: A History of Birth Defects and Prenatal Diagnosis* (Baltimore, MD: Johns Hopkins University Press, 2017), p. 5.

46. On the concept of the *corps relationnel* in kinship anthropology, see Porqueres i Gené, "Personne et parenté," p. 17; Porqueres i Gené, "Individu et parenté," p. 149, on the importance of not — or not only — thinking about the embryo's relationality in terms of kinship.

47. See Duden, "Zwischen 'wahrem Wissen' und Prophetie." On medicine's delegitimatization of the child's quickening, also Angus McLaren, "Policing Pregnancies: Changes in Nineteenth-Century Criminal and Canon Law," in *The Human Embryo: Aristotle and the Arabic and European Traditions*, ed. G. R. Dunstan (Exeter: University of Exeter Press, 1990).

48. I borrow the term from Oliver J. T. Harris and John Robb, "Multiple Ontologies and the Problem of the Body in History," *American Anthropologist* 114.4 (2012), pp. 668–79. Harris and Robb draw on the work of Sarah Tarlow. Duden, "Zwischen 'wahrem Wissen' und Prophetie," exemplifies the multimodal perspective. On the diversity of present-day ontological practices with regard to the unborn, see, for the embryo, Orland, "Labor-Reproduktion," and for fetal life, Luc Boltanski, *The Foetal Condition: A Sociology of Engendering and Abortion*, trans. Catherine Porter (Cambridge, UK: Polity, 2013).

49. On twentieth-century perinatal medicine's orientation on the threshold of birth, see, for example, Lorna Weir, *Pregnancy, Risk and Biopolitics: On the Threshold of the Living Subject* (London: Routledge, 2006). For an analysis of the practical conception of things as opposed to modes of "knowledge about," Annemarie Mol, *The Body Multiple: Ontology in Medical Practice* (Durham, NC: Duke University Press, 2002); Eduardo Viveiros de Castro, "The Relative Native," *HAU: Journal of Ethnographic Theory* 3.3 (2013), p. 484; Duden, "Zwischen 'wahrem Wissen' und Prophetie," p. 46.

50. Regarding my use of "ontology" see, in the history of science, Ian Hacking, *Historical Ontology* (Cambridge, MA: Harvard University Press, 2002). A key influence for me has been the debate in social anthropology around ontological grammars and the constitution of "things" through relations and practices, especially the work of Philippe Descola, Marilyn Strathern, and Eduardo Viveiros de Castro. For recent syntheses of this discussion, see Pierre Charbonnier, Gildas Salmon, and Peter Skafish (ed.), *Comparative Metaphysics: Ontology after Anthropology* (London: Rowman & Littlefield, 2017); Martin Holbraad and Morten Axel Pedersen, *The Ontological Turn: An Anthropological Exposition* (Cambridge, UK: Cambridge University Press, 2017).

51. A form of comparison oriented not on cultural pluralism but on ontological

heterogeneity is proposed by Strathern; see Martin Holbraad and Morten Axel Pedersen, "Planet M: The Intense Abstraction of Marilyn Strathern," *Anthropological Theory* 9.4 (2009), pp. 371–94. On the question of the composition of world versus the question of perspectives on or representations of world, see Steve Woolgar and Javier Lezaun, "The Wrong Bin Bag: A Turn to Ontology in Science and Technology Studies?," *Social Studies of Science* 43.3 (2013), p. 322.

52. Latour, *We Have Never Been Modern*, is foundational on the postulate of symmetry.

53. Foucault, *Order of Things*, p. 318.

54. On the internal Other and on heteronomy in naturalism, see Matei Candea and Lys Alcayna-Stevens, "Internal Others: Ethnographies of Naturalism," *Cambridge Journal of Anthropology* 30.2 (2012), pp. 36–47.

55. On other internal Others: Honegger, *Die Ordnung der Geschlechter*; Londa Schiebinger, *Nature's Body: Gender in the Making of Modern Science* (New Brunswick, NJ: Rutgers University Press, 1993). On procreation as a gap in Michel Foucault's foundations of a history of the human sciences, see Robert A. Nye, "Love and Reproductive Biology in Fin-de-Siècle France: A Foucauldian Lacuna?," in *Foucault and the Writing of History*, ed. Jan Goldstein (Oxford: Blackwell, 1994).

56. William T. Preyer, "Psychogenesis," in *Naturwissenschaftliche Thatsachen und Probleme: Populäre Vorträge* (Jena: Verlag von Gebrüder Paetel, 1880), p. 237.

57. Jean-Paul Galibert, "Le jeu des temps embryonnaires," in *L'embryon humain à travers l'histoire: Images, savoirs et rites*, ed. Véronique Dasen (Gollion: Infolio éditions, 2007), p. 257.

58. Johann Wolfgang von Goethe, *Elective Affinities*, trans. R. J. Hollingdale (London: Penguin, [1809] 1971), p. 163.

PART TWO: LIVING BEINGS

1. Michel Foucault, *The Order of Things: An Archaeology of the Human Sciences*, trans. from the French (London: Routledge, [1966] 1974), p. 276.

2. Hélène Rouch, "Le placenta comme tiers," *Langages* 21.85 (1987), p. 74.

3. See Tobias Cheung, "What is an 'Organism'? On the Occurrence of a New Term and Its Conceptual Transformations 1680–1850," *History and Philosophy of Life Sciences* 32.2–3 (2010), pp. 155–94; in more detail: François Jacob, *The Logic of Life: A History of Heredity*, trans. Betty E. Spillmann (Princeton, NJ: Princeton University Press, [1973] 1993).

4. See Foucault, *Order of Things*; Georges Canguilhem et al., *Du développement à l'évolution au XIXe siècle* (Paris: Presses universitaires de France, 1962).

5. See Barbara Duden, "Anatomie der guten Hoffnung" (unpublished typescript, 2003), pp. 265–88.

6. Barbara Duden, "Zwischen 'wahrem Wissen' und Prophetie: Konzeptionen des Ungeborenen," in *Geschichte des Ungeborenen: Zur Erfahrungs- und Wissenschaftsgeschichte der Schwangerschaft, 17.–20. Jahrhundert*, ed. Barbara Duden, Jürgen Schlumbohm, and Patrice Veit (Göttingen: Vandenhoeck & Ruprecht, 2002), p. 20. On the uterus in medieval anatomical dissection: Katharine Park, *Secrets of Women: Gender, Generation, and the Origins of Human Dissection* (New York: Zone Books, 2006), pp. 77–120.

7. Duden, "Zwischen 'wahrem Wissen' und Prophetie," p. 12.

8. See Duden, "Anatomie der guten Hoffnung," pp. 72–74. On Hunter's work on the uterus, see also Andrew Cunningham, *The Anatomist Anatomis'd: An Experimental Discipline in Enlightenment Europe* (London: Routledge, 2010), pp. 179–82.

9. Duden, "Anatomie der guten Hoffnung," p. 70, and for more detail on Hunter's interest in the uterus, pp. 31–76.

10. Preface to the *Icones embryonum humanorum* (1799), in Samuel Thomas Soemmerring, *Schriften zur Embryologie und Teratologie*, ed. Ulrike Enke (Basel: Schwabe, 2000), p. 173. On Soemmerring, see Enke, "Von der Schönheit der Embryonen: Samuel Thomas Soemmerrings Werk *Icones embryonum humanorum* (1799)," in Duden, Schlumbohm, and Veit, *Geschichte des Ungeborenen*. It is open to debate whether Soemmerring already introduced the changing of form into his images (and consequently whether his Latin word *metamorphosis* is correctly rendered by "development," *Entwicklung* in Enke's translation of the work). See Janina Wellmann, "Keine Ikone der Entwicklung: Die *Icones embryonum humanorum* von Samuel Thomas Soemmerring," in *Kulturen des Wissens im 18. Jahrhundert*, ed. Hans Ulrich Schneider (Berlin: De Gruyter, 2008), esp. pp. 593–94. Either way, he carried out the turn to the entitative embryo as described by Duden.

11. See Justin E. H. Smith (ed.), *The Problem of Animal Generation in Early Modern Philosophy* (Cambridge, UK: Cambridge University Press, 2006); Shirley Roe, *Matter, Life, and Generation: Eighteenth-Century Embryology and the Haller-Wolff Debate* (Cambridge, UK: Cambridge University Press, 1981).

12. See Jacob, *Logic of Life*, p. 125; Roe, *Matter, Life, and Generation*, pp. 148–56.

13. See Janina Wellmann, *The Form of Becoming: Embryology and the Epistemology of Rhythm, 1760–1830*, trans. Kate Sturge (New York: Zone Books, 2017), pp. 273–320.

14. Wilhelm His, *Anatomie menschlicher Embryonen*, 3 vols. (Leipzig: F. C. W. Vogel, 1880–1885).

15. See Nick Hopwood, "Producing Development: The Anatomy of Human Embryos and the Norms of Wilhelm His," *Bulletin of the History of Medicine* 74.1 (2000), pp. 29–79. For a survey, Hopwood, "Embryology," in *The Cambridge History of Science*, vol. 6: *The Modern Biological and Earth Sciences*, ed. Peter J. Bowler and John V. Pickstone (Cambridge, UK: Cambridge University Press, 2009). See also the online exhibition *Making Visible Embryos*, www.sites.hps.cam.ac.uk/visibleembryos/.

16. See Duden, "Anatomie der guten Hoffnung," pp. 268–69.

17. On microscopy and comparative anatomy, see Enke, "Von der Schönheit der Embryonen," p. 235; on measurement-based drawing techniques, Duden, "Anatomie der guten Hoffnung," pp. 273–81; on Wilhelm His's embryo research practices, Nick Hopwood, "'Giving Body' to Embryos: Modeling, Mechanism, and the Microtome in Late Nineteenth-Century Anatomy," *Isis* 90.3 (1999), pp. 462–96.

18. François Magendie, *Précis élémentaire de physiologie*, vol. 2 (Paris: Méquignon-Marvis, 1817), p. 436.

19. Cheung, "What is an 'Organism'?," p. 178.

20. Jacob, *Logic of Life*, p. 74. See also Foucault, *Order of Things*, pp. 226–32. On the "grand metaphor" of organization, also Claudia Honegger, *Die Ordnung der Geschlechter: Die Wissenschaften vom Menschen und das Weib, 1750–1850* (Frankfurt am Main: Campus, 1991), pp. 126–34.

21. See Alain Prochiantz, *Claude Bernard: La révolution physiologique* (Paris: Presses universitaires de France, 1990), p. 55. On the process of differentiation between anatomy and physiology, Richard L. Kremer, "Physiology," in Bowler and Pickstone, *Modern Biological and Earth Sciences*, p. 343; Karl E. Rothschuh, *Geschichte der Physiologie* (Berlin: Springer, 1953), pp. 92–93.

22. Albrecht von Haller, *La génération, ou Exposition des phénomènes relatifs a cette fonction naturelle*, vol. 2 (Paris: Ventes de la Doué, 1774), chap. 2. The proposition "Le fétus vit de bonne heure" is the summary of the volume's first section that Haller gives in the table of contents, p. 529.

23. On these visualizations, see Lynn M. Morgan, *Icons of Life: A Cultural History of Human Embryos* (Berkeley: University of California Press, 2009); Karen Newman, *Fetal Positions: Individualism, Science, Visuality* (Stanford, CA: Stanford University Press, 1996). For a critical discussion: Rebecca Wilkin, "Descartes, Individualism, and the Fetal Subject," *Differences* 19.1 (2008), p. 106.

24. On the biological objectification of the pregnant woman in parallel to that of the

unborn: Barbara Duden, *Disembodying Women: Perspectives on Pregnancy and the Unborn*, trans. Lee Hoinacki (Cambridge, MA: Harvard University Press, 1993).

25. Ludmilla Jordanova, "Gender, Generation and Science: William Hunter's Obstetrical Atlas," in *William Hunter and the Eighteenth-Century Medical World*, ed. W. F. Bynum and Roy Porter (Cambridge, UK: Cambridge University Press, 1985), p. 395. On related conjectures during the seventeenth century, for example in Harvey's work: Douglas M. Haynes, "The Human Placenta: Historical Considerations," in *The Human Placenta: Clinical Perspectives*, ed. J. Patrick Lavery (Rockville, MD: Aspen, 1987). On Harvey's attempts to think of the child-to-be as independent of the pregnant woman: Esther Fischer-Homberger, *Harvey's Troubles with the Egg* (Sheffield, UK: European Association for the History of Medicine and Health Publications, 2001).

26. Jordanova, "Gender, Generation and Science," p. 406. On the controversy over who discovered how the fetus's blood circulation and that of the mother are related, see Cunningham, *Anatomist Anatomis'd*, pp. 291–93.

27. Magendie, *Précis élémentaire de physiologie*, pp. 431–32. On the history of pregnant women's emotionality, see Lisa Malich, *Die Gefühle der Schwangeren: Eine Geschichte somatischer Emotionalität (1780–2010)* (Bielefeld: transcript-Verlag, 2017); Ziv Eisenberg, "Clear and Pregnant Danger: The Making of Prenatal Psychology in Mid-Twentieth-Century America," *Journal of Women's History* 22.3 (2010), pp. 112–35.

28. Jacob, *Logic of Life*, p. 155; see also Georges Canguilhem, "The Living and Its Milieu," in *Knowledge of Life*, trans. Stefanos Geroulanos and Daniela Ginsburg (New York: Fordham University Press, [1965] 2008). Also relevant for the present purposes is Laura Nuño de la Rosa, "Becoming Organisms: The Organisation of Development and the Development of Organisation," *History and Philosophy of Life Sciences* 32.2–3 (2010), p. 294.

29. Magendie, *Précis élémentaire de physiologie*, p. 450.

CHAPTER FOUR: THE LIFE OF THE FETUS

1. William T. Preyer, *Specielle Physiologie des Embryo: Untersuchungen über die Lebenserscheinungen vor der Geburt* (Leipzig: Grieben, 1883), quoted terms on pp. 1–3. Preyer is well known in developmental psychology — see the introduction to a modern reprint of another of his books: Georg Eckardt, "Einleitung," in William T. Preyer, *Die Seele des Kindes: Eingeleitet und mit Materialien zur Rezeptionsgeschichte versehen von Georg Eckardt* (Berlin: Springer, 1989) — but his work on developmental physiology has attracted very little attention from historians. An exception is Joseph Needham, *Chemical Embryology*, 3

vols. (Cambridge, UK: Cambridge University Press, 1931), vol. 1, p. 3. On the rediscovery of Preyer as the "true father of fetal studies," see Alessandra Piontelli, *Development of Normal Fetal Movements: The First 25 Weeks of Gestation* (Milan: Springer, 2010), p. 2; Jan G. Nijhuis (ed.), *Fetal Behaviour: Developmental and Perinatal Aspects* (Oxford: Oxford University Press, 1992).

2. Charles Féré, *La famille névropathique: Théorie tératologique de l'hérédité et de la prédisposition morbides et de la dégénérescence* (Paris: F. Alcan, [1894] 1898), p. 243. Selections from Preyer's book were also translated into English in 1937: William T. Preyer, "Embryonic Motility and Sensitivity: Translated from the Original German of *Specielle Physiologie des Embryo*," trans. G. E. Coghill and Wolfram K. Legner, *Monographs of the Society for Research in Child Development* 2.6 (1937), pp. 1–115.

3. See William Coleman, *Biology in the Nineteenth Century: Problems of Form, Function, and Transformation* (Cambridge, UK: Cambridge University Press, 1979), pp. 53–54. On His in more detail, Nick Hopwood, *Haeckel's Embryos: Images, Evolution, and Fraud* (Chicago: University of Chicago Press, 2015), chapters 6–8.

4. Preyer, *Specielle Physiologie des Embryo*, pp. 4, 16, 17. The phase that Preyer remitted to morphology corresponds approximately to what is today called the preembryonic phase (first to third week).

5. Ibid., p. 3.

6. Henri Beaunis, *Nouveaux éléments de physiologie humaine, comprenant les principes de la physiologie comparée et de la physiologie générale* (Paris: J.-B. Baillière et fils, editions of 1876, 1881, and 1888); F. Tourneux and G. Herrmann, "Embryon," in *Dictionnaire encyclopédique des sciences médicales*, ed. Amédée Dechambre, 1st series, vol. 33 (Paris: G. Masson, P. Asselin, 1886), pp. 657–730. On Preyer's role in French physiology, see Jean-Louis Fischer, "Embryogénése," in *Dictionnaire d'histoire et de philosophie des sciences*, ed. Dominique Lecourt (Paris: Presses universitaires de France, 1999), p. 393.

7. Preyer, *Specielle Physiologie des Embryo*, p. 17.

8. Barbara Duden, "Die 'Geheimnisse' der Schwangeren und das Öffentlichkeitsinteresse der Medizin: Zur sozialen Bedeutung der Kindsregung," in *Frauengeschichte—Geschlechtergeschichte*, ed. Karin Hausen and Heide Wunder (Frankfurt am Main: Campus, 1992), pp. 117 and 124. On the doctrine of *animatio* and its controversies, see Chapter 3, n3.

9. Nicolas Malebranche, *The Search after Truth*, ed. and trans. Thomas M. Lennon and Paul J. Olskamp (Cambridge, UK: Cambridge University Press, [1674–75] 1997), p. 112.

10. Malebranche, *The Search after Truth*, pp. 112–13.

11. On the rejection of "Hippocratic truths," see Nadia Maria Filippini, "Die 'erste Geburt': Eine neue Vorstellung vom Fötus und vom Mutterleib (Italien, 18. Jahrhundert)," in *Geschichte des Ungeborenen: Zur Erfahrungs- und Wissenschaftsgeschichte der Schwangerschaft, 17.–20. Jahrhundert*, ed. Barbara Duden, Jürgen Schlumbohm, and Patrice Veit (Göttingen: Vandenhoeck & Ruprecht, 2002), pp. 107–108; Daniela Watzke, "Embryologische Konzepte zur Entstehung von Missbildungen im 18. Jahrhundert," in *Imagination und Sexualität: Pathologien der Einbildungskraft im medizinischen Diskurs der frühen Neuzeit*, ed. Stefanie Zaun, Daniela Watzke, and Jörn Steigerwald (Frankfurt am Main: Klostermann, 2004), pp. 124–30; Ulrike Enke, "Einleitung," in Samuel Thomas Soemmerring, *Schriften zur Embryologie und Teratologie*, ed. Ulrike Enke (Basel: Schwabe, 2000), pp. 38–39. Needham's findings on the history of embryology indicate that research on this point proliferated in the seventeenth century. Joseph Needham, *A History of Embryology*, rev. ed. (Cambridge, UK: Cambridge University Press, 1959), pp. 115–227.

12. William Hunter, *An Anatomical Description of the Human Gravid Uterus, and Its Contents* (London: J. Johnson and G. Nicol, 1794), pp. 36–49, quotations p. 48.

13. It was not until Karl Ernst von Baer's 1828 studies of the vascular connections between mother and fetus in mammals that Hunter's postulate was conclusively confirmed. Foster de Witt, "An Historical Study on Theories of the Placenta to 1900," *Journal of the History of Medicine* 14.3 (1959), pp. 68–69. See also Lawrence D. Longo and Lawrence P. Reynolds, "Some Historical Aspects of Understanding Placental Development, Structure and Function," *International Journal of Developmental Biology* 54.2–3 (2010), pp. 242–44.

14. Xavier Bichat, *Physiological Researches on Life and Death*, trans. Tobias Watkins (Philadelphia: Smith & Maxwell, [1799] 1809), p. 120. On Bichat's multifarious and foundational significance, see Mohammadali M. Shoja et al., "Marie-François Xavier Bichat (1771–1802) and His Contributions to the Foundations of Pathological Anatomy and Modern Medicine," *Annals of Anatomy — Anatomischer Anzeiger* 190.5 (2008), pp. 413–20; in the context of the emergence of experimental physiology: John E. Lesch, *Science and Medicine in France: The Emergence of Experimental Physiology, 1790–1855* (Cambridge, MA: Harvard University Press, 1984), pp. 50–79.

15. Bichat, *Physiological Researches*, p. 94; Bichat, *Physiologische Untersuchungen über Leben und Tod*, trans. C. H. Pfaff (Copenhagen: Friedrich Brummer, 1802), p. 49.

16. Bichat, *Physiological Researches*, pp. 94 and 103; see also pp. 118–25.

17. Ibid., pp. 95 (in French, "entièrement anéantie," Bichat, *Recherches physiologiques sur la vie et la mort* [Paris: Brosson, Gabon, 1805], p. 108) and 94.

18. Bichat, *Physiological Researches*, pp. 97, 100, 103.

19. Ibid., p. 103. On the distinction between animated and nonanimated life, see Karl Figlio, "The Historiography of Scientific Medicine: An Invitation to the Human Sciences," *Comparative Studies in Society and History* 19.3 (1977), p. 268.

20. Bichat, *Physiological Researches*, pp. 104–107 (original emphasis).

21. Barbara Duden, "Zwischen 'wahrem Wissen' und Prophetie: Konzeptionen des Ungeborenen," in *Geschichte des Ungeborenen: Zur Erfahrungs- und Wissenschaftsgeschichte der Schwangerschaft, 17.–20. Jahrhundert*, ed. Barbara Duden, Jürgen Schlumbohm, and Patrice Veit (Göttingen: Vandenhoeck & Ruprecht, 2002), p. 27.

22. There was a risk of such confusion partly because the division into animal and organic life in Bichat's work did in fact draw upon the Aristotelian distinction between the nutritive and the sensitive soul. On the long-standing influence of that distinction, see Ilse Jahn, *Grundzüge der Biologiegeschichte* (Jena: G. Fischer, 1990), p. 73.

23. Bichat, *Physiological Researches*, p. 120.

24. Christian Friedrich Nasse, "Von der Beseelung des Kindes," *Zeitschrift für die Anthropologie* 2.1 (1824), pp. 9 and 8. On the potentiality argument, see Lynn M. Morgan, "The Potentiality Principle from Aristotle to Abortion," *Current Anthropology* 54.S7 (2013), pp. S15–S25.

25. Nasse, "Von der Beseelung des Kindes," pp. 8–9.

26. Ibid., pp. 1–6.

27. Ibid., pp. 9–10.

28. Ibid., pp. 9–10 and 2.

29. Ibid., p. 1.

30. Andrew Cunningham, *The Anatomist Anatomis'd: An Experimental Discipline in Enlightenment Europe* (London: Routledge, 2010), p. 317. Also on the process by which physiology became an autonomous discipline in the nineteenth century, see Georges Canguilhem, "La constitution de la physiologie comme science," in *Études d'histoire et de philosophie des sciences*, 2nd ed. (Paris: J. Vrin, 1970), pp. 227–30.

31. Magendie, quoted in Frank W. Stahnisch, "François Magendie (1783–1855)," *Journal of Neurology* 256.11 (2009), p. 1951. In more detail: Stahnisch, *Ideas in Action: Der Funktionsbegriff und seine methodologische Rolle im Forschungsprogramm des Experimentalphysiologen François Magendie (1783–1855)* (Münster: LIT, 2003); Lesch, *Science and Medicine in France*, pp. 89–124.

32. See François Magendie, *Précis élémentaire de physiologie*, vol. 1 (Paris: Méquignon-Marvis, 1816), pp. 108–109.

33. See ibid., pp. 119–20.

34. See ibid., p. 153.

35. See ibid., pp. 135–36 and 152–53.

36. See François Magendie, *Précis élémentaire de physiologie*, vol. 2 (Paris: Méquignon-Marvis, 1817), p. 436.

37. See ibid., pp. 435–36.

38. See ibid., pp. 437–42.

39. Ibid., p. 449.

40. See Canguilhem, "La constitution de la physiologie," p. 240.

41. See Stahnisch, *Ideas in Action*, p. 237; Timothy Lenoir, *The Strategy of Life: Teleology and Mechanics in Nineteenth-Century German Biology* (Dordrecht: D. Reidel, 1982), pp. 103–104.

42. Johannes Müller, "Zur Physiologie des Foetus," *Zeitschrift für die Anthropologie* 2.2 (1824), pp. 423–83; Müller, *De respiratione foetus, commentatio physiologica* (Leipzig: Cnobloch, 1823).

43. Müller, "Zur Physiologie des Foetus," p. 423. Müller did, however, also work embryologically. See Lenoir, *Strategy of Life*, pp. 54–111.

44. Müller, "Zur Physiologie des Foetus," p. 442.

45. This sketch would later be supplemented by the discussion in the chapter on development in Johannes Müller, *Handbuch der Physiologie des Menschen für Vorlesungen*, 3rd rev. ed., vol. 2 (Coblenz: Hölscher, 1840).

46. See the conclusion in Müller, "Zur Physiologie des Foetus," p. 465.

47. Ibid., p. 446.

48. Ibid., p. 447.

49. Ibid., pp. 461–62.

50. For anecdotal comments on these experiments, see Wilhelm Haberling, *Johannes Müller: Das Leben des Rheinischen Naturforschers* (Leipzig: Akademische Verlagsgesellschaft, 1924), pp. 36–38. On the biographical context: Laura Otis, *Müller's Lab* (Oxford: Oxford University Press, 2007), pp. 6–8.

51. See Richard L. Kremer, "Physiology," in *The Cambridge History of Science*, vol. 6: *The Modern Biological and Earth Sciences*, ed. Peter J. Bowler and John V. Pickstone (Cambridge, UK: Cambridge University Press, 2009), pp. 346–47. On the radicalism of Müller's experimentation: Canguilhem, "La constitution de la physiologie," p. 249. Generally on the experimentalization of physiology: William Coleman and Frederic L. Holmes (ed.), *The*

Investigative Enterprise: Experimental Physiology in Nineteenth-Century Medicine (Berkeley: University of California Press, 1988); Lesch, *Science and Medicine in France.*

52. From Karl Ernst von Baer's autobiography, quoted in Eckhard Struck, "Ignaz Döllinger 1770–1841: Ein Physiologe der Goethe-Zeit und der Entwicklungsgedanke in seinem Leben und Werk" (PhD diss., Ludwig Maximilian University, Munich, 1977), p. 121.

53. For more detail on this and the characteristics of Döllinger's work as relevant here, see Owsei Temkin, "German Concepts of Ontogeny and History around 1800," *Bulletin of the History of Medicine* 24.3 (1950), pp. 229–32; Janina Wellmann, *The Form of Becoming: Embryology and the Epistemology of Rhythm, 1760–1830*, trans. Kate Sturge (New York: Zone Books, 2017), pp. 144–54 and 267–320. On Döllinger, Pander, and von Baer, see also Hopwood, *Haeckel's Embryos*, pp. 16–24.

54. Ignaz Döllinger, *Grundzüge der Physiologie der Entwicklung des Zell-, Knochen- und Blutsystemes* (Regensburg: G. Joseph Manz, 1842), p. 3.

55. Ibid., p. 4.

56. On Döllinger's concept of time, see Temkin, "German Concepts of Ontogeny and History." In this respect, Temkin places Döllinger in a precise relationship to his era's philosophy of history. On temporality in Döllinger, see also Caroline Arni, "Traversing Birth: Continuity and Contingency in Research on Development in Nineteenth-Century Life and Human Sciences," *History and Philosophy of the Life Sciences* 37.1 (2015), pp. 50–67.

57. Döllinger, *Grundzüge der Physiologie der Entwicklung*, p. 70.

58. Ibid., p. 275.

59. Ibid., pp. 277–80. For Döllinger, incidentally, the human being as an organism is free only in death, which he therefore calls a "second birth" that leads the way to a "third sphere of life" (p. 279).

60. See the editor's preface in Döllinger, *Grundzüge der Physiologie der Entwicklung*, pp. iv–v. On the details of his illness: Struck, "Ignaz Döllinger," p. 184.

61. See Temkin, "German Concepts of Ontogeny and History," p. 236, including the reference to the 1841 obituary.

62. Müller, *Handbuch der Physiologie des Menschen*, pp. 760 and 764.

63. See Temkin, "German Concepts of Ontogeny and History," p. 231.

64. Döllinger, *Grundzüge der Physiologie der Entwicklung*, pp. 42–43.

65. Ibid., p. xi.

66. Ibid.

67. Müller, "Zur Physiologie des Foetus," p. 459.

68. Pierre-Jean-Georges Cabanis, *On the Relations between the Physical and Moral Aspects of Man*, vol. 1, ed. George Mora, trans. Margaret Duggan Saidi (Baltimore, MD: Johns Hopkins University Press, [1802] 1982), p. 50 (translation emended). On the significance of Cabanis for Johannes Müller's work, see Gerhard Scharbert, "*Psychologus nemo, nisi Physiologus*: Johannes Müller und die Perspektiven einer médecine philosophique; eine Entdeckung aus dem Universitätsarchiv," *Würzburger medizinhistorische Mitteilungen* 29 (2010), pp. 241–55. On Müller's psychophysiology: Michael Hagner, *Homo cerebralis: Der Wandel vom Seelenorgan zum Gehirn* (Frankfurt am Main: Suhrkamp, 2008), pp. 238–39.

69. Cabanis, *On the Relations*, p. 50. See Claudia Honegger, *Die Ordnung der Geschlechter: Die Wissenschaften vom Menschen und das Weib, 1750–1850* (Frankfurt am Main: Campus, 1991), p. 166. The second part of Honegger's book comments in detail on the significance of sensibility in a physiology that "was at once sociology, anthropology, and philosophy" (p. 164).

70. See Michael Gross, "The Lessened Locus of Feelings: A Transformation in French Physiology in the Early Nineteenth Century," *Journal of the History of Biology* 12 (1979), pp. 231–71.

71. Cabanis, *On the Relations*, p. 34.

72. Ibid., p. 562.

73. See Hagner, *Homo cerebralis*, pp. 195–201.

74. Müller to von Baer, January 20, 1828, in Michael Hagner, "Sieben Briefe von Johannes Müller an Karl Ernst von Baer," *Medizinhistorisches Journal* 27 (1992), p. 149.

75. Karl Friedrich Burdach (ed.), *Die Physiologie als Erfahrungswissenschaft*, vol. 2, with contributions by Karl E. von Baer et al., 2nd ed. (Leipzig: Leopold Voss, 1837), p. 779.

76. Ibid., p. 781.

77. Ibid., p. 785. This view of the individual living being as the realization of a continuing whole (such as a species) was crucial for the life sciences. On its emergence, see Ohad S. Parnes, "On the Shoulders of Generations: The New Epistemology of Heredity in the Nineteenth Century," in *Heredity Produced: At the Crossroads of Biology, Politics, and Culture, 1500–1870*, ed. Staffan Müller-Wille and Hans-Jörg Rheinberger (Cambridge, MA: MIT Press, 2007), pp. 319–23.

78. Burdach, *Die Physiologie als Erfahrungswissenschaft*, p. 781.

79. Ibid., p. 780.

80. See ibid., p. 779.

81. Ibid., p. 780.

82. Ibid., p. 781.

83. Ibid., p. 783.

84. Ibid., original emphasis. This, Burdach argues, can also be seen in the newborn, which physiologically resembles a "hibernating animal" (p. 784).

85. Ibid., p. 799. The term "appropriation" (*Aneignung*; in the French texts *appropriation*) occurs frequently, also in the work of Müller, who speaks of the child "appropriating for itself" the warmth and nourishment supplied by the mother (*Handbuch der Physiologie des Menschen*, p. 574), and that of Döllinger (*Grundzüge der Physiologie der Entwicklung*, p. 10).

86. Burdach, *Die Physiologie als Erfahrungswissenschaft*, pp. 802–803.

87. On this link in Aristotle, see Jane M. Oppenheimer, "When Sense and Life Begin: Background for a Remark in Aristotle's *Politics* (1335b24)," *Arethusa* 8.2 (1975), pp. 331–43.

88. Preyer, *Specielle Physiologie des Embryo*, pp. 43–44.

89. Ibid., p. 6.

90. Ibid.

91. Ibid.

92. William T. Preyer, *The Mind of the Child, Part I: The Senses and the Will*, trans. H. W. Brown (New York: Appleton, 1893), p. ix. First published in German, Preyer's book appeared in six languages and 180 editions between 1882 and 2010. On his importance in developmental psychology, see Eckardt, "Einleitung," pp. 41 and 43.

93. Preyer, *Mind of the Child*, p. xiv.

94. See Preyer, *Specielle Physiologie des Embryo*, p. 515.

95. See ibid., p. 486.

96. Ibid., p. 487.

97. Ibid., pp. 489 and 493.

98. Ibid., p. 547.

99. On the watershed of the 1880s, see Frederick B. Churchill, "From Heredity Theory to *Vererbung*: The Transmission Problem, 1850–1915," *Isis* 78.3 (1987), p. 337.

100. Needham, *Chemical Embryology*, vol. 1, p. 3.

101. C. S. Marshall, "Physiology of the Fetus: Origin and Extent of Function in Prenatal Life, by William Frederick Windle, W. B. Saunders Co., Philadelphia, 1940," *Yale Journal of Biology and Medicine* 13.3 (1941), pp. 425–26.

102. On this point and on embryologists' resistance to the definitive separation of heredity from development, see Jane Maienschein, "Heredity/Development in the United States, circa 1900," *History and Philosophy of Life Sciences* 9.1 (1987), pp. 79–93.

CHAPTER FIVE: PREGNANCY AS RELATION

1. On Prévost, see J. J. Dreifuss, "Un chercheur genevois insuffisamment connu: Jean-Louis Prevost (1790–1850)," *Revue médicale suisse* 51 (2006).

2. Jean-Louis Prévost, "Note sur le sang du foetus dans les animaux vertébrés," *Annales des sciences naturelles*, 1st series, 4 (1825), p. 499.

3. Ibid.

4. Jean-Louis Prévost and Antoine Morin, "Recherches physiologiques et chimiques sur la nutrition du foetus," *Mémoires de la Société Physique et d'Histoire naturelle de Genève* 9 (1842), p. 239.

5. Ibid., p. 245.

6. See Foster de Witt, *An Historical Study on Theories of the Placenta to 1900* (Bern: Arnaud, 1958), p. 20.

7. François Magendie, *Précis élémentaire de physiologie*, vol. 2 (Paris: Méquignon-Marvis, 1817), pp. 444 and 446.

8. Joseph Needham, *A History of Embryology*, rev. ed. (Cambridge, UK: Cambridge University Press, 1959), pp. 179–82.

9. Albrecht von Haller, *La génération, ou Exposition des phénomènes relatifs à cette fonction naturelle*, vol. 2 (Paris: Ventes de la Doué, 1774), p. 148.

10. See Needham, *History of Embryology*, p. 226.

11. August C. Mayer, "Ueber das Einsaugungsvermögen der Venen des großen und kleinen Kreislaufsystems," *Deutsches Archiv für die Physiologie* 3.4 (1817), p. 503.

12. Ibid., pp. 488–89.

13. Ibid., p. 503.

14. William T. Preyer, *Specielle Physiologie des Embryo: Untersuchungen über die Lebenserscheinungen vor der Geburt* (Leipzig: Grieben, 1883), p. 207.

15. Ibid.

16. Adolf S. Schauenstein and Josef Spaeth, "Uebergang von Medicamenten in die Milch der Säugenden und in den Fötus," *Froriep's Notizen aus dem Gebiet der Natur- und Heilkunde* 2.17 (1859), p. 266.

17. See ibid., pp. 266–67. Similar observations on the administration of chloroform during labor and of mercury to pregnant syphilis patients were made in 1862 by the physician Xavier Bourgeois in *De l'influence des maladies de la femme pendant la grossesse sur la santé et la constitution de l'enfant, Mémoire récompensé par l'Académie impériale de medicine, Séance du 17 décembre 1861* (Paris: J.-B. Baillière et fils, 1862), pp. 112–13.

18. Murat [probably Jean-Baptiste-Arnaud Murat], "Placenta," in *Dictionnaire des sciences médicales*, vol. 42, ed. Société de médecins et de chirurgiens (Paris: C. L. F. Panckoucke, 1820), pp. 517 and 531.

19. Ibid., pp. 534–35.

20. Ibid., pp. 538 and 540.

21. Xavier Delore, "Placenta," in *Dictionnaire encyclopédique des sciences médicales*, 2nd series, vol. 25, ed. Amédée Dechambre (Paris: G. Masson, P. Asselin, 1886), pp. 523–24. Delore mentions contradictory opinions regarding, for example, the connection between placenta and womb and the "material exchange" between amniotic fluid and the maternal blood. Ibid., pp. 516 and 521.

22. Ibid., pp. 522–23 and 527.

23. Preyer, *Specielle Physiologie des Embryo*, p. 205.

24. Louis Guinard and Henri Hochwelker, "Recherches sur le passage des substances solubles du foetus à la mère," *Comptes rendus hebdomadaires des séances et mémoires de la Société de biologie* 50 (1898), pp. 1183–84; Preyer, *Specielle Physiologie des Embryo*, pp. 218–28.

25. Adolphe Pinard, "Foetus," in *Dictionnaire encyclopédique des sciences médicales*, 4th series, vol. 2, ed. Amédée Dechambre (Paris: G. Masson, P. Asselin, 1878), p. 521.

26. Ignaz Döllinger, *Grundzüge der Physiologie der Entwicklung des Zell-, Knochen- und Blutsystemes* (Regensburg: G. Joseph Manz, 1842), pp. 378–79.

27. Ibid., p. 324.

28. See François Jacob, *The Logic of Life: A History of Heredity*, trans. Betty E. Spillmann (Princeton, NJ: Princeton University Press, [1973] 1993), p. 127, also see p. 86. On climate and nutrition in eighteenth-century natural history, and with respect to procreation, see Jean-Louis Fischer, "La callipédie, ou l'art d'avoir de beaux enfants," *Dix-huitième siècle* 23 (1991), p. 142.

29. Carl Gustav Carus, *Lehrbuch der Gynäkologie oder systematische Darstellung der Lehren von Erkenntniss und Behandlung eigenthümlicher gesunder und krankhafter Zustände sowohl der nicht schwangeren, schwangeren und gebärenden Frauen als der Wöchnerinnen und neugebornen Kinder: Zur Grundlage akademischer Vorlesungen und zum Gebrauche für practische Aerzte, Wundärzte und Geburtshelfer*, vol. 2 (Leipzig: Fleischer, 1820), pp. 53–54.

30. Gabriel Gustav Valentin, *Handbuch der Entwickelungsgeschichte des Menschen mit vergleichender Rücksicht der Entwickelung der Säugethiere und Vögel: nach fremden und eigenen Beobachtungen* (Berlin: A. Rücker, 1835), p. 651.

31. Karl Friedrich Burdach (ed.), *Die Physiologie als Erfahrungswissenschaft*, vol. 2, with

contributions by Karl E. von Baer et al., 2nd ed. (Leipzig: Leopold Voss, 1837), pp. 105–106.

32. Johannes Müller, *Handbuch der Physiologie des Menschen für Vorlesungen*, vol. 2, 3rd rev. ed. (Coblenz: Hölscher, 1840), quotations pp. 574 and 760.

33. See Alain Prochiantz, *Claude Bernard: La révolution physiologique* (Paris: Presses universitaires de France, 1990), p. 101, and on the entanglement of nutrition and development, p. 89.

34. Claude Bernard, *Leçons sur les phénomènes de la vie, communs aux animaux et aux végétaux*, vol. 2 (Paris: J.-B. Baillière et fils, 1879), pp. 58–59. Bernard firmly believed that no "corps solides" at all, and thus no microbes, were able to pass through the placenta.

35. Ibid., p. 59. Concerning Bernard's work on *glycogenie*, see Georges Canguilhem, "Théorie et technique de l'expérimentation chez Claude Bernard," in *Études d'histoire et de philosophie des sciences concernant les vivants et la vie*, 2nd ed. (Paris: J. Vrin, 1970), pp. 143–55.

36. Preyer, *Specielle Physiologie des Embryo*, p. 4 (original emphasis).

37. Ibid., p. 6, see also pp. 42–43. Others, too, dedicated much thought and many pages to this practical problem of research. See, for example, F. Tourneux and G. Herrmann, "Embryon," in *Dictionnaire encyclopédique des sciences médicales*, 1st series, vol. 33, ed. Amédée Dechambre (1886), pp. 726–27.

38. Preyer, *Specielle Physiologie des Embryo*, p. 4.

39. Ibid., pp. 5–4.

40. Ibid., p. 7.

41. Ibid., pp. 6–7. On the auscultation of the fetal heartbeat from the early nineteenth century onward, see Paule Herschkorn-Barnu, "Wie der Fötus einen klinischen Status erhielt: Bedingungen und Verfahren der Produktion eines medizinischen Fachwissens, Paris 1832–1848," in *Geschichte des Ungeborenen: Zur Erfahrungs- und Wissenschaftsgeschichte der Schwangerschaft, 17.–20. Jahrhundert*, ed. Barbara Duden, Jürgen Schlumbohm, and Patrice Veit (Göttingen: Vandenhoeck & Ruprecht, 2002), pp. 167–203.

42. Preyer, *Specielle Physiologie des Embryo*, p. 7 (original emphasis).

43. Ibid., p. 9.

44. Ibid., pp. 7–9.

45. Ibid., p. 5.

46. This was also a problem for embryology. See Jane M. Oppenheimer, *Essays in the History of Embryology and Biology* (Cambridge, MA: MIT Press, 1967), p. 185.

47. See Preyer, *Specielle Physiologie des Embryo*, p. 6.

48. See Jacob, *Logic of Life*, p. 100.

CHAPTER SIX: PRENATAL DANGER

1. William T. Preyer, *Specielle Physiologie des Embryo: Untersuchungen über die Lebenserscheinungen vor der Geburt* (Leipzig: Grieben, 1883), pp. 208, 215, 216, and 217.

2. Ibid., pp. 207 and 216.

3. Ibid., pp. 215, 217, and 207–208. Incidentally, the fact that chloroform adversely affects fetal life in small organisms meant Preyer "had to refrain from chloroforming pregnant animals for the purposes of vivisection" (ibid., p. 209).

4. Ibid., pp. 217–18.

5. François Jacob, *The Logic of Life: A History of Heredity*, trans. Betty E. Spillmann (Princeton, NJ: Princeton University Press, [1973] 1993), p. 123, also p. 184; on the experimental use of toxic agents as "one of the methods most favoured by physiologists for more than a century," see p. 186. On the reciprocal constitution of the normal and the pathological, Georges Canguilhem, *The Normal and the Pathological*, trans. Carolyn R. Fawcett and Robert S. Cohen (New York: Zone Books, [1966] 1991).

6. On the "single unit of health and disease," see J. Andrew Mendelsohn, "Medicine and the Making of Bodily Inequality in Twentieth-Century Europe," in *Heredity and Infection: The History of Disease Transmission*, ed. Jean-Paul Gaudillière and Ilana Löwy (London: Routledge, 2001), p. 42.

7. See Erna Lesky, *Die Zeugungs- und Vererbungslehren der Antike und ihr Nachwirken* (Wiesbaden: Franz Steiner, 1950), p. 1328.

8. Quoted in Joseph Needham, *A History of Embryology*, rev. ed. (Cambridge, UK: Cambridge University Press, 1959), pp. 228–29.

9. Christoph Wilhelm Hufeland, "Von den Krankheiten der Ungeborenen und der Vorsorge für das Leben und die Gesundheit des Menschen vor der Geburt" (1827), in *Sammlung auserlesener Abhandlungen* über *Kinder-Krankheiten*, vol. 5, ed. Franz Joseph von Mezler (Prague: Haase, 1836), p. 5.

10. Ibid., pp. 6–5.

11. Ibid., pp. 3–4.

12. Charles Billard, *Traité des maladies des enfans nouveaux-nés et à la mamelle, fondé sur de nouvelles observations cliniques et d'anatomie pathologique, faites à l'hôpital des enfans-trouvés de Paris, dans le service de M. Banon* (Paris: J.-B. Baillière et fils, 1828), pp. 2–3.

13. Ibid., pp. 3–4.

14. Xavier Bourgeois, *De l'influence des maladies de la femme pendant la grossesse sur la santé et la constitution de l'enfant, Mémoire récompensé par l'Académie impériale de médicine,*

Séance du 17 décembre 1861 (Paris: J.-B. Baillière et fils, 1862), p. 34.

15. Adolphe Pinard, "Foetus," in *Dictionnaire encyclopédique des sciences médicales*, ed. Amédée Dechambre, 4th series, vol. 2 (1878), pp. 535–36.

16. Ibid., p. 535. On Pinard's "puériculture intrauterine," see Paule Herschkorn, "Adolphe Pinard et l'enfant à naître: L'invention de la médecine fœtale," *Devenir* 8.3 (1996), pp. 77–87.

17. Pinard, "Foetus," p. 536.

18. Féré uses this term frequently, for example in Charles Féré, "Essai expérimental sur les rapports étiologiques, de l'infécondité, des monstruosités, de l'avortement, de la morti-natalité, du retard de développement et de la débilité congénitale," *Teratologia: A Quarterly Journal of Antenatal Pathology* 2.4 (1895), p. 245.

19. See Charles Féré, "Morphinisme et grossesse," *Comptes rendus des séances et mémoires de la Société de biologie* 35 (1883), pp. 526–28.

20. This characterization of Féré is found in the preface to *Index général des travaux de Charles Féré, médecin de Bicêtre, publié dans la "Normandie médicale" par un groupe d'amis de l'auteur* (Rouen: Girieud, 1909).

21. Féré published the results of the experiments discussed here in a series of articles in the proceedings of the Société de biologie, *Comptes rendus des séances de la Société de biologie et de ses filiales.* On this periodical's immense significance for biology in France, see Jacques Léonard, *La médecine entre les savoirs et les pouvoirs: Histoire intellectuelle et politique de la médecine française au XIXe siècle* (Paris: Aubier-Montaigne, 1981), p. 140. Here, I draw chiefly on two synthesizing articles by Charles Féré published in 1895 (Féré, "Essai expérimental sur les rapports étiologiques") and 1899 (Féré, "Tératogénie expérimentale et pathologie générale," in *Cinquantenaire de la Société de biologie*, ed. Société de biologie [Paris: Masson, 1899]). For a summary of the most important findings, see also Féré, *La famille névropathique: Théorie tératologique de l'hérédité et de la prédisposition morbides et de la dégénérescence* (Paris: F. Alcan, [1894] 1898), pp. 243–52.

22. Féré, "Essai expérimental sur les rapports étiologiques," pp. 245–46.

23. Féré, "Tératogénie expérimentale," p. 360, and on Dareste, p. 361.

24. See Michael Hagner, "Vom Naturalienkabinett zur Embryologie: Wandlungen des Monströsen und die Ordnung des Lebens," in *Der falsche Körper: Beiträge zu einer Geschichte der Monstrositäten*, ed. Michael Hagner (Göttingen: Wallstein, 1995), esp. p. 97. What was new in the nineteenth century was not the idea of contingent "accidents" in the events of procreation, but the idea that deviations followed regular laws. See Jacques Roger, *The Life Sciences in Eighteenth-Century French Thought*, ed. Keith R. Benson, trans. Robert Ellrich

(Stanford, CA: Stanford University Press, [1963] 1998), pp. 318–36. On teratology, see also Jean-Louis Fischer, *De la genèse fabuleuse à la morphogénèse des monstres* (Paris: Société française d'histoire des sciences et des techniques, 1986), p. 2; Urs Zürcher, *Monster oder Laune der Natur: Medizin und die Lehre von den Missbildungen 1780–1914* (Frankfurt am Main: Campus, 2004); Georges Canguilhem, "Monstrosity and the Monstrous," in *Knowledge of Life*, trans. Stefanos Geroulanos and Daniela Ginsburg (New York: Fordham University Press, [1965] 2008); Jane M. Oppenheimer, "Some Historical Relationships between Teratology and Experimental Embryology," *Bulletin of the History of Medicine* 42.2 (1968), pp. 145–59.

25. See Georges Canguilhem et al., *Du développement à l'évolution au XIXe siècle* (Paris: Presses universitaires de France, 1962), pp. 10–18, esp. p. 12; also Jacob, *Logic of Life*, p. 124. I return to Étienne Geoffroy Saint-Hilaire in Chapter 7.

26. Camille Dareste, *Recherches sur la production artificielle des monstruosités, ou Essais de tératogénie expérimentale* (Paris: Reinwald, 1877); Jean-Louis Fischer, "La vie et la carrière d'un biologiste du XIXe siècle, Camille Dareste, 1822–1899," 3 vols. (PhD diss., University of Paris 1 E.P.H.E., 1973; translated as Jean-Louis Fischer, *Leben und Werk von Camille Dareste, 1822–1899: Schöpfer der experimentellen Teratologie* [Halle an der Saale: Deutsche Akademie der Naturforscher Leopoldina, 1994]). On the terminology of teratology, teratogeny, and teratogenesis, see Fischer, *De la genèse fabuleuse à la morphogénèse des monstres.*

27. Féré, "Tératogénie expérimentale," pp. 360–61.

28. Ibid., p. 360.

29. See ibid., pp. 364–65.

30. Ibid., p. 365.

31. Ibid., pp. 364–68, quotations pp. 365–67; similarly in Féré, "Essai expérimental sur les rapports étiologiques," p. 247, and Féré, "Note sur les dégénérescences d'origine toxique ou infectueuse," *Comptes rendus hebdomadaires des séances et mémoires de la Société de biologie* 47 (1895), p. 569. On Dareste's conjecture: Camille Dareste, "Introduction," in *Recherches sur la production artificielle des monstruosités, ou Essais de tératogénie expérimentale* (Paris: Reinwald, 1876), p. 40.

32. See Féré, "Tératogénie expérimentale," p. 367.

33. Joseph Needham counts this among the "one or two interesting facts" discovered by Féré. Needham, *Chemical Embryology*, 3 vols. (Cambridge, UK: Cambridge University Press, 1931), vol. 3, pp. 1409 and 1429–30.

34. See Féré, "Tératogénie expérimentale," p. 367.

35. Charles Féré, "L'hérédité morbide," *Revue des deux mondes* 64.126 (November 15, 1894), p. 446.

36. See *Index général des travaux de Charles Féré.*

37. Féré, "Tératogénie expérimentale."

38. John William Ballantyne, *Manual of Antenatal Pathology and Hygiene*, vol. 2: *The Embryo* (Edinburgh: W. Green and Sons, 1904), p. 220. For more detail on Ballantyne: Salim Al-Gailani, "Teratology and the Clinic: Monsters, Obstetrics and the Making of Antenatal Life in Edinburgh, c. 1900" (PhD diss., University of Cambridge, 2010); Al-Gailani, "Pregnancy, Pathology and Public Morals: Making Antenatal Care in Early Twentieth-Century Edinburgh," in *Western Maternity and Medicine, 1880–1990*, ed. Janet Greenlees and Linda Bryder (London: Pickering & Chatto, 2013); Al-Gailani, "'Antenatal Affairs': Maternal Marking and the Medical Management of Pregnancy in Britain around 1900," in *Imaginationen des Ungeborenen: Kulturelle Konzepte pränataler Prägung von der Frühen Neuzeit zur Moderne*, ed. Urte Helduser and Burkhard Dohm (Heidelberg: Universitätsverlag Winter, 2018); H. E. Reiss, "Historical Insights: John William Ballantyne 1861–1923," *Human Reproduction Update* 5.4 (1999), pp. 386–89; G. S. Philip Alistair, "Perinatal Profiles: John William Ballantyne, Scottish Obstetrician and Prolific Writer," *NeoReviews* 9.11 (2008), pp. 503–505. Ballantyne is also considered the "founding father of antenatal care": Ann Oakley, *The Captured Womb: A History of the Medical Care of Pregnant Women* (Oxford: Blackwell, 1984), p. 20. For France, that honor probably goes to Adolphe Pinard. See Herschkorn, "Adolphe Pinard et l'enfant à naître."

39. John William Ballantyne, *Manual of Antenatal Pathology and Hygiene*, vol. 1: *The Foetus* (Edinburgh: W. Green and Sons, 1902), p. 1 and almost identically on p. 3.

40. See ibid., p. 3.

41. Ibid., pp. 5 and 2.

42. Ibid., p. 16.

43. Ibid., p. 2.

44. Georges Canguilhem, "The Development of the Concept of Biological Regulation in the Eighteenth and Nineteenth Centuries," in *Ideology and Rationality in the History of the Life Sciences*, trans. Arthur Goldhammer (Cambridge, MA: MIT Press, [1977] 1988), p. 100.

45. William Frederick Windle, *Physiology of the Fetus: Origin and Extent of Function in Prenatal Life* (Philadelphia: Saunders, 1940); Joseph Barcroft, *Researches on Pre-natal Life*, vol. 1 (Oxford: Blackwell, 1946).

46. Féré, "Essai expérimental sur les rapports étiologiques," p. 245.

47. See Carlos López-Beltrán, "The Medical Origins of Heredity," in *Heredity Produced:*

At the Crossroads of Biology, Politics, and Culture, 1500–1870, ed. Staffan Müller-Wille and Hans-Jörg Rheinberger (Cambridge, MA: MIT Press, 2007), esp. pp. 114–15. On knowledge about this distinction in the whole of the nineteenth century, see Mendelsohn, "Medicine and the Making of Bodily Inequality," p. 40.

48. Jacques A. Millot, *L'art d'améliorer et perfectionner les générations humaines*, 2nd rev. ed. (Paris: Migneret, 1803), p. 73.

49. Charles Féré, "La famille névropathique," *Archives de neurologie* 7.19–20 (1884), pp. 1–43 and 173–91. On the reception, see Patrice Pinell, "Degeneration Theory and Heredity Patterns between 1850 and 1900," in *Heredity and Infection: The History of Disease Transmission*, ed. Jean-Paul Gaudillière and Ilana Löwy (London: Routledge, 2001), p. 248.

50. Thus Féré's retrospective summary of his 1884 work. Féré, *La famille névropathique*, p. 2.

51. The book was translated into German in 1896, under the title *Nervenkrankheiten und ihre Vererbung* (Nervous illnesses and their heredity).

52. Féré, *La famille névropathique*, p. 2.

53. Ibid., p. 1.

54. See Frederick B. Churchill, "From Heredity Theory to *Vererbung*: The Transmission Problem, 1850–1915," *Isis* 78.3 (1987), p. 339. In more detail: Carlos López-Beltrán, "In the Cradle of Heredity: French Physicians and *L'Hérédité Naturelle* in the Early 19th Century," *Journal of the History of Biology* 37.1 (2004), pp. 39–72; López-Beltrán, "Medical Origins of Heredity."

55. See López-Beltrán, "Medical Origins of Heredity," p. 106. For more detail on this and on the circulation of the concept between law, life sciences, medicine, and discourses on society: Ohad S. Parnes, "On the Shoulders of Generations: The New Epistemology of Heredity in the Nineteenth Century," in Müller-Wille and Rheinberger, *Heredity Produced*. On the interface with historiography: Zrinka Stahuljak, "History as a Medical Category: Heredity, Positivism, and the Study of the Past in Ninenteenth-Century France," *History of the Present* 3.2 (2013), pp. 140–59.

56. See Laure Cartron, "Degeneration and 'Alienism' in Early Nineteenth-Century France," in Müller-Wille and Rheinberger, *Heredity Produced*, pp. 155–57.

57. See Hans-Jörg Rheinberger and Staffan Müller-Wille, *Vererbung: Geschichte und Kultur eines biologischen Konzepts* (Frankfurt am Main: Fischer, 2009), p. 39; Staffan Müller-Wille, "Figures of Inheritance, 1650–1850," in Müller-Wille and Rheinberger, *Heredity Produced*.

58. See Churchill, "From Heredity Theory to *Vererbung*," p. 262. On the switch from a

concept of act to a concept of substance in procreation: Caroline Arni, "Menschen machen aus Akt und Substanz: Prokreation und Vaterschaft im reproduktionsmedizinischen und im literarischen Experiment," *Gesnerus: Swiss Journal of the History of Medicine and Science* 65 (2008), pp. 196–224. On semen in this context: Florence Vienne, "Eggs and Sperm as Germ Cells," in *Reproduction: Antiquity to the Present Day*, ed. Nick Hopwood, Rebecca Flemming, and Lauren Kassell (Cambridge, UK: Cambridge University Press, 2018).

59. See Rheinberger and Müller-Wille, *Vererbung*, p. 124; also Ohad Parnes, Ulrike Vedder, and Stefan Willer, *Das Konzept der Generation: Eine Wissenschafts- und Kulturgeschichte* (Frankfurt am Main: Suhrkamp, 2008), pp. 291–92.

60. Féré, *La famille névropathique*, pp. 2–3.

61. Ibid., pp. 3–4 (original emphasis).

62. Churchill, "From Heredity Theory to *Vererbung*," p. 337; see August Weismann, *The Germ-Plasm: A Theory of Heredity*, trans. W. Newton Parker and Harriet Rönnfeldt (New York: Charles Scribner's Sons, [1892] 1893), xi. Rheinberger and Müller-Wille consider Weismann's one of the "first syntheses": Rheinberger and Müller-Wille, *Vererbung*, pp. 115–29. For more detail on Weismann, see Frederick B. Churchill, *August Weismann: Development, Heredity, and Evolution* (Cambridge, MA: Harvard University Press, 2015).

63. See Churchill, "From Heredity Theory to *Vererbung*," p. 354.

64. See ibid., pp. 360–61.

65. See ibid., p. 362.

66. Féré, *La famille névropathique*, pp. 3–4.

67. Ibid., p. 4.

68. Ibid., p. 5.

69. On the Lamarckian note in Féré's work, see Pinell, "Degeneration Theory and Heredity Patterns," p. 247.

70. August Weismann, *Vorträge* über *Descendenztheorie, gehalten an der Universität zu Freiburg im Breisgau*, vol. 2 (Jena: G. Fischer, 1902), p. 77 (original emphasis).

71. See Rasmus Winther, "August Weismann on Germ-Plasm Variation," *Journal of the History of Biology* 34 (2001), esp. p. 522.

72. Prosper Lucas, *Traité philosophique et physiologique de l'hérédité naturelle dans les états de santé et de maladie du système nerveux*, 2 vols. (Paris: J.-B. Baillière et fils, 1847–1850). On Lucas, see Churchill, "From Heredity Theory to *Vererbung*," pp. 342–43; Annemarie Wettley, "Zur Problemgeschichte der 'dégénérescence,'" *Sudhoffs Archiv für Geschichte der Medizin und der Naturwissenschaften* 43.3 (1959), p. 194; López-Beltrán, "In the Cradle of Heredity," p. 48.

73. See Churchill, "From Heredity Theory to *Vererbung*," pp. 342–43; Rheinberger and Müller-Wille, *Vererbung*, p. 107.

74. Regarding predisposition, incidentally, Féré also refers to the eighteenth-century work of John Hunter: Féré, *La famille névropathique*, p. 8. On the significance of predisposition since the seventeenth century, see López-Beltrán, "Medical Origins of Heredity," p. 109; on the early nineteenth century: Cartron, "Degeneration and 'Alienism,'" pp. 160–61.

75. On this concept of degeneration, see Bénédict Augustin Morel, *Traité des dégénérescences physiques, intellectuelles et morales de l'espèce humaine et des causes qui produisent ces variétés maladives* (Paris: J.-B. Baillière et fils, 1857).

76. See Féré, *La famille névropathique*, p. 153.

77. Ibid., p. 7–8.

78. Ibid., pp. 7 and 18.

79. See chaps. 2 and 7 of Féré, *La famille névropathique.*

80. Féré, *La famille névropathique*, preface and p. 18.

81. Ibid., p. 208. On the relationship between hereditarily and congenitally transmitted dispositions: ibid., p. 210.

82. Féré, "Tératogénie expérimentale," p. 368.

83. See also Jean-Paul Gaudillière and Ilana Löwy, "Introduction: Horizontal and Vertical Transmission of Diseases, The Impossible Separation," in *Heredity and Infection: The History of Disease Transmission*, ed. Jean-Paul Gaudillière and Ilana Löwy (London: Routledge, 2001).

84. See Bourgeois, *De l'influence des maladies de la femme pendant la grossesse.*

85. See also Jane Maienschein, "Heredity/Development in the United States, circa 1900," *History and Philosophy of Life Sciences* 9.1 (1987), pp. 82–84.

86. The quotations are from the 1876 edition, Dareste, *Recherches sur la production artificielle des monstruosités: Introduction*, p. 40, and the revised edition of 1891, Camille Dareste, *Recherches sur la production artificielle des monstruosités, ou Essais de tératogénie expérimentale*, 2nd rev. ed. (Paris: Reinwald, 1891), p. 76. See also Canguilhem, "Monstrosity and the Monstrous," esp. p. 143. On the broader context: Staffan Müller-Wille, "Reproducing Difference: Race and Heredity from a *longue durée* Perspective," in *Reproduction, Race, and Gender in Philosophy and the Early Life Sciences*, ed. Susanne Lettow (Albany: SUNY Press, 2014).

87. Ballantyne, *Manual of Antenatal Pathology and Hygiene*, vol. 1, p. 14.

88. Émile Apert, *Traité des maladies familiales et des maladies congenitales* (Paris: J.-B. Baillière et fils, 1907), p. 208. On Apert: Pinell, "Degeneration Theory and Heredity

Patterns," esp. pp. 256–58. On Vignes and Apert: Anne Carol, *Histoire de l'eugénisme en France: Les médecins et la procréation XIXe–XXe siècle* (Paris: Éditions du Seuil, 1995). On the continuity of the "inseparable mix" of heredity, the quality of the seed, and development: Jean-Paul Gaudillière and Ilana Löwy, "The Hereditary Transmission of Human Pathologies between 1900 and 1940: The Good Reasons Not to Become 'Mendelian,'" in *Heredity Explored: Between Public Domain and Experimental Science, 1850–1930*, ed. Staffan Müller-Wille and Christina Brandt (Cambridge, MA: MIT Press, 2016); Jean-Paul Gaudillière, "Le syndrome nataliste: Hérédité, médecine et eugénisme en France et en Grande-Bretagne, 1920–1965," in *L'éternel retour de l'eugénisme*, ed. Jean Gayon and Daniel Jacobi (Paris: Presses universitaires de France, 2006).

89. Féré, *La famille névropathique*, p. 208.

90. Henri Legrand du Saulle, "Influence des événements politiques sur les caractères du délire et anomalies physiques et intellectuelles que l'on observe chez les enfants conçus pendant le siège de Paris," *Le praticien* 7 (1884), p. 185.

91. On alcohol, which along with syphilis was firmly established in the canon of prenatal pathogenic factors, see Elizabeth M. Armstrong, *Conceiving Risk, Bearing Responsibility: Fetal Alcohol Syndrome and the Diagnosis of Moral Disorder* (Baltimore, MD: Johns Hopkins University Press, 2003), esp. p. 23–67; on paternal transmission of disease: Antje Kampf, "Times of Danger: Embryos, Sperm and Precarious Reproduction ca. 1870s–1910s," *History and Philosophy of the Life Sciences* 37.1 (2015), pp. 68–86.

92. Féré, "L'hérédité morbide," p. 443; Charles Féré, "Nerve Troubles as Foreshadowed in the Child," *Brain: A Journal of Neurology* 8 (1886), p. 230.

93. See Féré, "Essai expérimental sur les rapports étiologiques," p. 245.

94. Bourgeois, *De l'influence des maladies de la femme pendant la grossesse*, pp. 113–22.

95. Karl Friedrich Burdach (ed.), *Die Physiologie als Erfahrungswissenschaft*, vol. 2, with contributions by Karl E. von Baer et al., 2nd ed. (Leipzig: Leopold Voss, 1837), pp. 119–21 (original emphasis).

96. Hufeland, "Von den Krankheiten der Ungeborenen," pp. 10 and 16–17.

97. Théodule A. Ribot, *Heredity: A Psychological Study of Its Phenomena, Laws, Causes, and Consequences*, trans. from the French (New York: Appleton, [1873] 1875), p. 203.

98. Ibid., p. 204.

99. Yves Delage, *La structure du protoplasma et les théories sur l'hérédité et les grands problèmes de la biologie générale* (Paris: Reinwald, 1895), p. 228.

100. Féré, *La famille névropathique*, p. 252.

101. See Needham, *Chemical Embryology*, vol. 3, pp. 1429–30.

102. Ibid., vol. 1, p. 100.

PART THREE: INNER LIFE

1. Georges Canguilhem, "Monstrosity and the Monstrous," in *Knowledge of Life*, trans. Stefanos Geroulanos and Daniela Ginsburg (New York: Fordham University Press, [1965] 2008), p. 145.

2. William T. Preyer, "Psychogenesis," in *Naturwissenschaftliche Thatsachen und Probleme: Populäre Vorträge* (Jena: Verlag von Gebrüder Paetel, 1880), p. 202.

3. Gerhard von Welsenburg [Iwan Bloch], *Das Versehen der Frauen in Vergangenheit und Gegenwart und die Anschauungen der Aerzte, Naturforscher und Philosophen darüber* (Leipzig: H. Barsdorf, 1899), p. 2 (original emphasis). In his history of ideas, Bloch closely follows the work of Julius Preuss, *Vom Versehen der Schwangeren: Eine historisch-kritische Studie* (Berlin: Fischer, 1892).

4. On antiquity, see Véronique Dasen, "Empreintes maternelles," *Micrologus: Natura, scienze e società medievali* 17 (2009), pp. 35–54. On the Middle Ages, see, for example, John William Ballantyne, *Manual of Antenatal Pathology and Hygiene*, vol. 2: *The Embryo* (Edinburgh: W. Green and Sons, 1904), p. 108; von Welsenburg [Bloch], *Das Versehen der Frauen*, pp. 20–21.

5. Charles Féré, *La famille névropathique: Théorie tératologique de l'hérédité et de la prédisposition morbides et de la dégénérescence* (Paris: F. Alcan, [1894] 1898), p. 17.

6. Ibid. Discussing the theory of emotions, Féré also cites stories of the "violent emotions" of mothers during pregnancy. See the chapter "The Pathological Effects of the Emotions" in Charles Féré, *The Pathology of Emotions: Physiological and Clinical Studies*, trans. Robert Park (London: University Press, [1892] 1899), p. 234.

7. Jean-Louis Fischer, "Défense et critiques de la thèse 'imaginationiste' à l'époque de Spallanzani," in *Lazzaro Spallanzani e la biologia del settecento*, ed. Giuseppe Montalenti and Paolo Rossi (Florence: L. S. Olschki, 1982), p. 413.

8. Marie-Hélène Huet, *Monstrous Imagination* (Cambridge, MA: Harvard University Press, 1993), p. 6.

9. Ballantyne, *Manual of Antenatal Pathology and Hygiene*, vol. 2, p. 119.

10. Among others, Ernest Martin, *Histoire des monstres depuis l'antiquité jusqu'à nos jours* (Paris: Reinwald, 1880), esp. pp. 266–94; Preuss, *Vom Versehen der Schwangeren*; Ballantyne, *Manual of Antenatal Pathology and Hygiene*, pp. 105–27; von Welsenburg [Bloch], *Das Versehen der Frauen*.

11. Prosper Lucas, *Traité philosophique et physiologique de l'hérédité naturelle dans les états de santé et de maladie du système nerveux*, vol. 2 (Paris: J.-B. Baillière et fils, 1850), p. 504.

12. On continuity, see Lisa Malich, "Zeitpfeile, Zeitfaltungen und Diskursanalyse: Zu Kontinuitäten der Imaginationslehre," *Berichte zur Wissenschaftsgeschichte* 34 (2011), pp. 363–78; Malich, *Die Gefühle der Schwangeren: Eine Geschichte somatischer Emotionalität (1780–2010)* (Bielefeld: transcript-Verlag, 2017), pp. 94–108. See also Jean-Louis Fischer, "La callipédie, ou l'art d'avoir de beaux enfants," *Dix-huitième siècle* 23 (1991), pp. 141–58; Françoise Loux, *Le jeune enfant et son corps dans la médecine traditionelle* (Paris: Flammarion, 1978). On the trope of maternal imagination in literary history: Urte Helduser, *Imaginationen des Monströsen: Wissen, Literatur und Poetik der "Missgeburt," 1600–1835* (Göttingen: Wallstein, 2016); on imagination in aesthetics: Huet, *Monstrous Imagination*, esp. chap. 3. On medicine and the imagination: Esther Fischer-Homberger, *Krankheit Frau und andere Arbeiten zur Medizingeschichte der Frau* (Bern: H. Huber, 1979), pp. 125–26.

13. For example, Anke Bennholdt-Thomsen and Alfredo Guzzoni, "Zur Theorie des Versehens im 18. Jahrhundert: Ansätze einer pränatalen Psychologie," in *Klio und Psyche*, ed. Thomas Kornbichler (Pfaffenweiler: Centaurus, 1990); Julia Epstein, "The Pregnant Imagination, Women's Bodies, and Fetal Rights," in *Inventing Maternity: Politics, Science, and Literature, 1650–1865*, ed. Susan C. Greenfield and Carol Barash (Lexington: University Press of Kentucky, 1999); Jane M. Oppenheimer, "Some Historical Relationships between Teratology and Experimental Embryology," *Bulletin of the History of Medicine* 42.2 (1968), pp. 145–59. Currently the most comprehensive collection on the doctrine of maternal imagination and its transformations is Urte Helduser and Burkhard Dohm (ed.), *Imaginationen des Ungeborenen/Imaginations of the Unborn: Kulturelle Konzepte pränataler Prägung von der Frühen Neuzeit zur Moderne/Cultural Concepts of Prenatal Imprinting from the Early Modern Period to the Present* (Heidelberg: Universitätsverlag Winter, 2018).

CHAPTER SEVEN: HOW THE PREGNANT WOMAN FEELS

1. See Anke Bennholdt-Thomsen and Alfredo Guzzoni, "Zur Theorie des Versehens im 18. Jahrhundert: Ansätze einer pränatalen Psychologie," in *Klio und Psyche*, ed. Thomas Kornbichler (Pfaffenweiler: Centaurus, 1990), pp. 113–14.

2. On Descartes and the mechanist view, see Justin E. H. Smith, "Imagination and the Problem of Heredity in Mechanist Embryology," in *The Problem of Animal Generation in Early Modern Philosophy*, ed. Justin E. H. Smith (Cambridge, UK: Cambridge University Press, 2006), pp. 86 and 95; on physiological perspectives, Bennholdt-Thomsen and Guzzoni, "Zur

Theorie des Versehens," p. 114; Daniela Watzke, "Embryologische Konzepte zur Entstehung von Missbildungen im 18. Jahrhundert," in *Imagination und Sexualität: Pathologien der Einbildungskraft im medizinischen Diskurs der frühen Neuzeit*, ed. Stefanie Zaun, Daniela Watzke, and Jörn Steigerwald (Frankfurt am Main: Klostermann, 2004), pp. 125–27.

3. Nicolas Malebranche, *The Search after Truth*, ed. and trans. Thomas M. Lennon and Paul J. Olskamp (Cambridge, UK: Cambridge University Press, [1674–1675] 1997), p. 113. Malebranche's significance for the doctrine of imagination in the eighteenth century has been noted widely, recently by Urte Helduser, *Imaginationen des Monströsen: Wissen, Literatur und Poetik der "Missgeburt," 1600–1835* (Göttingen: Wallstein, 2016), pp. 64–70.

4. Malebranche, *Search after Truth*, pp. 115–17.

5. Ibid., p. 117. On the principle of resemblance in premodern natural philosophy, see Hans-Jörg Rheinberger and Staffan Müller-Wille, *Vererbung: Geschichte und Kultur eines biologischen Konzepts* (Frankfurt am Main: Fischer, 2009), pp. 35–36. See also Smith, "Imagination and the Problem of Heredity"; Silvia de Renzi, "Resemblance, Paternity, and Imagination in Early Modern Courts," in *Heredity Produced: At the Crossroads of Biology, Politics, and Culture, 1500–1870*, edited by Staffan Müller-Wille and Hans-Jörg Rheinberger (Cambridge, MA: MIT Press, 2007); Esther Fischer-Homberger, *Medizin vor Gericht: Zur Sozialgeschichte der Gerichtsmedizin* (Darmstadt: Luchterhand, 1988), pp. 248–61. On antiquity: Erna Lesky, *Die Zeugungs- und Vererbungslehren der Antike und ihr Nachwirken* (Wiesbaden: Franz Steiner, 1950), pp. 103–104.

6. See Philip K. Wilson, "Out of Sight, Out of Mind? The Daniel Turner–James Blondel Dispute Over the Power of the Maternal Imagination," *Annals of Science: The History of Science and Technology* 49.1 (1992), pp. 63–85. On Blondel's emphasis on fetal individuality, see Jean-Louis Fischer, "Défense et critiques de la thèse 'imaginationiste' à l'époque de Spallanzani," in *Lazzaro Spallanzani e la biologia del settecento*, ed. Giuseppe Montalenti and Paolo Rossi (Florence: L. S. Olschki, 1982), p. 419; Nadia Maria Filippini, "Die 'erste Geburt': Eine neue Vorstellung vom Fötus und vom Mutterleib (Italien, 18. Jahrhundert)," in *Geschichte des Ungeborenen: Zur Erfahrungs- und Wissenschaftsgeschichte der Schwangerschaft, 17.–20. Jahrhundert*, ed. Barbara Duden, Jürgen Schlumbohm, and Patrice Veit (Göttingen: Vandenhoeck & Ruprecht, 2002), p. 117. For more detail on the eighteenth-century controversies: Helduser, *Imaginationen des Monströsen*.

7. Albrecht von Haller, *La génération, ou Exposition des phénomènes relatifs à cette fonction naturelle*, vol. 1 (Paris: Ventes de la Doué, 1774), pp. 548–49. Malebranche is mentioned directly on p. 567.

8. See ibid., p. 550.

9. See Watzke, "Embryologische Konzepte," p. 127.

10. Bennholdt-Thomsen and Guzzoni, "Zur Theorie des Versehens," p. 113.

11. See Ulrike Enke, "Einleitung," in Samuel Thomas Soemmerring, *Schriften zur Embryologie und Teratologie*, ed. Ulrike Enke (Basel: Schwabe, 2000), pp. 39–40.

12. Blondel, paraphrased after Enke, "Einleitung," p. 37. See also Smith, "Imagination and the Problem of Heredity," p. 86.

13. Jean-Baptiste Demangeon, *De l'imagination considérée dans ses effets directs sur l'homme et les animaux, et dans ses effets indirects sur les produits de la gestation; avec une notice sur la génération et les causes les plus probables des difformités de naissance. Ouvrage ou l'on fait la part de l'imagination dans les phénomènes du magnétisme, de l'exorcisme, de l'ascétisme et d'autres prestiges*, 2nd ed. (Paris: Rouen, 1829), pp. v–vi. A few years later, a third edition appeared, with the title *Du pouvoir de l'imagination sur le physique et le moral de l'homme, nouvelle édition* (Paris: J. Rouvier & E. Le Bouvier, 1834). All three editions were reprinted numerous times. It is possible that Demangeon was responding to the 1807 defense of a medical dissertation that indeed remained somewhat undecided on the turn from imagination to feeling: P. Courby, *Des effets généraux des passions dans l'économie animale, et de leur influence chez les femmes grosses* (Paris: Didot Jeune, 1807).

14. Jean-Baptiste Demangeon, *Considérations physiologiques sur le pouvoir de l'imagination maternelle durant la grossesse et sur les autres causes, prétendues ou réelles, des difformités et des variétés naturelles* (Paris: Chez l'auteur, 1807), p. 2.

15. Ibid., p. 11.

16. Ibid., pp. 17 and 19.

17. Ibid., p. 20.

18. Ibid., pp. 21–22.

19. Antoine J. L. Jourdan, "Imagination," in *Dictionnaire des sciences médicales*, vol. 24 (Paris: C. L. F. Panckoucke, 1818), pp. 78–80.

20. Jacques A. Millot, *Médecine perfective, ou Code des bonnes mères* (Paris: Collin & Millot, 1809), pp. 92–93.

21. Christoph Wilhelm Hufeland, "Von den Krankheiten der Ungeborenen und der Vorsorge für das Leben und die Gesundheit des Menschen vor der Geburt" (1827), in *Sammlung auserlesener Abhandlungen* über *Kinder-Krankheiten*, vol. 5, ed. Franz Joseph von Mezler (Prague: Haase, 1836), p. 6.

22. Ibid., pp. 7 and 8.

23. François Magendie, *Précis élémentaire de physiologie*, vol. 2 (Paris: Méquignon-Marvis, 1817), p. 450.

24. Johannes Müller, *Handbuch der Physiologie des Menschen für Vorlesungen*, vol. 2, 3rd rev. ed. (Coblenz: Hölscher, 1840), p. 575 (emphasis added).

25. Ibid., p. 574. See Chapter 5.

26. Theodor Bischoff, "Entwicklungsgeschichte, mit besonderer Berücksichtigung der Missbildungen (Nachtrag)," in *Handwörterbuch der Physiologie, mit Rücksicht auf die physiologische Pathologie*, ed. Rudolph Wagner, vol. 1 (Braunschweig: Vieweg, 1842), pp. 886–87.

27. See Everard Home, "II. The Croonian Lecture: On the Existence of Nerves in the Placenta," *Philosophical Transactions of the Royal Society* 115 (1825), pp. 73–78.

28. Carl Gustav Carus, *Lehrbuch der Gynäkologie oder systematische Darstellung der Lehren von Erkenntniss und Behandlung eigenthümlicher gesunder und krankhafter Zustände sowohl der nicht schwangeren, schwangeren und gebärenden Frauen als der Wöchnerinnen und neugebornen Kinder: Zur Grundlage akademischer Vorlesungen und zum Gebrauche für practische Aerzte, Wundärzte und Geburtshelfer*, vol. 2 (Leipzig: Fleischer, 1820), p. 280. For this reason, Carus also regards "mother" and "child" as "*one organism*" (ibid.; original emphasis).

29. Ibid., p. 61.

30. See Karl Friedrich Burdach (ed.), *Die Physiologie als Erfahrungswissenschaft*, vol. 2, with contributions by Karl E. von Baer et al., 2nd ed. (Leipzig: Leopold Voss, 1832–40), p. 784.

31. Bischoff, "Entwicklungsgeschichte," pp. 887–88.

32. Ibid., p. 889.

33. Ibid.

34. Ibid., p. 894.

35. On this point, see the persuasive reflections by Georges Canguilhem in *Knowledge of Life*, trans. Stefanos Geroulanos and Daniela Ginsburg (New York: Fordham University Press, 2008), especially the chapter "Monstrosity and the Monstrous." On the naturalization of the monstrous, also Michael Hagner, "Vom Naturalienkabinett zur Embryologie: Wandlungen des Monströsen und die Ordnung des Lebens," in *Der falsche Körper: Beiträge zu einer Geschichte der Monstrositäten*, ed. Michael Hagner (Göttingen: Wallstein, 1995); Jean-Louis Fischer, *De la genèse fabuleuse à la morphogénèse des monstres* (Paris: Société française d'histoire des sciences et des techniques, 1986).

36. See Georges Canguilhem et al., *Du développement à l'évolution au XIXe siècle* (Paris: Presses universitaires de France, 1962), pp. 10–18. The term *retardement de développement* is

found in Étienne Geoffroy Saint-Hilaire, *Philosophie anatomique: Des monstruosités humaines* (Paris: Chez l'auteur, 1822), p. 508. He later credited Meckel with having discovered the mechanism. Étienne Geoffroy Saint-Hilaire, *Considérations générales sur les monstres, comprenant une théorie des phénomènes de la monstruosité* (Paris: J. Tastu, 1826), p. 9. On Étienne Geoffroy Saint-Hilaire's teratology, see Bernard Duhamel, "L'œuvre tératologique d'Étienne Geoffroy Saint-Hilaire," *Revue d'histoire des sciences* 25.4 (1972), pp. 337–46; Jean-Louis Fischer, "Le concept expérimental dans l'œuvre tératologique d'Étienne Geoffroy Saint-Hilaire," *Revue d'histoire des sciences* 25.4 (1972), pp. 347–64. The link between teratology and emotional influence in Étienne Geoffroy Saint-Hilaire's work as discussed here was pointed out by Jane M. Oppenheimer, "Some Historical Relationships between Teratology and Experimental Embryology," *Bulletin of the History of Medicine* 42.2 (1968), pp. 145–59.

37. Étienne Geoffroy Saint-Hilaire, *Philosophie anatomique*, p. 501.

38. Ibid., pp. 505–507.

39. Ibid., quotations on pp. 522 and 519.

40. Étienne Geoffroy Saint-Hilaire, "Considérations générales sur la monstruosité, et Description d'un genre nouveau observé dans l'espèce humain, et nommé Aspalosome," *Annales des sciences naturelles* 4 (1825), p. 463.

41. Étienne Geoffroy Saint-Hilaire, *Considérations générales sur les monstres*, p. 41.

42. Ibid., p. 42.

43. Incidentally, Theodor Bischoff underlined this point. Even if the theorem of retarded development explains the genesis of malformations, he argued, that "certainly does not answer the question of what causes this check." Bischoff, "Entwicklungsgeschichte," p. 894.

44. Isidore Geoffroy Saint-Hilaire, *Histoire générale et particulière des anomalies de l'organisation chez l'homme et les animaux: Ouvrage comprenant des recherches sur les caractères, la classification, l'influence physiologique et pathologique, les rapports généraux, les lois et les causes des monstruosités, des variétés et vices de conformation, ou Traité de tératologie*, vol. 3 (Paris: J.-B. Baillière et fils, 1836), p. 542.

45. Ibid.

46. See ibid., p. 542 and 548. The formula can be found, for example, in Burdach, *Die Physiologie als Erfahrungswissenschaft*, p. 125.

47. Isidore Geoffroy Saint-Hilaire, *Histoire générale et particulière des anomalies*, p. 548.

48. Ibid., p. 542.

49. Étienne Geoffroy Saint-Hilaire, *Considérations générales sur les monstres*, p. 41; see

also Étienne Geoffroy Saint-Hilaire, *Philosophie anatomique*, p. 507.

50. On statistics in a medicine shaped by hospital practices, see Esther Fischer-Homberger, *Geschichte der Medizin*, 2nd rev. ed. (Berlin: Springer, 1977), pp. 83–84. More comprehensively: J. Rosser Matthews, *Quantification and the Quest for Medical Certainty* (Princeton, NJ: Princeton University Press, 1995). On statistics in the study of congenital illness and "degeneration," Laure Cartron, "Degeneration and 'Alienism' in Early Nineteenth-Century France," in *Heredity Produced: At the Crossroads of Biology, Politics, and Culture, 1500–1870*, ed. Staffan Müller-Wille and Hans-Jörg Rheinberger (Cambridge, MA: MIT Press, 2007), p. 168; Bernd Gausemeier, "Pedigrees of Madness: The Study of Heredity in Nineteenth and Early Twentieth Century Psychiatry," *History and Philosophy of the Life Sciences* 36.4 (2015), pp. 467–83.

51. See Elizabeth M. Armstrong, *Conceiving Risk, Bearing Responsibility: Fetal Alcohol Syndrome & the Diagnosis of Moral Disorder* (Baltimore, MD: Johns Hopkins University Press, 2003), p. 31.

52. Henri Legrand du Saulle, "Influence des événements politiques sur les caractères du délire et anomalies physiques et intellectuelles que l'on observe chez les enfants conçus pendant le siège de Paris," *Le praticien* 7 (1884), pp. 160–61.

53. For further anecdotes concerning maternal "émotions violentes" during pregnancy, see, for example, Charles Féré, *The Pathology of Emotions: Physiological and Clinical Studies*, trans. Robert Park (London: University Press, [1892] 1899), pp. 234–35.

54. See Charles Féré, "Les enfants du Siège," *Progrès médical* 12.13 (March 29, 1884), p. 245.

55. Charles Féré, "L'hérédité morbide," *Revue des deux mondes* 64.126 (November 15, 1894), p. 443.

56. Charles Féré, "Les stigmates fonctionnels de la dégénérescence," *Journal des connaissances médicales pratiques et de pharmacologie* 64.1 (January 2, 1896), p. 187.

57. See Michel Foucault, *Abnormal: Lectures at the Collège de France, 1974–1975*, trans. Graham Burchell (London: Verso, 2003), p. 299.

58. Paul Moreau (de Tours), *La folie chez les enfants* (Paris: J.-B. Baillière et fils, 1888), p. 37. The German translation is *Der Irrsinn im Kindesalter*, trans. Demetrio Galatti (Stuttgart: Enke, 1889).

59. Gabriel Compayré, *The Intellectual and Moral Development of the Child*, vol. 2, trans. Mary E. Wilson (New York: Appleton, [1893] 1902), p. 256.

60. Ibid., pp. 256–57.

61. Adolphe Combe, *La nervosité de l'enfant: Quatre conférences*, 2nd. ed. (Lausanne: H. Mignot, 1903), pp. 73–74.

62. John William Ballantyne, *Manual of Antenatal Pathology and Hygiene*, vol. 2: *The Embryo* (Edinburgh: W. Green and Sons, 1904), p. 127. Ballantyne too adopts Geoffroy Saint-Hilaire's maxim regarding the enduring and/or intense mental states of the mother. On the debate around maternal impressions in Ballantyne's immediate circle, and for detail on Ballantyne's first publications on the theme, see Salim Al-Gailani, "Teratology and the Clinic: Monsters, Obstetrics and the Making of Antenatal Life in Edinburgh, c. 1900" (PhD diss., University of Cambridge, 2010), pp. 115–26.

63. For example, in Ernest Martin, *Histoire des monstres depuis l'antiquité jusqu'à nos jours* (Paris: Reinwald, 1880), pp. 293–94.

CHAPTER EIGHT: IF THE FETUS SENSES

1. Charles Féré, "Sensation et mouvement: Contribution à la psychologie du foetus," *Revue philosophique de la France et de l'étranger* 21 (1886), p. 261.

2. On the experimentalization of psychophysiology, see Otniel E. Dror, "The Affect of Experiment: The Turn to Emotions in Anglo-American Physiology, 1900–1940," *Isis* 90.2 (1999), p. 212.

3. On Féré's materialism, see Laurent Mucchielli, "Aux origines de la psychologie universitaire en France (1870–1900): Enjeux intellectuels, contexte politique, réseaux et stratégies d'alliance autour de la *Revue philosophique* de Théodule Ribot," *Annals of Science: The History of Science and Technology* 55.3 (1998), p. 281. Also Roger Courtin, *Charles Féré (1852–1907), médecin de la Bicêtre, et la "Néo-psychologie"* (Paris: Connaissances et Savoirs, 2007), p. 72; Jacqueline Carroy, Annick Ohayon, and Régine Pas, *Histoire de la psychologie en France, XIXe–XXe siècles* (Paris: La Découverte, 2006), pp. 95–96.

4. See Féré, "Sensation et mouvement." The text is also found in a shorter and slightly revised form as the chapter "Sur la psychologie du foetus" in Charles Féré, *Sensation et mouvement: Études expérimentales de psycho-mécanique* (Paris: F. Alcan, 1887).

5. On the *Revue*, see Mucchielli, "Aux origines de la psychologie universitaire en France"; Jacqueline Thirard, "La fondation de la *Revue philosophique*," *Revue philosophique de la France et de l'étranger* 166.4 (1976), pp. 401–13; Serge Nicolas, *Histoire de la psychologie française: Naissance d'une nouvelle science* (Paris: In Press, 2002), pp. 113–18; Carroy, Ohayon, and Pas, *Histoire de la psychologie en France*, p. 40. Before the *Revue philosophique*, there was no French journal specializing in psychology; from 1894, the *Année psychologique* appeared, under the direction of Alfred Binet.

6. See Féré, "Sensation et mouvement," p. 255, and on the measurability of muscle

activity, p. 248.

7. Ibid., p. 256; see also p. 258.

8. Ibid., p. 257.

9. Ibid., pp. 257–58.

10. Ibid., pp. 258–59.

11. Ibid., p. 259. Féré makes the same point succinctly in Charles Féré, *The Pathology of Emotions: Physiological and Clinical Studies*, trans. Robert Park (London: University Press, [1892] 1899), p. 234.

12. On the concept of reflex in the nineteenth century, see Michael Hagner, *Homo cerebralis: Der Wandel vom Seelenorgan zum Gehirn* (Frankfurt am Main: Suhrkamp, 2008), pp. 238–46; Georges Canguilhem, "La constitution de la physiologie comme science," in *Études d'histoire et de philosophie des sciences concernant les vivants et la vie*, 2nd ed. (Paris: J. Vrin, 1970), p. 266.

13. In 1757, the Göttingen obstetrician Johann Georg Roederer examined fetal movement in 135 unmarried pregnant women. Barbara Duden, "Die 'Geheimnisse' der Schwangeren und das Öffentlichkeitsinteresse der Medizin: Zur sozialen Bedeutung der Kindsregung," in *Frauengeschichte—Geschlechtergeschichte*, ed. Karin Hausen and Heide Wunder (Frankfurt am Main: Campus, 1992); on quickening, see Cathy McClive, "The Hidden Truths of the Belly: The Uncertainties of Pregnancy in Early Modern Europe," *Social History of Medicine* 15.2 (2002), pp. 209–27.

14. Féré, "Sensation et mouvement," p. 256.

15. Jean-Marie Jacquemier, *Manuel des accouchements et des maladies des femmes grosses et accouchées, contenant les soins à donner aux nouveaux-nés*, 2 vols. (Paris: J.-B. Baillière et fils, 1846), vol. 1, pp. 323–26.

16. See ibid., vol. 1, pp. 316, 324, 431.

17. Ibid., vol. 1, p. 324. On Dubois's and Jacquemier's interest in fetal physiology with regard to the determination of heartbeats, see Paule Herschkorn-Barnu, "Wie der Fötus einen klinischen Status erhielt: Bedingungen und Verfahren der Produktion eines medizinischen Fachwissens, Paris 1832–1848," in *Geschichte des Ungeborenen: Zur Erfahrungs- und Wissenschaftsgeschichte der Schwangerschaft, 17.–20. Jahrhundert*, ed. Barbara Duden, Jürgen Schlumbohm, and Patrice Veit (Göttingen: Vandenhoeck & Ruprecht, 2002).

18. Adolphe Pinard, "Foetus," in *Dictionnaire encyclopédique des sciences médicales*, ed. Amédée Dechambre, 4th series, vol. 2 (Paris: G. Masson, P. Asselin, 1878), p. 522. On this experiment, see Jacquemier, *Manuel des accouchements*, vol. 1, p. 324.

19. Jacquemier, *Manuel des accouchements*, vol. 1, p. 323.

20. Pinard, "Foetus," p. 522; see also p. 497.

21. See the section "A Psychophysiological Prelude" in Chapter 4.

22. Jules Bernard Luys, *Le cerveau et ses fonctions*, 6th ed. (Paris: F. Alcan, [1873] 1888), p. 100.

23. Théodule Ribot, *Heredity: A Psychological Study of Its Phenomena, Laws, Causes, and Consequences*, translated from the French (New York: Appleton, 1875), p. 218.

24. Ibid., p. 227.

25. Xavier Bichat, *Physiological Researches on Life and Death*, trans. Tobias Watkins (Philadelphia: Smith & Watkins, [1799] 1809), p. 95; Pierre-Jean-Georges Cabanis, *On the Relations between the Physical and Moral Aspects of Man*, vol. 2, ed. George Mora, trans. Margaret Duggan Saidi (Baltimore, MD: Johns Hopkins University Press, [1802] 1982), p. 562.

26. William T. Preyer, *Specielle Physiologie des Embryo: Untersuchungen über die Lebenserscheinungen vor der Geburt* (Leipzig: Grieben, 1883), p. 445.

27. Ibid., quotations on pp. 431–32.

28. Ibid., p. 430.

29. See Jacquemier, *Manuel des accouchements*, vol. 1, pp. 325–26.

30. Féré, "Sensation et mouvement," p. 260 (emphasis added).

31. This is made explicit in Charles Féré, "Un fait pour servir à l'histoire des bouffées de chaleur et des rougeurs morbides," *Comptes rendus hebdomadaires des séances et mémoires de la Société de biologie* 46 (1894), p. 645.

32. Julius Preuss, *Vom Versehen der Schwangeren: Eine historisch-kritische Studie* (Berlin: Fischer, 1892), pp. 48–49. Similarly Joseph Drzewiecki, "Einfluss der Eindrücke der Mutter auf den Fötus oder das sogenannte 'Versehen' der Schwangeren," *Wiener Medizinische Wochenschrift* 41.46 (1891), p. 1858, also printed in *Revue de l'hypnotisme et de la psychologie physiologique* 6.2 (1892), pp. 196–202.

33. Pierre Janet, "Charles Féré: Sensations et mouvements," *Revue philosophique de la France et de l'étranger* 24 (1887), p. 200.

34. Bernard Perez, "Les facultés de l'enfant à l'époque de la naissance," *Revue philosophique de la France et de l'étranger* 13.1 (1882), esp. p. 133.

35. Bernard Perez, *La psychologie de l'enfant: Les trois premières années* (Paris: J.-B. Baillière et fils, 1882). Translated, with significant amendments, as Bernard Perez, *The First Three Years of Childhood*, ed. and trans. Alice M. Christie (London: Sonnenschein, [1882] 1885). A German translation followed in 1894.

36. Perez, *First Three Years of Childhood*, p. 1.

37. Ibid.

38. On the discussion at the beginning of the century, see the section "A Psychophysiological Prelude" in Chapter 4.

39. Perez, *La psychologie de l'enfant*, p. 3 (this sentence is not included in the English translation of the chapter).

40. Perez, *First Three Years of Childhood*, p. 6.

41. Ibid., p. 3.

42. Perez, *La psychologie de l'enfant*, p. 4.

43. Perez, "Les facultés de l'enfant," pp. 133–34.

44. Perez, *La psychologie de l'enfant*, p. 8.

45. Perez, *First Three Years of Childhood*, p. 3. On Perez's work as a classic of French developmental psychology, see Dominique Ottavi, *De Darwin à Piaget: Pour une histoire de la psychologie de l'enfant* (Paris: CNRS Éditions, 2001), pp. 143–51; Robert B. Cairns and Beverley D. Cairns, "The Making of Developmental Psychology," in *Handbook of Child Psychology*, ed. Richard M. Lerner, vol. 1, 6th ed. (Hoboken, NJ: Wiley, 2006), p. 100; Guenther Reinert, "History of Life-Span Development Psychology," in *Life-Span Development and Behavior*, ed. Paul B. Baltes and Orville G. Brim, vol. 2 (New York: Academic Press, 1979), p. 218.

46. Bernard Perez, "L'âme de l'embryon et l'âme de l'enfant," *Revue philosophique de la France et de l'étranger* 23 (1887), pp. 585–86.

47. William T. Preyer, "Psychogenesis," in *Naturwissenschaftliche Thatsachen und Probleme: Populäre Vorträge* (Jena: Verlag von Gebrüder Paetel, 1880), pp. 201–202.

48. Ibid., p. 202 (original emphasis).

49. See ibid., pp. 228–29.

50. William T. Preyer, *The Mind of the Child, Part I: The Senses and the Will*, trans. H. W. Brown (New York: Appleton, [1882] 1893), p. xii.

51. Preyer, *Specielle Physiologie des Embryo*, p. 18.

52. See Cairns and Cairns, "Making of Developmental Psychology," pp. 96–100; Georg Eckardt, "Einleitung," in William T. Preyer, *Die Seele des Kindes: Eingeleitet und mit Materialien zur Rezeptionsgeschichte versehen von Georg Eckardt* (Berlin: Springer, 1989), pp. 40–45.

53. Preyer, *Mind of the Child*, p. ix. On the purely practical division and substantive interlock of these volumes, see Chapter 4, "Traversing Birth."

54. Gabriel Compayré, *The Intellectual and Moral Development of the Child*, 2 vols., trans. Mary E. Wilson (New York: Appleton, [1893] 1900–2), vol. 1, p. 28. On the history of such

statements, see Nick Hopwood, "'Not Birth, Marriage or Death, but Gastrulation': The Life of a Quotation in Biology," *The British Journal for the History of Science* 55.1 (2022).

55. Compayré, *Intellectual and Moral Development of the Child*, p. 30.

56. Ibid., pp. 35 and 39.

57. Ibid., p. 29.

58. Preyer, *Specielle Physiologie des Embryo*, p. 5.

59. On Kussmaul as a follower of Müller, see Andrew I. Schafer, "History of the Physician as Scientist," in *The Vanishing Physician-Scientist?*, ed. Andrew I. Schafer (Ithaca, NY: Cornell University Press, 2009). On Kussmaul in the context of developmental psychology, Günther Reinert, "Grundzüge einer Geschichte der Human-Entwicklungspsychologie," in *Die Psychologie des 20. Jahrhunderts. Band 1: Die europäische Tradition: Tendenzen, Schulen, Entwicklungslinien*, ed. Heinrich Balmer (Zurich: Kindler, 1976), p. 873.

60. Adolf Kussmaul, *Untersuchungen über das Seelenleben des neugeborenen Menschen* (Leipzig: C. F. Winter, 1859), p. 36; Kussmaul expressly criticizes Bichat on p. 10. He names Hegel, Herholdt, and Nasse as further advocates of the plant idea (p. 36). On Bichat's fetal vegetative life, see Chapter 4, "Vegetative Life and the Human Child."

61. Kussmaul, *Untersuchungen*, p. 8.

62. Ibid., p. 6.

63. Ibid., pp. 35–36.

64. Ibid., pp. 39–38.

65. Ibid., p. 36.

66. Silvio Canestrini, *Über das Sinnesleben des Neugeborenen (nach physiologischen Experimenten)* (Berlin: J. Springer, 1913), preface.

67. See ibid., pp. 7–8. Canestrini devotes particular attention to Compayré, who had compiled the available literature very comprehensively around the same time.

68. Ibid., p. 9.

69. Ibid., p. 8.

70. Ibid., pp. 15–16.

71. Ibid., p. 17 and preface.

72. Ibid., pp. 100–101.

73. Albrecht Peiper, "Sinnesempfindungen des Kindes vor seiner Geburt," *Monatsschrift für Kinderheilkunde* 29 (1925), p. 238.

74. Ibid., p. 240.

75. See Reinhart Koselleck, *Futures Past: On the Semantics of Historical Time*, trans. Keith

Tribe (New York: Columbia University Press, [1979] 2004), pp. 103–104.

76. Sigmund Freud, "Three Essays on the Theory of Sexuality" (1905), in vol. 7 of *The Standard Edition of the Complete Psychological Works of Sigmund Freud*, ed. and trans. James Strachey (London: The Hogarth Press, 1953), pp. 173–74 n. 2.

77. Sigmund Freud, "Inhibitions, Symptoms and Anxiety" (1926), in vol. 20 of *The Standard Edition of the Complete Psychological Works of Sigmund Freud*, ed. and trans. James Strachey (London: The Hogarth Press, 1959), p. 135.

78. Freud to Ferenczi, July 18, 1915, in *The Correspondence of Sigmund Freud and Sándor Ferenczi*, 3 vols., ed. Ernst Falzeder and Eva Brabant, trans. Peter T. Hoffer (Cambridge, MA: Belknap Press, 1993–2000), vol. 2, p. 68.

79. See Freud to Ferenczi, July 12, 1915, in ibid., pp. 65–66.

80. See Sigmund Freud, "The Ego and the Id" (1923), in vol. 19 of *The Standard Edition of the Complete Psychological Works of Sigmund Freud*, ed. and trans. James Strachey (London: The Hogarth Press, 1961), p. 38. On heredity in psychoanalysis, see Laura Otis, *Organic Memory: History and the Body in the Late Nineteenth and Early Twentieth Centuries* (Lincoln: University of Nebraska Press, 1994), pp. 181–205; Ilse Grubrich-Simitis, "Metapsychology and Metabiology," in Sigmund Freud, *A Phylogenetic Fantasy: Overview of the Transference Neuroses*, ed. Grubrich-Simitis, trans. Axel Hoffer and Peter T. Hoffer (Cambridge, MA: Belknap Press, 1987), pp. 97–100; Ohad Parnes, Ulrike Vedder, and Stefan Willer, *Das Konzept der Generation: Eine Wissenschafts- und Kulturgeschichte* (Frankfurt am Main: Suhrkamp, 2008), pp. 293–99; Georges Canguilhem et al., *Du développement à l'évolution au XIXe siècle* (Paris: Presses universitaires de France, 1962), p. 46.

81. Freud to Ferenczi, December 22, 1916, in *Correspondence of Sigmund Freud and Sándor Ferenczi*, vol. 2, p. 166.

82. Ferenczi to Freud, July 24, 1915, in ibid., p. 70 (original emphasis).

83. Sándor Ferenczi, "Stages in the Development of the Sense of Reality" (1913), in *First Contributions to Psycho-Analysis*, trans. Ernest Jones (London: The Hogarth Press, 1952), p. 217.

84. Ibid., p. 218.

85. Ibid., p. 219.

86. In pathological cases, where the mother is injured or the umbilical cord is damaged, this confrontation may begin even before birth. Ibid.

87. Ibid.

88. Ibid., pp. 219–20.

89. For more detail on the idea of recapitulation, see Steven Jay Gould, *Ontogeny and Phylogeny* (Cambridge, MA: Belknap Press, 1977). On the embryological controversy around Haeckel, Nick Hopwood, *Haeckel's Embryos: Images, Evolution, and Fraud* (Chicago: University of Chicago Press, 2015). On Freud's engagement with evolution, also Gerhard Scharbert, "Freud and Evolution," *History and Philosophy of the Life Sciences* 31.2 (2009), pp. 295–312.

90. See Canguilhem et al., *Du développement à l'évolution*, pp. 46–48. On psychology's reception of recapitulation theory, also Gould, *Ontogeny and Phylogeny*, pp. 155–64.

91. Preyer, *Mind of the Child*, p. xiv.

92. On the systematic significance for theories of development, see John C. Cavanaugh, "Early Developmental Theories: A Brief Review of Attempts to Organise Developmental Data prior to 1925," *Journal of the History of Behavioural Sciences* 17.1 (1981), pp. 38–47.

93. Sigmund Freud, *The Interpretation of Dreams*, in vol. 5 of *The Standard Edition of the Complete Psychological Works of Sigmund Freud*, ed. and trans. James Strachey (London: The Hogarth Press, 1958), pp. 400–401 n.3. See Marina Leitner, *Freud, Rank und die Folgen: Ein Schlüsselkonflikt für die Psychoanalyse* (Vienna: Turia & Kant, 1998), p. 391.

94. Sigmund Freud, *Introductory Lectures on Psycho-Analysis (Part III)* (1916–17), vol. 16 of *The Standard Edition of the Complete Psychological Works of Sigmund Freud*, ed. and trans. James Strachey (London: The Hogarth Press, 1963), p. 396 (original emphasis).

95. Ibid., pp. 396–97.

96. See Hermine Hug-Hellmuth, *A Study of the Mental Life of the Child*, trans. James J. Putnam and Mabel Stevens (Washington, DC: Nervous and Mental Diseases Publication Co., [1913] 1919), p. xiii. The book appeared as part of the series Papers on Applied Psychology, edited by Freud. On Hug-Hellmuth, see Angela Graf-Nold: *Der Fall Hermine Hug-Hellmuth: Eine Geschichte der frühen Kinder-Psychoanalyse* (Munich: Verlag Internationale Psychoanalyse, 1988).

97. See Ferenczi, "Stages in the Development of the Sense of Reality," p. 220 n.10; Ernest Jones, "Angstaffekt und Geburtsakt," *Internationale Zeitschrift für Psychoanalyse* 9.1 (1923), p. 79.

98. Freud, *Introductory Lectures*, p. 397.

99. Dorothy Garley, "Über den Schock des Geborenwerdens und seine möglichen Nachwirkungen," *Internationale Zeitschrift für Psychoanalyse* 10.2 (1924), pp. 135–36.

100. Ibid., pp. 139 and 162–63.

101. Ibid., pp. 136–38.

102. Freud, *Introductory Lectures*, p. 397.

103. Jones, "Angstaffekt und Geburtsakt"; Ernest Jones, "Cold, Disease, and Birth," in *Papers on Psycho-Analysis*, 3rd ed. (London: Bailliere, Tindall and Cox, 1923).

104. Gustav Hans Graber, *Die Ambivalenz des Kindes* (Leipzig: Internationaler psychoanalytischer Verlag, 1924); Sándor Ferenczi, *Thalassa: A Theory of Genitality*, trans. Henry Alden Bunker (New York: The Psychoanalytic Quarterly, [1924] 1938), p. 43.

105. Freud to Rank, August 27, 1924, in *The Letters of Sigmund Freud and Otto Rank: Inside Psychoanalysis*, ed. E. James Lieberman and Robert Kramer, letters translated by Gregory C. Richter (Baltimore, MD: Johns Hopkins University Press, 2012), p. 215.

106. See Leitner, *Freud, Rank und die Folgen*, p. 16.

107. Otto Rank, *The Trauma of Birth* (London: Kegan Paul, [1924] 1929), p. 11.

108. Ibid., p. xiii.

109. Ibid., p. 187.

110. Ibid. (original emphasis).

111. Ibid., pp. 194–95 (original emphasis).

112. Ibid., p. 3. On Ferenczi and Rank as originators of the concept of the pre-Oedipal, see Leitner, *Freud, Rank und die Folgen*, p. 238.

113. Freud to Rank, December 1, 1923, in *Letters of Sigmund Freud and Otto Rank*, p. 179.

114. Sigmund Freud, "The Dissolution of the Oedipus Complex" (1924), in vol. 19 of *The Standard Edition of the Complete Psychological Works of Sigmund Freud*, ed. and trans. James Strachey (London: The Hogarth Press, 1961), p. 179.

115. Freud to Ferenczi, February 4, 1924, in *Correspondence of Sigmund Freud and Sándor Ferenczi*, vol. 3, p. 123.

116. Freud to Ferenczi, March 26, 1924, in ibid., p. 135.

117. Freud to Rank, July 23, 1924, in *Letters of Sigmund Freud and Otto Rank*, p. 208.

118. Freud to Ferenczi, August 14, 1925, in *Correspondence of Sigmund Freud and Sándor Ferenczi*, vol. 3, p. 222. On this conflict, see Leitner, *Freud, Rank und die Folgen*, pp. 58–154.

119. Freud, "Inhibitions, Symptoms and Anxiety," pp. 135–36.

120. Ibid., p. 138.

121. Ibid., p. 138.

122. Ibid., p. 135.

123. See John Forrester, *Dispatches from the Freud Wars: Psychoanalysis and Its Passions* (Cambridge, MA: Harvard University Press, 1997), p. 185. Incidentally, Ferenczi only just escaped Freud's anathema regarding a psychology that commenced too early; in this relationship, the rift came later. A brief comment on this point is found in Henri F.

Ellenberger, *The Discovery of the Unconscious: The History and Evolution of Dynamic Psychiatry* (New York: Basic Books, 1970), pp. 844–45.

124. See Siegfried Bernfeld, *The Psychology of the Infant*, trans. Rosetta Hurwitz (London: Kegan Paul, [1925] 1929), p. 211.

125. Ibid., p. 212.

126. Ibid.; see also pp. 1–2.

127. Ibid., p. 214.

128. See ibid., pp. 222–24.

129. See ibid., p. 30.

130. Ibid.

131. Ibid. On such impressions, see also Gerrit Breeuwsma, "The Nephew of an Experimentalist: Ambivalences in Developmental Thinking," in *Beyond the Century of the Child: Cultural History and Developmental Psychology*, ed. Willem Koops and Michael Zuckerman (Philadelphia: University of Pennsylvania Press, 2003), esp. p. 184.

132. On prenatal psychology in the American context, see Sara Dubow, *Ourselves Unborn: A History of the Fetus in Modern America* (Oxford: Oxford University Press, 2011), pp. 46–49; Ziv Eisenberg, "Clear and Pregnant Danger: The Making of Prenatal Psychology in Mid-Twentieth-Century America," *Journal of Women's History* 22.3 (2010), pp. 112–35.

133. In the version of January 2022, the sentence reads: "The ISPPM [International Society for Prenatal and Perinatal Psychology and Medicine] comprehends this prenatal and perinatal phase of life as the first ecological situation as a human being, being inseparable with its mother and its enviroment in form of a continuous dialogue" ("About," *International Society for Prenatal and Perinatal Psychology and Medicine*, n.d., https://isppm.ngo/about-the-international-society-for-prenatal-and-perinatal-psychology-and-medicine/?lang=en). See also Peter G. Hepper, "Fetal Psychology: An Embryonic Science," in *Fetal Behaviour: Developmental and Perinatal Aspects*, ed. Jan G. Nijhuis (Oxford: Oxford University Press, 1992).

CHAPTER NINE: WHAT THE ENVIRONMENT DOES

1. See Chapter 6, "Experimenting with Development."

2. Charles Féré, "Note relative aux réactions du foetus aux émotions de la mère," *Comptes rendus hebdomadaires des séances et mémoires de la Société de biologie* 55 (1903), pp. 74–75.

3. See Charles Féré, *La famille névropathique: Théorie tératologique de l'hérédité et de la*

prédisposition morbides et de la dégénérescence (Paris: F. Alcan, [1894] 1898), p. 252.

4. Josef Halban, "Ueber Schwangerschaftsreaktionen fötaler Organe und ihre puerperale Involution," *Münchener medizinische Wochenschrift* 51.49 (1904), p. 2210.

5. See Roy Porter, *The Greatest Benefit to Mankind: A Medical History of Humanity* (New York: W. W. Norton, 1999), pp. 562–70; Georges Canguilhem, "La constitution de la physiologie comme science," in *Études d'histoire et de philosophie des sciences concernant les vivants et la vie*, 2nd ed. (Paris: J. Vrin, 1970), pp. 262–65. For a history of the hormonal body beginning in the nineteenth century, see Chandak Sengoopta, *The Most Secret Quintessence of Life: Sex, Glands, and Hormones, 1850–1950* (Chicago: University of Chicago Press, 2006).

6. On Halban's work on the placenta, see Victor C. Medvei, *The History of Clinical Endocrinology* (Carnforth, UK: Parthenon, 1993), pp. 199–200 and 429. On the hormonization of pregnancy: Lisa Malich, "Die hormonelle Natur und ihre Technologien: Zur Hormonisierung der Schwangerschaft im zwanzigsten Jahrhundert," *L'Homme* 25.2 (2014), pp. 69–84. On the role of endocrinology for the reproductive sciences, Adele E. Clarke, *Disciplining Reproduction: Modernity, American Life Sciencs, and "the Problems of Sex"* (Berkeley: University of California Press, 1998), pp. 121–62.

7. Hugo Sellheim, "Mutter-Kinds-Beziehungen auf Grund innersekretorischer Verknüpfung," *Münchener Medizinische Wochenschrift* 71.38 (1924), pp. 1305–1306.

8. Ibid., pp. 1306–1307 (original emphasis).

9. Ibid., p. 1307. On the conceptual significance of the "chemical correlation," see Canguilhem, "La constitution de la physiologie," p. 264.

10. Sellheim, "Mutter-Kinds-Beziehungen," p. 1307.

11. Lester W. Sontag, "War and the Fetal-Maternal Relationship," *Marriage and Family Living* 6.1 (1944), p. 3.

12. Ibid.

13. Ibid., p. 4.

14. Ibid., p. 16.

15. See Lester W. Sontag, "The Significance of Fetal Environmental Differences," *American Journal of Obstetrics and Gynecology* 42 (July–December 1941), p. 996.

16. The pathology he examined was reduced learning performance, which Sontag was able to demonstrate by sending the young animals through mazes. See William D. Thompson and Lester W. Sontag, "Behavioral Effects in the Offspring of Rats Subjected to Audiogenic Seizure during the Gestational Period," *Journal of Comparative and Physiological Psychology* 49.5 (1956), pp. 454–56. On the experiments on pregnant women, for which

Sontag drew on Peiper's work, Lester W. Sontag and Robert F. Wallace, "The Movement Response of the Human Fetus to Sound Stimuli," *Child Development* 6.4 (1935), pp. 253–58. On Sontag in the context of further changes to maternal-fetal relations, Tatjana Buklijaš, "Transformations of the Maternal-Fetal Relationship in the Twentieth Century: From Maternal Impressions to Epigenetic States," in *Imaginationen des Ungeborenen/Imaginations of the Unborn: Kulturelle Konzepte pränataler Prägung von der Frühen Neuzeit zur Moderne/ Cultural Concepts of Prenatal Imprinting from the Early Modern Period to the Present*, ed. Urte Helduser and Burkhard Dohm (Heidelberg: Universitätsverlag Winter, 2018).

17. "History: Fels Longitudinal Study Collection," Wright State University, Boonshoft School of Medicine, n.d., www.med.wright.edu/lhrc/fels.

18. "On Organizing for Continuity," *Child Development* 42.4 (1971), p. 983.

19. Lester W. Sontag, "The History of Longitudinal Research: Implications for the Future," *Child Development* 42.4 (1971), p. 988.

20. Jerome Kagan, "American Longitudinal Research on Psychological Development," *Child Development* 35.1 (1964), pp. 1–2. For a brief treatment of the American longitudinal studies in the context of developmental psychology, see Günther Reinert, "Grundzüge einer Geschichte der Human-Entwicklungspsychologie," in *Die Psychologie des 20. Jahrhunderts. Band 1: Die europäische Tradition: Tendenzen, Schulen, Entwicklungslinien*, ed. Heinrich Balmer (Zurich: Kindler, 1976), pp. 882–83. Equally briefly on the British and WHO longitudinal studies, Nikolas Rose, *Governing the Soul: The Shaping of the Private Self* (London: Free Association Books, [1989] 1999), pp. 187–91.

21. Sontag, "History of Longitudinal Research," p. 988.

22. "On Organizing for Continuity," *Child Development* 42.4 (1971), pp. 983–85.

23. Sontag, "Significance of Fetal Environmental Differences," p. 1002.

24. On this concept of stress and its spread in the twentieth century, see Mark Jackson, *The Age of Stress: Science and the Search for Stability* (Oxford: Oxford University Press, 2013).

25. Martha E. Rogers, Abrahahm M. Lilienfeld, and Benjamin Pasamanick, *Prenatal and Paranatal Factors in the Development of Childhood Behavior Disorders* (Copenhagen: Ejnar Munksgaard, 1955), p. 20. The authors were particularly interested in differences between Black and white families.

26. Ibid., pp. 17–20.

27. Rogers, Lilienfeld, and Pasamanick quote the term "reproductive casualty" in ibid., p. 81.

28. Abraham M. Lilienfeld and Elizabeth Parkhurst, "A Study of the Association of

Factors of Pregnancy and Parturition with the Development of Cerebral Palsy: A Preliminary Report," *American Journal of Hygiene* 53.3 (1951), p. 278.

29. Elizabeth K. Turner, "The Syndrome in the Infant Resulting from Maternal Emotional Tension during Pregnancy," *Medical Journal of Australia* 43.6 (1956), pp. 221–22.

30. Denis H. Stott, "Physical and Mental Handicaps Following a Disturbed Pregnancy," *Lancet* (May 1957), pp. 1007, 1008, 1011.

31. Antonio J. Ferreira, "The Pregnant Woman's Emotional Attitude and Its Reflection on the Newborn," *American Journal of Orthopsychiatry: A Journal of Human Behavior* 30.3 (1960), pp. 553 and 559 (original emphasis).

32. Anthony Davids, Spencer de Vault, and Max Talmadge, "Psychological Study of Emotional Factors in Pregnancy: A Preliminary Report," *Psychosomatic Medicine* 23.2 (1961), p. 94.

33. Jane M. Oppenheimer, "Some Historical Relationships between Teratology and Experimental Embryology," *Bulletin of the History of Medicine* 42.2 (1968), p. 147.

34. Michel Foucault, *The Archaeology of Knowledge*, trans. A. M. Sheridan Smith (New York: Pantheon, [1969] 1972), p. 4.

35. In historical epistemology, this is described as "recurrence," the assessment of past knowledge through the lens of new knowledge. Hans-Jörg Rheinberger took the notion further with the concept of historiality. Georges Canguilhem, "L'histoire des sciences dans l'œuvre épistémologique de Gaston Bachelard," in *Études d'histoire et de philosophie des sciences concernant les vivants et la vie*, 2nd ed. (Paris: J. Vrin, 1970), pp. 173–86; Hans-Jörg Rheinberger, *Toward a History of Epistemic Things* (Stanford, CA: Stanford University Press, 1997), pp. 176–86.

36. On the polythetic category, see Gianna Pomata, "Die Geschichte der Frauen zwischen Anthropologie und Biologie," *Feministische Studien* 2.2 (1983), pp. 113–27.

CHAPTER TEN: THRESHOLDS IN TIME

1. Dr. Dupouy, "Les enfants du siège," *Le médecin: Moniteur de la policlinique* 11.18 (May 3, 1885), p. 3.

2. Henri Legrand du Saulle, "Influence des événements politiques sur les caractères du délire et anomalies physiques et intellectuelles que l'on observe chez les enfants conçus pendant le siège de Paris," *Le praticien* 7 (1884), p. 184.

3. Dupouy, "Les enfants du siège," p. 3.

4. Paul Moreau (de Tours), *La folie chez les enfants* (Paris: J.-B. Baillière et fils, 1888), p. 40.

5. Legrand du Saulle, "Influence des événements politiques," p. 185.

6. See Wolfgang Schivelbusch, *Die Kultur der Niederlage: Der amerikanische Süden 1865, Frankreich 1871, Deutschland 1918* (Berlin: A. Fest, 2001); Colette E. Wilson, *Paris and the Commune, 1871–78: The Politics of Forgetting* (Manchester: Manchester University Press, 2007).

7. See Charles Féré, "Les enfants du Siège," *Progrès médical* 12.13 (March 29, 1884), p. 245.

8. See Michel Foucault, *Abnormal: Lectures at the Collège de France, 1974–1975*, trans. Graham Burchell (London: Verso, 2003), p. 151.

9. On the belief that the number of the mentally ill had grown since 1789, see Laure Cartron, "Degeneration and 'Alienism' in Early Nineteenth-Century France," in *Heredity Produced: At the Crossroads of Biology, Politics, and Culture, 1500–1870*, ed. Staffan Müller-Wille and Hans-Jörg Rheinberger (Cambridge, MA: MIT Press, 2007), p. 158.

10. On the theorization of "the masses," see Michael Gamper, *Masse lesen, Masse schreiben: Eine Diskurs- und Imaginationsgeschichte der Menschenmenge 1765–1930* (Munich: W. Fink, 2007), pp. 355–72; Daniel Pick, *Faces of Degeneration: A European Disorder, c. 1848–c. 1918* (Cambridge, UK: Cambridge University Press, 1989), pp. 87–96.

11. Jean-Vincent Laborde, *Les hommes et les actes de l'insurrection de Paris devant la psychologie morbide* (Paris: Germer Baillière, 1872), pp. 3–10, quotations pp. iii, 3, 6.

12. Henri Legrand du Saulle, *Le délire des persecutions* (Paris: H. Plon, 1871), pp. 483–84. An abridged form of the appendix appeared in 1896 as a three-part series titled "L'état mental des Parisiens pendant le Siège de Paris (1871)," in the section "Actualités médicales rétrospectives" of *Chronique médicale* 3 (1896), pp. 77–80, 119–21, and 147–51.

13. Legrand du Saulle, *Le délire des persecutions*, pp. 484–85.

14. Ibid., pp. 515 and 505.

15. On the controversy around revolutionary madness, see Pick, *Faces of Degeneration*, p. 70.

16. Legrand du Saulle, *Le délire des persecutions*, p. 504.

17. Ibid., p. 498.

18. Ibid., pp. 501–502.

19. Féré, "Les enfants du Siège," p. 245.

20. See Fréderic Carbonel, "Un oublié normand de la psychologie française: Le Docteur Féré (1852–1907)," *Bulletin de la Société libre d'Emulation de la Seine-Maritime* (2002), p. 38; Roger Courtin, *Charles Féré (1852–1907), médecin de la Bicêtre, et la "Néo-psychologie"* (Paris: Connaissances et Savoirs, 2007), p. 59.

21. Charles Féré, "L'hérédité morbide," *Revue des deux mondes* 64.126 (November 15, 1894), p. 440.

22. Legrand du Saulle, *Le délire des persecutions*, p. 504.

23. Ibid., pp. 506–10.

24. Ibid., pp. 512–13.

25. Ibid., p. 513.

26. Legrand du Saulle, "Influence des événements politiques," p. 184.

27. See Ohad S. Parnes, "On the Shoulders of Generations: The New Epistemology of Heredity in the Nineteenth Century," in Müller-Wille and Rheinberger, *Heredity Produced*, esp. p. 317. The following remarks also draw on the study of the generation concept presented by the historians of science Ohad Parnes, Ulrike Vedder, and Stefan Willer, *Das Konzept der Generation: Eine Wissenschafts- und Kulturgeschichte* (Frankfurt am Main: Suhrkamp, 2008).

28. See Parnes, "On the Shoulders of Generations," pp. 316–19.

29. See François Jacob, *The Logic of Life: A History of Heredity*, trans. Betty E. Spillmann (Princeton, NJ: Princeton University Press, [1973] 1993), esp. pp. 19–20; Parnes, Vedder, and Willer, *Das Konzept der Generation*, pp. 207–208; Parnes, "On the Shoulders of Generations," pp. 314–15 and in detail on pp. 319–23.

30. See Parnes, "On the Shoulders of Generations," pp. 323–26.

31. See Laura Otis, *Organic Memory: History and the Body in the Late Nineteenth and Early Twentieth Centuries* (Lincoln: University of Nebraska Press, 1994), pp. 23–25; Ilana Löwy, "On Guinea Pigs, Dogs and Men: Anaphylaxis and the Study of Biological Individuality, 1902–1939," *Studies in History and Philosophy of Biological and Biomedical Sciences* 34.3 (2003), p. 402. Specifically on organic memory's role in remembering the Commune, see Wilson, *Paris and the Commune*, pp. 10–11.

32. See the section "Pregnancy Accidents and Heredity" in Chapter 6.

33. Féré, "L'hérédité morbide," p. 443.

34. Legrand du Saulle, "Influence des événements politiques," pp. 185–86. He expressly framed this as an exhortation to "pacifism" (p. 186). For Legrand du Saulle, "race" is nationally specified, but also refers to humanity as such.

35. References to stigma can be found, among other locations, in Charles Féré, *Sensation et mouvement: Études expérimentales de psycho-mécanique* (Paris: F. Alcan, 1887), p. 96. On Féré's distinction between morphological and functional stigma, see Féré, "Les stigmates fonctionnels de la dégénérescence," *Journal des connaissances médicales pratiques et de*

pharmacologie 64.1 (January 2, 1896), pp. 186–88. On the concept of stigma more generally, Anne Carol, *Histoire de l'eugénisme en France: Les médecins et la procréation XIXe–XXe siècle* (Paris: Éditions du Seuil, 1995), p. 94. On the child in debates on degeneration during the first half of the century: Cartron, "Degeneration and 'Alienism' in Early Nineteenth-Century France," p. 168; on the second half of the century: Patrice Pinell, "Degeneration Theory and Heredity Patterns between 1850 and 1900," in *Heredity and Infection: The History of Disease Transmission*, ed. Jean-Paul Gaudillière and Ilana Löwy (London: Routledge, 2001), p. 249.

36. See Cartron, "Degeneration and 'Alienism' in Early Nineteenth-Century France," pp. 156–57.

37. Bénédict Augustin Morel, *Traité des dégénérescences physiques, intellectuelles et morales de l'espèce humaine et des causes qui produisent ces variétés maladives* (Paris: J.-B. Baillière et fils, 1857); see Carol, *Histoire de l'eugénisme en France*, p. 92 and, for more on Morel, pp. 87–114. See also Pick, *Faces of Degeneration*, pp. 50–51; Pinell, "Degeneration Theory and Hereditary Patterns," p. 246.

38. On the role of teratology in Morel's notion of degeneration, see Jacques Léonard, *La médecine entre les savoirs et les pouvoirs: Histoire intellectuelle et politique de la médecine française au XIXe siècle* (Paris: Aubier-Montaigne, 1981), p. 184.

39. Morel, *Traité des dégénérescences physiques, intellectuelles et morales*, pp. 47–63, quotation p. 60.

40. Ibid., p. 47. On milieu-based notions of species variation, see Staffan Müller-Wille, "Reproducing Difference: Race and Heredity from a *longue durée* Perspective," in *Reproduction, Race, and Gender in Philosophy and the Early Life Sciences*, ed. Susanne Lettow (Albany: SUNY Press, 2014). On the idea of bodies in chemical milieus as an overarching concept in nineteenth-century French medicine and bioscience, J. Andrew Mendelsohn,"Medicine and the Making of Bodily Inequality in Twentieth-Century Europe," in Gaudillière and Löwy, *Heredity and Infection*, pp. 43–44. On the accentuation of *influences occasionelles* in Morel's work, also Annemarie Wettley, "Zur Problemgeschichte der 'dégénérescence,'" *Sudhoffs Archiv für Geschichte der Medizin und der Naturwissenschaften* 43.3 (1959), pp. 195–96.

41. See Carol, *Histoire de l'eugénisme en France*, p. 92.

42. Morel, *Traité des dégénérescences physiques, intellectuelles et morales*, p. 50.

43. Legrand du Saulle, *Le délire des persecutions*, pp. 485–86.

44. See Pinell, "Degeneration Theory and Hereditary Patterns," p. 249; Paule Herschkorn-Barnu, "Wie der Fötus einen klinischen Status erhielt: Bedingungen und Verfahren

der Produktion eines medizinischen Fachwissens, Paris 1832–1848," in *Geschichte des Ungeborenen: Zur Erfahrungs- und Wissenschaftsgeschichte der Schwangerschaft, 17.–20. Jahrhundert*, ed. Barbara Duden, Jürgen Schlumbohm, and Patrice Veit (Göttingen: Vandenhoeck & Ruprecht, 2002), p. 202.

45. See Carol, *Histoire de l'eugénisme en France*, pp. 97–109. On the environmental emphasis in the French concept of degeneration: Mendelsohn, "Medicine and the Making of Bodily Inequality"; on the environmental aspect in the French discourse on population: Joshua Cole, *The Power of Large Numbers: Population, Politics, and Gender in Nineteenth-Century France* (Ithaca, NY: Cornell University Press, 2000), p. 152.

46. See Féré, "L'hérédité morbide." On Féré's use of Morel's work: Pinell, "Degeneration Theory and Hereditary Patterns," pp. 246–50.

47. See Féré, "L'hérédité morbide," pp. 436–38, 448.

48. Ibid., pp. 444 and 448.

49. Ibid., p. 452.

50. Féré, *Sensation et mouvement*, p. 96.

51. On these iterations of the degeneration concept, see Pick, *Faces of Degeneration*, p. 52; Carol, *Histoire de l'eugénisme en France*, pp. 92–93.

52. Féré, "L'hérédité morbide," p. 452.

53. A chapter on this issue was included in the new edition of *La famille névropathique* in 1896. On the optimistic note in Morel's work, and on hygiene and *métissage* as "regenerative" measures, see Carol, *Histoire de l'eugénisme en France*, p. 92. On the history of these ideas: Susanne Lettow, "Improving Reproduction: Articulations of Breeding and 'Race-Mixing' in French and German Discourse (1750–1800)," in *The Secrets of Generation: Reproduction in the Long Eighteenth Century*, ed. Raymond Stephanson and Darren N. Wagner (Toronto: University of Toronto Press, 2015).

54. Féré, "Les stigmates fonctionnels de la dégénérescene," p. 188.

55. Féré, "L'hérédité morbide," p. 450.

56. Foucault, *Abnormal*, p. 313. See also Zrinka Stahuljak,"History as a Medical Category: Heredity, Positivism, and the Study of the Past in Ninenteenth-Century France," *History of the Present* 3.2 (2013), pp. 140–59.

57. Eugene S. Talbot, *Degeneracy: Its Causes, Signs and Results* (New York: C. Scribner's Sons, 1898), p. 60. The work appeared in thirty-eight editions until 2011. On Talbot, see Pick, *Faces of Degeneration*, p. 23. In 1911, Talbot published a "developmental pathology" that followed Féré in thinking of degeneration as something that happened before birth: Eugene S. Talbot,

Developmental Pathology: A Study in Degenerative Evolution (Boston: R. G. Badger, 1911), p. v.

58. See Pick, *Faces of Degeneration*, pp. 69–70.

59. See Peter J. Bowler, *Life's Splendid Drama: Evolutionary Biology and the Reconstruction of Life's Ancestry, 1860–1940* (Chicago: University of Chicago Press, 1996).

60. Legrand du Saulle, "Influence des événements politiques," p. 184.

61. See Mark S. Micale and Paul Lerner, "Trauma, Psychiatry, and History: A Conceptual and Historiographical Introduction," in *Traumatic Pasts: History, Psychiatry, and Trauma in the Modern Age, 1870–1930*, ed. Mark S. Micale and Paul Lerner (Cambridge, UK: Cambridge University Press, 2001), p. 20.

62. See Mark S. Micale, "Jean-Martin Charcot and *les névroses traumatiques*: From Medicine to Culture in French Trauma Theory of the Late Nineteenth Century," in Micale and Lerner, *Traumatic Pasts*, pp. 136–39.

63. On ideas regarding the transmission of trauma in the form of heredity or a constantly repeated traumatization suffered by successor generations through cultural memory, see Parnes, Vedder, and Willer, *Das Konzept der Generation*, pp. 291–313.

64. On the generalization of trauma in the twentieth century, the beginnings of which can be identified here, see Michael S. Roth, *Memory, Trauma, and History: Essays on Living with the Past* (New York: Columbia University Press, 2012); Didier Fassin and Richard Rechtman, *The Empire of Trauma: An Inquiry into the Condition of Victimhood*, trans. Rachel Gomme (Princeton, NJ: Princeton University Press, 2009); Ruth Leys, *Trauma. A Genealogy* (Chicago: University of Chicago Press, 2000).

65. Dupouy, "Les enfants du siège," p. 3. On the futurization of generation and heredity at the end of the nineteenth century, see also Parnes, Vedder, and Willer, *Das Konzept der Generation*, pp. 82–119; Hans-Jörg Rheinberger and Staffan Müller-Wille, *Vererbung: Geschichte und Kultur eines biologischen Konzepts* (Frankfurt am Main: Fischer, 2009), p. 129.

66. Camille Dareste, "Introduction," in *La production artificielle des monstruosités, ou Essais de tératogénie expérimentale* (Paris: Reinwald, 1876), p. 42.

67. See Chapter 7, "A Natural Experiment on Mental Pain." On the genesis of statistical rationality: Alain Desrosières, *The Politics of Large Numbers: A History of Statistical Reasoning*, trans. Camille Naish (Cambridge, MA: Harvard University Press, 1998).

68. Michel Foucault, *Security, Territory, Population: Lectures at the Collège de France, 1977–78*, ed. Michel Sellenart, trans. Graham Burchell (Basingstoke: Palgrave Macmillan, 2009), p. 473.

69. Léonard, *La médecine entre les savoirs et les pouvoirs*, p. 302; see Foucault, *Security, Territory, Population*. With a focus on the aspects relevant here, also Susanne Lettow,

"Population, Race and Gender: On the Genealogy of the Modern Politics of Reproduction," *Distinktion: Scandinavian Journal of Social Theory* 16.3 (2015), pp. 267–82. On hygiene: Philipp Sarasin, *Reizbare Maschinen: Eine Geschichte des Körpers 1765–1914* (Frankfurt am Main: Suhrkamp, 2001).

70. On the medicalization of obstetrics, see Ann Oakley, *The Captured Womb: A History of the Medical Care of Pregnant Women* (Oxford: Blackwell, 1984); Jacques Gélis, *La sage-femme ou le médecin: Une nouvelle conception de la vie* (Paris: Fayard, 1988); Jürgen Schlumbohm, *Lebendige Phantome: Ein Entbindungshospital und seine Patientinnen, 1751–1830* (Göttingen: Wallstein, 2012); Lucia Aschauer, *Gebärende unter Beobachtung: Die Etablierung der männlichen Geburtshilfe in Frankreich (1750–1830)* (Frankfurt am Main: Campus, 2020). On the coexistence of different models of pregnancy: Clare Hanson, *A Cultural History of Pregnancy: Pregnancy, Medicine and Culture, 1750–2000* (Basingstoke: Palgrave Macmillan, 2004). On statistical rationality in obstetrics: Herschkorn-Barnu, "Wie der Fötus einen klinischen Status erhielt"; on the beginnings in the eighteenth century: Barbara Duden, "Die 'Geheimnisse' der Schwangeren und das Öffentlichkeitsinteresse der Medizin: Zur sozialen Bedeutung der Kindsregung," in *Frauengeschichte — Geschlechtergeschichte*, ed. Karin Hausen and Heide Wunder (Frankfurt am Main: Campus, 1992).

71. Jacques A. Millot, *L'art d'améliorer et perfectionner les générations humaines*, 2nd rev. ed. (Paris: Migneret, 1803), pp. v and 59–61.

72. Ibid., pp. 79 and v.

73. See Pierre Darmon, *Le mythe de la procréation à l'âge baroque* (Paris: Éditions du Seuil, 1981). Also Jean-Louis Fischer, "La callipédie, ou l'art d'avoir de beaux enfants," *Dix-huitième siècle* 23 (1991), p. 143.

74. On the politicization of pregnancy in the eighteenth century, see Carol, *Histoire de l'eugénisme en France*, pp. 17–26; in detail, also Elsa Dorlin, *La matrice de la race: Généalogie sexuelle et coloniale de la Nation française*, 2nd ed. (Paris: La Découverte, 2009), pp. 97–189.

75. See Carol, *Histoire de l'eugénisme en France*, pp. 38–45.

76. See Adolphe Pinard, "Note pour servir à l'histoire de la puériculture intra-uterine," *Bulletin de l'Académie de médecine* 59.47 (1895), pp. 593–97. On Pinard: Carol, *Histoire de l'eugénisme en France*, pp. 45–51; Paule Herschkorn, "Adolphe Pinard et l'enfant à naître: L'invention de la médecine fœtale," *Devenir* 8.3 (1996), pp. 77–87.

77. See Adolphe Pinard, *Traité du palper abdominal, au point de vue obstétrical, et de la version par manœuvres externes* (Paris: Lauwereyns, 1878).

78. Adolphe Pinard, "Foetus," in *Dictionnaire encyclopédique des sciences médicales*, ed.

Amédée Dechambre, 4th series, vol. 2 (1878), pp. 472–556.

79. Pinard, "Note pour servir à l'histoire de la puériculture intra-uterine."

80. Quoted in Carol, *Histoire de l'eugénisme en France*, pp. 47–48.

81. On Ballantyne's motif of prevention, see Salim Al-Gailani, "Pregnancy, Pathology and Public Morals: Making Antenatal Care in Early Twentieth-Century Edinburgh," in *Western Maternity and Medicine, 1880–1990*, ed. Janet Greenlees and Linda Bryder (London: Pickering & Chatto, 2013).

CHAPTER ELEVEN: OF HUMAN BORN

1. Johannes Müller, "Zur Physiologie des Foetus," *Zeitschrift für die Anthropologie* 2.2 (1824), p. 446.

2. Hans-Jörg Rheinberger, *Toward a History of Epistemic Things* (Stanford, CA: Stanford University Press, 1997), 28; on the intersections, Rheinberger, *An Epistemology of the Concrete: Twentieth-Century Histories of Life*, trans. G. M. Goshgarian (Durham, NC: Duke University Press, 2010), pp. 217–32.

3. William T. Preyer, *Specielle Physiologie des Embryo: Untersuchungen über die Lebenserscheinungen vor der Geburt* (Leipzig: Grieben, 1883), p. 6 (original emphasis).

4. Ibid., p. 5.

5. Gabriel Compayré, *The Intellectual and Moral Development of the Child*, vol. 1, trans. Mary E. Wilson (New York: Appleton, [1893] 1900), p. 30.

6. William T. Preyer, "Psychogenesis," in *Naturwissenschaftliche Thatsachen und Probleme: Populäre Vorträge* (Jena: Verlag von Gebrüder Paetel, 1880), p. 237.

7. Siegfried Bernfeld, *The Psychology of the Infant*, trans. Rosetta Hurwitz (London: Kegan Paul, [1925] 1929), p. 30.

8. Preyer, "Psychogenesis," p. 237; Bernfeld, *Psychology of the Infant*, p. 30.

9. Bernfeld, *Psychology of the Infant*, pp. 30 and 33.

10. Friedrich Balke coins the term "non-knowledge" (*Nicht-Wissen*) in his work on Michel Foucault's discussion of invalid speech. Balke, "Der Riesenmaulwurf: Zur Rolle des Nicht-Wissens bei Kafka und Foucault," *Nach Feierabend: Zürcher Jahrbuch für Wissensgeschichte* 5 (2009), p. 21.

11. Foucault describes invalid speech — speech that is "null and void" — as part of the distinction between reason and madness. Michel Foucault, "The Order of Discourse" (1970), trans. Ian McLeod, in *Untying the Text: A Post-Structuralist Reader*, ed. Robert Young (London: Routledge & Kegan Paul, 1981), p. 53.

12. James Sully, "Babies and Science," *Cornhill Magazine* 43 (January-June 1881), p. 546.

13. Quoted in Charles Billard, *Traité des maladies des enfans nouveaux-nés et à la mamelle, fondé sur de nouvelles observations cliniques et d'anatomie pathologique, faites à l'Hôpital des enfans-trouvés de Paris* (Paris: J.-B. Baillière et fils, 1828), p. vii.

14. Ibid., p. vii.

15. Preyer, *Specielle Physiologie des Embryo*, p. 5.

16. See Silvio Canestrini, *Über das Sinnesleben des Neugeborenen (nach physiologischen Experimenten)* (Berlin: J. Springer, 1913), preface.

17. Sully, "Babies and Science," p. 546.

18. Preyer, "Psychogenesis," p. 205.

19. Ibid.

20. Ibid., pp. 202–203.

21. Foucault, "Order of Discourse," p. 60, quoting Canguilhem.

22. Sully, "Babies and Science," p. 547.

23. William T. Preyer, *The Mind of the Child, Part I: The Senses and the Will*, trans. H. W. Brown (New York: Appleton, 1893), p. x.

24. Ibid., pp. x–xi.

25. Preyer, "Psychogenesis," p. 204.

26. On women's contributions to observing children, see Christine von Oertzen, "Science in the Cradle: Milicent Shinn and Her Home-Based Network of Baby Observers, 1890–1910," *Centaurus* 55.2 (2013), pp. 175–95.

27. Preyer, "Psychogenesis," p. 237.

28. Preyer, *Mind of the Child*, p. 2; Preyer, "Psychogenesis," p. 237.

29. Sully, "Babies and Science," p. 543.

30. Preyer, "Psychogenesis," p. 237.

31. Philippe Descola, *Beyond Nature and Culture*, trans. Janet Lloyd (Chicago: University of Chicago Press, 2014), p. 62.

Bibliography

SOURCES

Apert, Émile. *Traité des maladies familiales et des maladies congenitales.* Paris: J.-B. Baillière et fils, 1907.

Ballantyne, John William. *Manual of Antenatal Pathology and Hygiene*, vol. 1: *The Foetus*; vol. 2: *The Embryo.* Edinburgh: W. Green and Sons, 1902–1904.

Barcroft, Joseph. *Researches on Pre-natal Life.* Vol. 1. Oxford: Blackwell, 1946.

Beaunis, Henri. *Nouveaux éléments de physiologie humaine, comprenant les principes de la physiologie comparée et de la physiologie générale.* 3 vols. Paris: J.-B. Baillière et fils, 1876, 1881, 1888.

Bernard, Claude. "Cours de physiologie générale de la Faculté des sciences: Leçon d'ouverture—Exposition de la méthode." *Le moniteur des hôpitaux: Journal des progrès de la médecine et de la chirurgie pratique* 1–2.54 (1854), pp. 409–12.

———. *Leçons sur la physiologie et la pathologie du système nerveux.* Paris: J.-B. Baillière et fils, 1858.

———. *Leçons sur les phénomènes de la vie, communs aux animaux et aux végétaux.* Vol. 2. Paris: J.-B. Baillière et fils, 1879.

Bernfeld, Siegfried. *The Psychology of the Infant.* Trans. Rosetta Hurwitz. London: Kegan Paul, [1925] 1929.

Bichat, Xavier. *Physiological Researches on Life and Death.* Trans. Tobias Watkins. Philadelphia: Smith & Maxwell, [1799] 1809.

———. *Physiologische Untersuchungen über Leben und Tod.* Trans. C. H. Pfaff. Copenhagen: Friedrich Brummer, 1802.

———. *Recherches physiologiques sur la vie et la mort*. Paris: Brosson, Gabon, 1805.

Billard, Charles. *Traité des maladies des enfans nouveaux-nés et à la mamelle, fondé sur de nouvelles observations cliniques et d'anatomie pathologique, faites à l'Hôpital des enfans-trouvés de Paris*. Paris: J.-B. Baillière et fils, 1828.

Bischoff, Theodor. "Entwicklungsgeschichte, mit besonderer Berücksichtigung der Missbildungen (Nachtrag)." In *Handwörterbuch der Physiologie, mit Rücksicht auf die physiologische Pathologie*, edited by Rudolph Wagner, vol. 1, pp. 860–928. Braunschweig: Vieweg, 1842.

Bourgeois, Xavier. *De l'influence des maladies de la femme pendant la grossesse sur la santé et la constitution de l'enfant, Mémoire récompensé par l'Académie impériale de médicine, Séance du 17 décembre 1861*. Paris: J.-B. Baillière et fils, 1862.

Bourneville, Désiré-Magloire. *Recueil de photographies sur les aliénés de Bicêtre, Photographies datées de 1880–1881*. N. p., [1880–1881].

———, et al. *Recherches cliniques et thérapeutiques sur l'épilepsie, l'hystérie et l'idiotie: Compte-rendu du Service des enfants idiots, épileptiques et arriérés de Bicêtre*. Paris: Progrès médical, 1881–1908.

Burdach, Karl Friedrich (ed.). *Die Physiologie als Erfahrungswissenschaft*. 6 vols. With contributions by Karl E. von Baer et al. Leipzig: Leopold Voss, 1832–1840.

Cabanis, Pierre-Jean-Georges. *On the Relations between the Physical and Moral Aspects of Man*. 2 vols. Edited by George Mora and trans. Margaret Duggan Saidi. Baltimore, MD: Johns Hopkins University Press, [1802] 1982.

———. *Rapports du physique et du moral de l'homme et lettres sur les causes premières*. 2 vols. Paris: Crapart, Caille et Ravier, 1802.

Canestrini, Silvio. *Über das Sinnesleben des Neugeborenen (nach physiologischen Experimenten)*. Berlin: J. Springer, 1913.

Carus, Carl Gustav. *Lehrbuch der Gynäkologie oder systematische Darstellung der Lehren von Erkenntniss und Behandlung eigenthümlicher gesunder und krankhafter Zustände sowohl der nicht schwangeren, schwangeren und gebärenden Frauen als der Wöchnerinnen und neugebornen Kinder: Zur Grundlage akademischer Vorlesungen und zum Gebrauche für practische Aerzte, Wundärzte und Geburtshelfer*. 2 vols. Leipzig: Fleischer, 1820.

Charcot, Jean-Martin. *Leçons du mardi à la Salpêtrière: Policlinique 1888–1889*. Paris: E. Lecrosnier & Babé, 1889.

Combe, Adolphe. *La nervosité de l'enfant: Quatre conferences*, 2nd ed. Lausanne: H. Mignot, 1903.

Compayré, Gabriel. *The Intellectual and Moral Development of the Child.* 2 vols. Trans. Mary E. Wilson. New York: Appleton, [1893] 1900–1902.

Courby, P. *Des effets généraux des passions dans l'économie animale, et de leur influence chez les femmes grosses.* Paris: Didot Jeune, 1807.

Dareste, Camille. "Introduction." In *Recherches sur la production artificielle des monstruosités, ou Essais de tératogénie expérimentale.* Paris: Reinwald, 1876.

———. *Recherches sur la production artificielle des monstruosités, ou Essais de tératogénie expérimentale.* Paris: Reinwald, 1877.

———. *Recherches sur la production artificielle des monstruosités, ou Essais de tératogénie expérimentale*, 2nd revised ed. Paris: Reinwald, 1891.

Darwin, Charles. "A Biographical Sketch of an Infant." *Mind* 2.7 (1877), pp. 285–94.

———. *The Expression of the Emotions in Man and Animals.* London: John Murray, 1872.

Davids, Anthony, Spencer de Vault, and Max Talmadge. "Psychological Study of Emotional Factors in Pregnancy: A Preliminary Report." *Psychosomatic Medicine* 23.2 (1961), pp. 93–103.

Dechambre, Amédée (ed.). *Dictionnaire encyclopédique des sciences médicales.* Paris: G. Masson, P. Asselin, 1864–1889.

Delage, Yves. *La structure du protoplasma et les théories sur l'*hérédité *et les grands problèmes de la biologie générale.* Paris: Reinwald, 1895.

Delore, Xavier. "Placenta." In *Dictionnaire encyclopédique des sciences médicales*, 2nd series, vol. 25, edited by Amédée Dechambre, pp. 489–541. Paris: G. Masson, P. Asselin, 1886.

Demangeon, Jean-Baptiste. *Considérations physiologiques sur le pouvoir de l'imagination maternelle durant la grossesse et sur les autres causes, prétendues ou réelles, des difformités et des variétés naturelles.* Paris: Chez l'auteur, 1807.

———. *De l'imagination considérée dans ses effets directs sur l'homme et les animaux, et dans ses effets indirects sur les produits de la gestation; avec une notice sur la génération et les causes les plus probables des difformités de naissance. Ouvrage ou l'on fait la part de l'imagination dans les phénomènes du magnétisme, de l'exorcisme, de l'ascétisme et d'autres prestiges*, 2nd ed. Paris: Rouen, 1829.

———. *Du pouvoir de l'imagination sur le physique et le moral de l'homme*, new ed. Paris: J. Rouvier & E. Le Bouvier, 1834.

Döllinger, Ignaz. *Grundzüge der Physiologie der Entwicklung des Zell-, Knochen- und Blutsystemes.* Regensburg: G. Joseph Manz, 1842.

Drzewiecki, Joseph. "Einfluss der Eindrücke der Mutter auf den Fötus oder das sogenannte 'Versehen' der Schwangeren." *Wiener Medizinische Wochenschrift* 41.46 (1891), pp. 1856–59.

Dupouy, Dr. "Les enfants du siège." *Le médecin: Moniteur de la policlinique* 11.18 (May 3, 1885), pp. 2–3.

Féré, Charles. "Contribution à l'histoire du choc moral chez les enfants." *Bulletin de la Société de médecine mentale de Belgique* 74 (1894), pp. 333–40.

______. "Les enfants du Siège." *Progrès médical* 12.13 (March 1884), pp. 245–46.

______. "Essai expérimental sur les rapports étiologiques, de l'infécondité, des monstruosités, de l'avortement, de la morti-natalité, du retard de développement et de la débilité congénitale." *Teratologia: A Quarterly Journal of Antenatal Pathology* 2.4 (1895), pp. 245–55.

______. "Un fait pour servir à l'histoire des bouffées de chaleur et des rougeurs morbides." *Comptes rendus hebdomadaires des séances et mémoires de la Société de biologie* 46 (1894), pp. 643–45.

______. "La famille névropathique." *Archives de neurologie* 7.19–20 (1884), pp. 1–43 and 173–91.

______. *La famille névropathique: Théorie tératologique de l'hérédité et de la prédisposition morbides et de la dégénérescence.* Paris: F. Alcan, [1894] 1898.

______. "L'hérédité morbide." *Revue des deux mondes* 64.126 (November 15, 1894), pp. 437–52.

______. "Morphinisme et grossesse." *Comptes rendus des séances et mémoires de la Société de biologie* 35 (1883), pp. 526–28.

______. "Nerve Troubles as Foreshadowed in the Child." *Brain: A Journal of Neurology* 8 (1886), pp. 230–38.

______. "Note relative aux réactions du foetus aux émotions de la mère." *Comptes rendus hebdomadaires des séances et mémoires de la Société de biologie* 55 (1903), pp. 74–75.

______. "Note sur les dégénérescences d'origine toxique ou infectueuse." *Comptes rendus hebdomadaires des séances et mémoires de la Société de biologie* 47 (1895), pp. 568–69.

______. *The Pathology of Emotions: Physiological and Clinical Studies.* Trans. Robert Park. London: University Press, [1892] 1899.

______. "Sensation et mouvement: Contribution à la psychologie du foetus." *Revue philosophique de la France et de l'étranger* 21 (1886), pp. 247–64.

______. *Sensation et mouvement: Études expérimentales de psycho-mécanique.* Paris: F. Alcan, 1887.

______. "Les stigmates fonctionnels de la dégénérescence." *Journal des connaissances médicales pratiques et de pharmacologie* 64.1 (January 2, 1896), pp. 186–88.

______. "Tératogénie expérimentale et pathologie générale." In *Cinquantenaire de la Société de biologie*, edited by Société de biologie, pp. 360–69. Paris: Masson, 1899.

Ferenczi, Sándor. "Stages in the Development of the Sense of Reality" (1913). In *First Contributions to Psycho-Analysis*. Trans. Ernest Jones, pp. 213–39. London: The Hogarth Press, 1952.

———. *Thalassa: A Theory of Genitality*. Trans. Henry Alden Bunker. New York: The Psychoanalytic Quarterly, [1924] 1938.

Ferreira, Antonio J. "The Pregnant Woman's Emotional Attitude and Its Reflection on the Newborn." *American Journal of Orthopsychiatry: A Journal of Human Behavior* 30.3 (1960), pp. 553–61.

Forgue, Émile. "Traumatisme." In *Dictionnaire encyclopédique des sciences médicales*, 3rd series, vol. 18, edited by Amédée Dechambre, pp. 39–52. Paris: G. Masson, P. Asselin, 1885.

Freud, Sigmund. "The Dissolution of the Oedipus Complex" (1924). In vol. 19 of *The Standard Edition of the Complete Psychological Works of Sigmund Freud*. Edited and trans. James Strachey, pp. 171–80. London: The Hogarth Press, 1961.

———. "The Ego and the Id" (1923). In vol. 19 of *The Standard Edition of the Complete Psychological Works of Sigmund Freud*. Edited and trans. James Strachey, pp. 12–66. London: The Hogarth Press, 1961.

———. "Inhibitions, Symptoms and Anxiety" (1926). In vol. 20 of *The Standard Edition of the Complete Psychological Works of Sigmund Freud*. Edited and trans. James Strachey, pp. 75–176. London: The Hogarth Press, 1959.

———. *The Interpretation of Dreams (II)* (1900). In vol. 5 of *The Standard Edition of the Complete Psychological Works of Sigmund Freud*. Edited and trans. James Strachey, pp. 339–627. London: The Hogarth Press, 1958.

———. *Introductory Lectures on Psycho-Analysis (Part III)* (1916–17). Vol. 16 of *The Standard Edition of the Complete Psychological Works of Sigmund Freud*. Edited and trans. James Strachey. London: The Hogarth Press, 1963.

———. *A Phylogenetic Fantasy: Overview of the Transference Neuroses*. Edited by Ilse Grubrich-Simitis and trans. Axel Hoffer and Peter T. Hoffer. Cambridge, MA: Belknap Press, 1987.

———. "Three Essays on the Theory of Sexuality" (1905). In vol. 7 of *The Standard Edition of the Complete Psychological Works of Sigmund Freud*. Edited and trans. James Strachey, pp. 123–246. London: The Hogarth Press, 1953.

Freud, Sigmund, and Sándor Ferenczi. *The Correspondence of Sigmund Freud and Sándor*

Ferenczi. 3 vols. Edited by Ernst Falzeder and Eva Brabant and trans. Peter T. Hoffer. Cambridge, MA: Belknap Press, 1993–2000.

Garley, Dorothy. "Über den Schock des Geborenwerdens und seine möglichen Nachwirkungen." *Internationale Zeitschrift für Psychoanalyse* 10.2 (1924), pp. 134–63.

Geoffroy Saint-Hilaire, Étienne. "Considérations générales sur la monstruosité, et Description d'un genre nouveau observé dans l'espèce humain, et nommé Aspalosome." *Annales des sciences naturelles* 4 (1825), pp. 451–68.

———. *Considérations générales sur les monstres, comprenant une théorie des phénomènes de la monstruosité*. Paris: J. Tastu, 1826.

———. *Philosophie anatomique: Des monstruosités humaines*. Paris: Chez l'auteur, 1822.

Geoffroy Saint-Hilaire, Isidore. *Histoire générale et particulière des anomalies de l'organisation chez l'homme et les animaux: Ouvrage comprenant des recherches sur les caractères, la classification, l'influence physiologique et pathologique, les rapports généraux, les lois et les causes des monstruosités, des variétés et vices de conformation, ou Traité de tératologie*. 3 vols. Paris: J.-B. Baillière, 1832–1836.

Goethe, Johann Wolfgang von. *Elective Affinities*. Trans. R. J. Hollingdale. London: Penguin, [1809] 1971.

Graber, Gustav Hans. *Die Ambivalenz des Kindes*. Leipzig: Internationaler psychoanalytischer Verlag, 1924.

Guinard, Louis, and Henri Hochwelker. "Recherches sur le passage des substances solubles du foetus à la mère." *Comptes rendus hebdomadaires des séances et mémoires de la Société de biologie* 50 (1898), pp. 1183–85.

Halban, Josef. "Ueber Schwangerschaftsreaktionen fötaler Organe und ihre puerperale Involution." *Münchener medizinische Wochenschrift* 51.49 (1904), p. 2210.

Haller, Albrecht von. *La génération, ou exposition des phénomènes relatifs à cette fonction naturelle*. 2 vols. Paris: Ventes de la Doué, 1774.

His, Wilhelm. *Anatomie menschlicher Embryonen*. 3 vols. Leipzig: F. C. W. Vogel, 1880–1885.

Home, Everard. "II. The Croonian Lecture: On the Existence of Nerves in the Placenta." *Philosophical Transactions of the Royal Society* 115 (1825), pp. 66–80.

Hufeland, Christoph Wilhelm. "Von den Krankheiten der Ungeborenen und der Vorsorge für das Leben und die Gesundheit des Menschen vor der Geburt" (1827). In *Sammlung auserlesener Abhandlungen über Kinder-Krankheiten*, vol. 5, edited by Franz Joseph von Mezler, pp. 3–34. Prague: Haase, 1836.

Hug-Hellmuth, Hermine. *A Study of the Mental Life of the Child*. Trans. James J. Putnam and Mabel Stevens. Washington, DC: Nervous and Mental Diseases Publication Co., [1913] 1919.

Hunter, William. *An Anatomical Description of the Human Gravid Uterus, and Its Contents.* London: J. Johnson and G. Nicol, 1794.

Index général des travaux de Charles Féré, médecin de Bicêtre, publié dans la "Normandie médicale" par un groupe d'amis de l'auteur. Rouen: Girieud, 1909.

Jacquemier, Jean-Marie. *Manuel des accouchements et des maladies des femmes grosses et accouchées, contenant les soins à donner aux nouveaux-nés.* 2 vols. Paris: Germer Baillière, 1846.

Janet, Pierre. "Charles Féré: Sensations et mouvements." *Revue philosophique de la France et de l'étranger* 24 (1887), pp. 198–202.

Jones, Ernest. "Angstaffekt und Geburtsakt." *Internationale Zeitschrift für Psychoanalyse* 9.1 (1923), p. 79.

———. "Cold, Disease, and Birth" (1923). In *Papers on Psycho-Analysis*, 3rd ed., pp. 595–600. London: Bailliere, Tindall and Cox, 1923.

Jourdan, Antoine J. L. "Imagination." In *Dictionnaire des sciences médicales*, vol. 24, pp. 15–85. Paris: C. L. F. Panckoucke, 1818.

Kagan, Jerome. "American Longitudinal Research on Psychological Development." *Child Development* 35.1 (1964), pp. 1–32.

Kussmaul, Adolf. *Untersuchungen über das Seelenleben des neugeborenen Menschen.* Leipzig: C. F. Winter, 1859.

Laborde, Jean-Vincent. *Les hommes et les actes de l'insurrection de Paris devant la psychologie morbide. Lettres à M. le Docteur Moreau (de Tours).* Paris: Germer Baillière, 1872.

Legrand du Saulle, Henri. *Le délire des persecutions.* Paris: H. Plon, 1871.

———. "L'état mental des Parisiens pendant le Siège de Paris (1871)." *Chronique médicale* 3 (1896), pp. 77–80, 119–21, 147–51.

———. *Étude médico-légale sur l'interdiction des aliénés et sur le conseil judiciaire: Suivie de recherches sur la situation juridique des fous et des incapables, à l'époque romaine.* Paris: Delahaye & Lecrosnier, 1881.

———. "Influence des événements politiques sur les caractères du délire et anomalies physiques et intellectuelles que l'on observe chez les enfants conçus pendant le siège de Paris." *Le praticien* 7 (1884), pp. 160–63 and 184–86.

Lieberman, E. James, and Robert Kramer (eds). *The Letters of Sigmund Freud and Otto Rank: Inside Psychoanalysis.* Letters translated by Gregory C. Richter. Baltimore, MD: Johns Hopkins University Press, 2012.

Lilienfeld, Abraham M., and Elizabeth Parkhurst. "A Study of the Association of Factors of Pregnancy and Parturition with the Development of Cerebral Palsy: A Preliminary Report." *American Journal of Hygiene* 53.3 (1951), pp. 262–82.

Lucas, Prosper. *Traité philosophique et physiologique de l'hérédité naturelle dans les états de santé et de maladie du système nerveux*. 2 vols. Paris: J.-B. Baillière, 1847–50.

Luys, Jules Bernard. *Le cerveau et ses fonctions*, 6th ed. Paris: F. Alcan, [1873] 1888.

Magendie, François. *Précis élémentaire de physiologie*. 2 vols. Paris: Méquignon-Marvis, 1816–1817.

Malebranche, Nicolas. *The Search after Truth*. Edited and trans. Thomas M. Lennon and Paul J. Olskamp. Cambridge, UK: Cambridge University Press, [1674–75] 1997.

Marshall, C. S. "Physiology of the Fetus: Origin and Extent of Function in Prenatal Life, by William Frederick Windle, W. B. Saunders Co., Philadelphia, 1940." *Yale Journal of Biology and Medicine* 13.3 (1941), pp. 425–26.

Martin, Ernest. *Histoire des monstres depuis l'antiquité jusqu'à nos jours*. Paris: Reinwald, 1880.

Mayer, August C. "Ueber das Einsaugungsvermögen der Venen des großen und kleinen Kreislaufsystems." *Deutsches Archiv für die Physiologie* 3.4 (1817), pp. 485–503.

Meckel, Johann Friedrich. *Handbuch der pathologischen Anatomie*. 2 vols. Leipzig: Reclam, 1812–1818.

Millot, Jacques A. *L'art d'améliorer et perfectionner les générations humaines*, 2nd revised ed. Paris: Migneret, 1803.

———. *Médecine perfective, ou Code des bonnes mères*. Paris: Collin & Millot, 1809.

Moreau, Paul (de Tours). *La folie chez les enfants*. Paris: J.-B. Baillière, 1888.

———. *Der Irrsinn im Kindesalter*. Trans. Demetrio Galatti. Stuttgart: Enke, 1889.

Morel, Bénédict Augustin. *Traité des dégénérescences physiques, intellectuelles et morales de l'espèce humaine et des causes qui produisent ces variétés maladives*. Paris: J.-B. Baillière, 1857.

Müller, Johannes. *De respiratione foetus, commentatio physiologica*. Leipzig: Cnobloch, 1823.

———. *Handbuch der Physiologie des Menschen für Vorlesungen*. 2 vols., 3rd revised ed. Coblenz: Hölscher, 1837–1840.

———. "Zur Physiologie des Foetus." *Zeitschrift für die Anthropologie* 2.2 (1824), pp. 423–83.

Murat [probably Jean-Baptiste-Arnaud Murat]. "Placenta." In *Dictionnaire des sciences médicales*, vol. 42, edited by Société de médecins et de chirurgiens, pp. 516–50. Paris: C. L. F. Panckoucke, 1820.

Nasse, Christian Friedrich. "Von der Beseelung des Kindes." *Zeitschrift für die Anthropologie* 2.1 (1824), pp. 1–22.

"On Organizing for Continuity." *Child Development* 42.4 (1971), pp. 983–85.

Peiper, Albrecht. "Sinnesempfindungen des Kindes vor seiner Geburt." *Monatsschrift für Kinderheilkunde* 29 (1925), pp. 236–41.

Perez, Bernard. "L'âme de l'embryon et l'âme de l'enfant." *Revue philosophique de la France et de l'étranger* 23 (1887), pp. 582–602.

———. "Les facultés de l'enfant à l'époque de la naissance." *Revue philosophique de la France et de l'étranger* 13.1 (1882), pp. 133–45.

———. *The First Three Years of Childhood.* Edited and trans. Alice M. Christie. London: Sonnenschein, [1882] 1885.

———. *La psychologie de l'enfant: Les trois premières années.* Paris: Germer Baillière, 1882.

Pinard, Adolphe. "Foetus." In *Dictionnaire encyclopédique des sciences médicales*, 4th series, vol. 2, edited by Amédée Dechambre, pp. 472–556. Paris: G. Masson, P. Asselin, 1878.

———. "Note pour servir à l'histoire de la puériculture intra-uterine." *Bulletin de l'Académie de médecine* 59.47 (1895), pp. 593–97.

———. *Traité du palper abdominal, au point de vue obstétrical, et de la version par manoeuvres externes.* Paris: Lauwereyns, 1878.

Preuss, Julius. *Vom Versehen der Schwangeren: Eine historisch-kritische Studie.* Berlin: Fischer, 1892.

Prévost, Jean-Louis. "Note sur le sang du foetus dans les animaux vertébrés." *Annales des sciences naturelles*, 1st series, 4 (1825), p. 499.

Prévost, Jean-Louis, and Antoine Morin. "Recherches physiologiques et chimiques sur la nutrition du foetus." *Mémoires de la Société Physique et d'Histoire naturelle de Genève* 9 (1842), pp. 235–45.

Preyer, William T. "Embryonic Motility and Sensitivity: Translated from the Original German of *Specielle Physiologie des Embryo*." Trans. G. E. Coghill and Wolfram K. Legner. *Monographs of the Society for Research in Child Development* 2.6 (1937), pp. 1–115.

———. *The Mind of the Child, Part I: The Senses and the Will.* Trans. H. W. Brown. New York: Appleton, [1882] 1893.

———."Psychogenesis." In *Naturwissenschaftliche Thatsachen und Probleme: Populäre Vorträge*, pp. 199–237. Jena: Verlag von Gebrüder Paetel, 1880.

———. *Specielle Physiologie des Embryo: Untersuchungen über die Lebenserscheinungen vor der Geburt.* Leipzig: Grieben, 1883.

Rank, Otto. *The Trauma of Birth.* London: Kegan Paul, [1924] 1929.

Ribot, Théodule. *Heredity: A Psychological Study of Its Phenomena, Laws, Causes, and*

Consequences. Translated from the French. New York: Appleton, [1873] 1875.

Rogers, Martha E., Abrahahm M. Lilienfeld, and Benjamin Pasamanick. *Prenatal and Paranatal Factors in the Development of Childhood Behavior Disorders*. Copenhagen: Ejnar Munksgaard, 1955.

Schauenstein, Adolf S., and Josef Spaeth. "Uebergang von Medicamenten in die Milch der Säugenden und in den Fötus." *Froriep's Notizen aus dem Gebiet der Natur- und Heilkunde* 2.17 (1859), pp. 266–71.

Sellheim, Hugo. "Mutter-Kinds-Beziehungen auf Grund innersekretorischer Verknüpfung." *Münchener Medizinische Wochenschrift* 71.38 (1924), pp. 1304–1307.

Soemmerring, Samuel Thomas. *Schriften zur Embryologie und Teratologie*. Edited by Ulrike Enke. Basel: Schwabe, 2000.

Sontag, Lester W. "The History of Longitudinal Research: Implications for the Future." *Child Development* 42.4 (1971), pp. 987–1002.

———. "The Significance of Fetal Environmental Differences." *American Journal of Obstetrics and Gynecology* 42 (July–December 1941), pp. 996–1003.

———. "War and the Fetal-Maternal Relationship." *Marriage and Family Living* 6.1 (1944), pp. 3–4 and 16.

Sontag, Lester W., and Robert F. Wallace. "The Movement Response of the Human Fetus to Sound Stimuli." *Child Development* 6.4 (1935), pp. 253–58.

Stott, Denis H. "Physical and Mental Handicaps Following a Disturbed Pregnancy." *The Lancet* (May 18, 1957), pp. 1006–1012.

Sully, James. "Babies and Science." *The Cornhill Magazine* 43 (January–June 1881), pp. 539–54.

Talbot, Eugene S. *Degeneracy: Its Causes, Signs and Results*. New York: Charles Scribner's Sons, 1898.

———. *Developmental Pathology: A Study in Degenerative Evolution*. Boston: R. G. Badger, 1911.

Thompson, William D., and Lester W. Sontag. "Behavioral Effects in the Offspring of Rats Subjected to Audiogenic Seizure during the Gestational Period." *Journal of Comparative and Physiological Psychology* 49.5 (1956), pp. 454–56.

Tourneux, F., and G. Herrmann. "Embryon." In *Dictionnaire encyclopédique des sciences médicales*, 1st series, vol. 33, edited by Amédée Dechambre, pp. 657–730. Paris: G. Masson, P. Asselin, 1886.

Turner, Elizabeth K. "The Syndrome in the Infant Resulting from Maternal Emotional Tension during Pregnancy." *Medical Journal of Australia* 43.6 (February 11, 1956), pp. 221–22.

Valentin, Gabriel Gustav. *Handbuch der Entwickelungsgeschichte des Menschen mit vergleichender Rücksicht der Entwickelung der Säugethiere und Vögel: nach fremden und eigenen Beobachtungen*. Berlin: A. Rücker, 1835.

Weismann, August. *The Germ-Plasm: A Theory of Heredity*. Trans. W. Newton Parker and Harriet Rönnfeldt. New York: Charles Scribner's Sons, [1892] 1893.

———. *Vorträge über Descendenztheorie, gehalten an der Universität zu Freiburg im Breisgau*. 2 vols. Jena: G. Fischer, 1902.

Welsenburg, Gerhard von [Iwan Bloch]. *Das Versehen der Frauen in Vergangenheit und Gegenwart und die Anschauungen der Aerzte, Naturforscher und Philosophen darüber*. Leipzig: H. Barsdorf, 1899.

Windle, William Frederick. *Physiology of the Fetus: Origin and Extent of Function in Prenatal Life*. Philadelphia: Saunders, 1940.

SECONDARY WORKS

Alcayna-Stevens, Lys, and Matei Candea (ed.). "Internal Others: Ethnographies of Naturalism." Special issue, *Cambridge Journal of Anthropology* 30.2 (2012).

Al-Gailani, Salim. "'Antenatal Affairs': Maternal Marking and the Medical Management of Pregnancy in Britain around 1900." In *Imaginationen des Ungeborenen/Imaginations of the Unborn: Kulturelle Konzepte pränataler Prägung von der Frühen Neuzeit zur Moderne/Cultural Concepts of Prenatal Imprinting from the Early Modern Period to the Present*, edited by Urte Helduser and Burkhard Dohm, pp. 153–72. Heidelberg: Universitätsverlag Winter, 2018.

———. "Pregnancy, Pathology and Public Morals: Making Antenatal Care in Early Twentieth-Century Edinburgh." In *Western Maternity and Medicine, 1880–1990*, edited by Janet Greenlees and Linda Bryder, pp. 31–46. London: Pickering & Chatto, 2013.

———. "Teratology and the Clinic: Monsters, Obstetrics and the Making of Antenatal Life in Edinburgh, c. 1900." PhD diss., University of Cambridge, 2010.

Alistair, G. S. Philip. "Perinatal Profiles: John William Ballantyne, Scottish Obstetrician and Prolific Writer." *NeoReviews* 9.11 (2008), pp. 503–505.

Armstrong, Elizabeth M. *Conceiving Risk, Bearing Responsibility: Fetal Alcohol Syndrome and the Diagnosis of Moral Disorder*. Baltimore, MD: Johns Hopkins University Press, 2003.

Arni, Caroline. "Menschen machen aus Akt und Substanz: Prokreation und Vaterschaft im reproduktionsmedizinischen und im literarischen Experiment." *Gesnerus: Swiss Journal of the History of Medicine and Science* 65 (2008), pp. 196–224.

———. "The Prenatal: Contingencies of Procreation and Transmission in the Nineteenth

Century." In *Heredity Explored: Between Public Domain and Experimental Science, 1850–1930*, edited by Staffan Müller-Wille and Christina Brandt, pp. 285–309. Cambridge, MA: MIT Press, 2016.

———. "Psychischer Einfluss und generationelles Trauma: Pränatale Prägung als Problem der Transmission, oder: Die Kinder des Année terrible 1870/71." In *Imaginationen des Ungeborenen/Imaginations of the Unborn: Kulturelle Konzepte pränataler Prägung von der Frühen Neuzeit zur Moderne/Cultural Concepts of Prenatal Imprinting from the Early Modern Period to the Present*, edited by Urte Helduser and Burkhard Dohm, pp. 133–52. Heidelberg: Universitätsverlag Winter, 2018.

———. "Traversing Birth: Continuity and Contingency in Research on Development in Nineteenth-Century Life and Human Sciences." *History and Philosophy of the Life Sciences* 37.1 (2015), pp. 50–67.

———. "Vom Unglück des mütterlichen 'Versehens' zur Biopolitik des Pränatalen: Aspekte einer Wissensgeschichte der maternal-fötalen Beziehung." In *Biopolitik und Geschlecht: Zur Regulierung des Lebendigen*, edited by Eva Sänger and Malaika Rödel, pp. 44–66. Münster: Westfälisches Dampfboot, 2012.

Aschauer, Lucia. *Gebärende unter Beobachtung: Die Etablierung der männlichen Geburtshilfe in Frankreich (1750–1830)*. Frankfurt am Main: Campus, 2020.

Asdal, Kristin, and Ingunn Moser. "Experiments in Context and Contexting." *Science, Technology & Human Values* 37.4 (2012), pp. 291–306.

Balke, Friedrich. "Der Riesenmaulwurf: Zur Rolle des Nicht-Wissens bei Kafka und Foucault." *Nach Feierabend: Zürcher Jahrbuch für Wissensgeschichte* 5 (2009), pp. 13–35.

Bennholdt-Thomsen, Anke, and Alfredo Guzzoni. "Zur Theorie des Versehens im 18. Jahrhundert: Ansätze einer pränatalen Psychologie." In *Klio und Psyche*, edited by Thomas Kornbichler, pp. 112–25. Pfaffenweiler: Centaurus, 1990.

Blanckaert, Claude. "L'histoire générale des sciences de l'homme: Principes et périodisation." In *L'histoire des sciences de l'homme: Trajectoire, enjeux et questions vives*, edited by Claude Blanckaert, Loïc Blondiaux, Laurent Loty, Marc Renneville, and Nathalie Richard, pp. 23–60. Paris: L'Harmattan, 1999.

Blanckaert, Claude, Loïc Blondiaux, Laurent Loty, Marc Renneville, and Nathalie Richard (ed.). *L'histoire des sciences de l'homme: Trajectoire, enjeux et questions vives*. Paris: L'Harmattan, 1999.

Bock von Wülfingen, Bettina, Christina Brandt, Susanne Lettow, and Florence Vienne (ed.). "Temporalities of Reproduction: Practices and Concepts from the Eighteenth to the Early Twentieth-First Century." Special issue, *History and Philosophy of the Life Sciences* 37.1 (2015), pp. 1–16.

Bogousslavsky, Julien (ed.). *Following Charcot: A Forgotten History of Neurology and Psychiatry*. Basel: Karger, 2011.

Boltanski, Luc. *The Foetal Condition: A Sociology of Engendering and Abortion*. Trans. Catherine Porter. Cambridge, UK: Polity, 2013.

Bowler, Peter J. "Biology and Human Nature." In *The Cambridge History of Science*, vol. 6: *The Modern Biological and Earth Sciences*, edited by Peter J. Bowler and John V. Pickstone, pp. 563–82. Cambridge, UK: Cambridge University Press, 2009.

———. *Life's Splendid Drama: Evolutionary Biology and the Reconstruction of Life's Ancestry, 1860–1940*. Chicago: University of Chicago Press, 1996.

Breeuwsma, Gerrit. "The Nephew of an Experimentalist: Ambivalences in Developmental Thinking." In *Beyond the Century of the Child: Cultural History and Developmental Psychology*, edited by Willem Koops and Michael Zuckerman, pp. 183–203. Philadelphia: University of Pennsylvania Press, 2003.

Brisson, Luc, Marie-Hélène Congourdeau, and Jean-Luc Solère (ed.). *L'embryon: Formation et animation; Antiquité grecque et latine, tradition hébraïque, chrétienne et islamique*. Paris: J. Vrin, 2008.

———. "Préface." In *L'embryon: Formation et animation; Antiquité grecque et latine, tradition hébraïque, chrétienne et islamique*, edited by Luc Brisson, Marie-Hélène Congourdeau, and Jean-Luc Solère, pp. 9–14. Paris: J. Vrin, 2008.

Buklijaš, Tatjana. "Food, Growth and Time: Elsie Widdowson's and Robert McCance's Research into Prenatal and Early Postnatal Growth." *Studies in History and Philosophy of Biological and Biomedical Sciences* 47, Part B (2014), pp. 267–77.

———. "Transformations of the Maternal-Fetal Relationship in the Twentieth Century: From Maternal Impressions to Epigenetic States." In *Imaginationen des Ungeborenen/Imaginations of the Unborn: Kulturelle Konzepte pränataler Prägung von der Frühen Neuzeit zur Moderne/Cultural Concepts of Prenatal Imprinting from the Early Modern Period to the Present*, edited by Urte Helduser and Burkhard Dohm, pp. 213–33. Heidelberg: Universitätsverlag Winter, 2018.

Cairns, Robert B., and Beverley D. Cairns. "The Making of Developmental Psychology." In *Handbook of Child Psychology*, vol. 1, 6th ed., edited by Richard M. Lerner, pp. 89–165. Hoboken, NJ: Wiley, 2006.

Candea, Matei, and Lys Alcayna-Stevens. "Internal Others: Ethnographies of Naturalism."

Cambridge Journal of Anthropology 30.2 (2012), pp. 36–47.

Canguilhem, Georges. "La constitution de la physiologie comme science." *In Études d'histoire et de philosophie des sciences concernant les vivants et la vie*, 2nd ed., pp. 226–73. Paris: J. Vrin, 1970.

———. "The Development of the Concept of Biological Regulation in the Eighteenth and Nineteenth Centuries." In *Ideology and Rationality in the History of the Life Sciences.* Trans. Arthur Goldhammer, pp. 81–102. Cambridge, MA: MIT Press, [1977] 1988.

———. "L'histoire des sciences dans l'œuvre épistémologique de Gaston Bachelard." In *Études d'histoire et de philosophie des sciences concernant les vivants et la vie*, 2nd ed., pp. 173–86. Paris: J. Vrin, 1970.

———. "The Living and Its Milieu." In *Knowledge of Life.* Trans. Stefanos Geroulanos and Daniela Ginsburg, pp. 98–120. New York: Fordham University Press, [1965] 2008.

———. "Monstrosity and the Monstrous." In *Knowledge of Life.* Trans. Stefanos Geroulanos and Daniela Ginsburg, pp. 134–48. New York: Fordham University Press, [1965] 2008.

———. *The Normal and the Pathological.* Trans. Carolyn R. Fawcett and Robert S. Cohen. New York: Zone Books, [1966] 1991.

———. "The Object of the History of Sciences" (1966). Trans. Mary Tiles. In *Continental Philosophy of Science*, edited by Gary Gutting, pp. 198–207. Oxford: Blackwell, 2005.

———. "On the History of the Life Sciences since Darwin." In *Ideology and Rationality in the History of the Life Sciences.* Trans. Arthur Goldhammer, pp. 103–23. Cambridge, MA: MIT Press, [1977] 1988.

———. "Théorie et technique de l'expérimentation chez Claude Bernard." In *Études d'histoire et de philosophie des sciences concernant les vivants et la vie*, 2nd ed., pp. 143–55. Paris: J. Vrin, 1970.

Canguilhem, Georges, Georges Lapassade, Jacques Piquemal, and Jacques Ulmann. *Du développement à l'évolution au XIXe siècle.* Paris: Presses universitaires de France, 1962.

Carbonel, Fréderic. "Le Docteur Féré (1852–1907): Une vie, une œuvre, de la médecine aux sciences sociales." *L'information psychiatrique* 82 (2006), pp. 59–69.

———. "Un oublié normand de la psychologie française: Le Docteur Féré (1852–1907)." *Bulletin de la Société libre d'Emulation de la Seine-Maritime* (2002), pp. 29–51.

Carol, Anne. *Histoire de l'eugénisme en France: Les médecins et la procréation XIXe–XXe siècle.* Paris: Éditions du Seuil, 1995.

Carroy, Jacqueline, Annick Ohayon, and Régine Pas. *Histoire de la psychologie en France,*

XIXe–XXe siècles. Paris: La Découverte, 2006.

Cartron, Laure. "Degeneration and 'Alienism' in Early Nineteenth-Century France." In *Heredity Produced: At the Crossroads of Biology, Politics, and Culture, 1500–1870*, edited by Staffan Müller-Wille and Hans-Jörg Rheinberger, pp. 155–74. Cambridge, MA: MIT Press, 2007.

Cavanaugh, John C. "Early Developmental Theories: A Brief Review of Attempts to Organise Developmental Data prior to 1925." *Journal of the History of Behavioural Sciences* 17.1 (1981), pp. 38–47.

Chappey, Jean-Luc. *La Société des Observateurs de l'homme (1799–1804): Des anthropologues au temps de Bonaparte*. Paris: Société des études robespierristes, 2002.

Charbonnier, Pierre, Gildas Salmon, and Peter Skafish (ed.). *Comparative Metaphysics: Ontology after Anthropology*. London: Rowman & Littlefield, 2017.

Cheung, Tobias. "What is an 'Organism'? On the Occurrence of a New Term and Its Conceptual Transformations 1680–1850." *History and Philosophy of Life Sciences* 32.2–3 (2010), pp. 155–94.

Chiapperino, Luca, Francesco Panese, and Umberto Simeoni. "L'epigénétique et le concept DOHaD: Vers de nouvelles temporalités de la médecine 'personnalisée'?" *Revue Médicale Suisse* 13 (2017), pp. 334–36.

Churchill, Frederick B. *August Weismann: Development, Heredity, and Evolution*. Cambridge, MA: Harvard University Press, 2015.

_____. "From Heredity Theory to *Vererbung*: The Transmission Problem, 1850–1915." *Isis* 78.3 (1987), pp. 336–64.

Clarke, Adele E. *Disciplining Reproduction: Modernity, American Life Sciences, and "the Problems of Sex"*. Berkeley: University of California Press, 1998.

Cole, Joshua. *The Power of Large Numbers: Population, Politics, and Gender in Nineteenth-Century France*. Ithaca, NY: Cornell University Press, 2000.

Coleman, William. *Biology in the Nineteenth Century: Problems of Form, Function, and Transformation*. Cambridge, UK: Cambridge University Press, 1979.

Coleman, William, and Frederic L. Holmes (ed.). *The Investigative Enterprise: Experimental Physiology in Nineteenth-Century Medicine*. Berkeley: University of California Press, 1988.

Courtin, Roger. *Charles Féré (1852–1907), médecin de la Bicêtre, et la "Néo-psychologie."* Paris: Connaissances et Savoirs, 2007.

Cunningham, Andrew. *The Anatomist Anatomis'd: An Experimental Discipline in*

Enlightenment Europe. London: Routledge, 2010.

Darmon, Pierre. *Le mythe de la procréation à l'âge baroque*. Paris: Éditions du Seuil, 1981.

Dasen, Véronique. "Empreintes maternelles." *Micrologus: Natura, scienze e società medievali* 17 (2009), pp. 35–54.

Daston, Lorraine (ed.). *Biographies of Scientific Objects*. Chicago: University of Chicago Press, 2000.

Daston, Lorraine, and Peter Galison. *Objectivity*. New York: Zone Books, 2007.

Démier, Francis. *La France du XIXe siècle, 1814–1914*. Paris: Éditions du Seuil, 2000.

Descola, Philippe. *Beyond Nature and Culture*. Trans. Janet Lloyd. Chicago: University of Chicago Press, 2014.

Desrosières, Alain. *The Politics of Large Numbers: A History of Statistical Reasoning*. Trans. Camille Naish. Cambridge, MA: Harvard University Press, 1998.

De Witt, Foster. *An Historical Study on Theories of the Placenta to 1900*. Bern: Arnaud, 1958.

———. "An Historical Study on Theories of the Placenta to 1900." *Journal of the History of Medicine* 14.3 (1959), pp. 360–74.

Dorlin, Elsa. *La matrice de la race: Généalogie sexuelle et coloniale de la Nation française*, 2nd ed. Paris: La Découverte, 2009.

Dreifuss, J. J. "Un chercheur genevois insuffisamment connu: Jean-Louis Prévost (1790–1850)." *Revue médicale suisse* 51 (2006).

Dror, Otniel E. "The Affect of Experiment: The Turn to Emotions in Anglo-American Physiology, 1900–1940." *Isis* 90.2 (1999), pp. 205–37.

Dubow, Sara. *Ourselves Unborn: A History of the Fetus in Modern America*. Oxford: Oxford University Press, 2011.

Duden, Barbara. "Anatomie der guten Hoffnung." Unpublished typescript, 2003.

———. *Disembodying Women: Perspectives on Pregnancy and the Unborn*. Trans. Lee Hoinacki. Cambridge, MA: Harvard University Press, 1993.

———. "Die 'Geheimnisse' der Schwangeren und das Öffentlichkeitsinteresse der Medizin: Zur sozialen Bedeutung der Kindsregung." In *Frauengeschichte—Geschlechtergeschichte*, edited by Karin Hausen and Heide Wunder, pp. 117–28. Frankfurt am Main: Campus, 1992.

———. *Die Gene im Kopf—der Fötus im Bauch: Historisches zum Frauenkörper*. Hannover: Offizin, 2002.

———. "Zwischen 'wahrem Wissen' und Prophetie: Konzeptionen des Ungeborenen." In *Geschichte des Ungeborenen: Zur Erfahrungs- und Wissenschaftsgeschichte der*

Schwangerschaft, 17.–20. Jahrhundert, edited by Barbara Duden, Jürgen Schlumbohm, and Patrice Veit, pp. 11–48. Göttingen: Vandenhoeck & Ruprecht, 2002.

Duden, Barbara, Jürgen Schlumbohm, and Patrice Veit (ed.). *Geschichte des Ungeborenen: Zur Erfahrungs- und Wissenschaftsgeschichte der Schwangerschaft, 17.–20. Jahrhundert.* Göttingen: Vandenhoeck & Ruprecht, 2002.

———. "Vorwort." In *Geschichte des Ungeborenen: Zur Erfahrungs- und Wissenschaftsgeschichte der Schwangerschaft, 17.–20. Jahrhundert*, edited by Barbara Duden, Jürgen Schlumbohm, and Patrice Veit, pp. 7–9. Göttingen: Vandenhoeck & Ruprecht, 2002.

Duhamel, Bernard. "L'œuvre tératologique d'Étienne Geoffroy Saint-Hilaire." *Revue d'histoire des sciences* 25.4 (1972), pp. 337–46.

Dunstan, G. R (ed.). *The Human Embryo: Aristotle and the Arabic and European Traditions.* Exeter: University of Exeter Press, 1990.

———. "Introduction: Text and Context." In *The Human Embryo: Aristotle and the Arabic and European Traditions*, edited by G. R. Dunstan, pp. 1–9. Exeter: University of Exeter Press, 1990.

Dupont, Jean-Claude. "Un autre embryon? Quelques relectures classiques de l'embryologie antique." In *L'embryon: Formation et animation; Antiquité grecque et latine, tradition hébraïque, chrétienne et islamique*, edited by Luc Brisson, Marie-Hélène Congourdeau, and Jean-Luc Solère, pp. 255–69. Paris: J. Vrin, 2008.

Eckardt, Georg. "Einleitung." In William T. Preyer, *Die Seele des Kindes: Eingeleitet und mit Materialien zur Rezeptionsgeschichte versehen von Georg Eckardt*, pp. 11–52. Berlin: Springer, 1989.

Eisenberg, Ziv. "Clear and Pregnant Danger: The Making of Prenatal Psychology in Mid-Twentieth-Century America." *Journal of Women's History* 22.3 (2010), pp. 112–35.

Ellenberger, Henri F. *The Discovery of the Unconscious: The History and Evolution of Dynamic Psychiatry.* New York: Basic Books, 1970.

Enke, Ulrike. "Einleitung." In Samuel Thomas Soemmerring, *Schriften zur Embryologie und Teratologie.* Edited by Ulrike Enke, pp. 1–110. Basel: Schwabe, 2000.

———. "Von der Schönheit der Embryonen: Samuel Thomas Soemmerrings Werk *Icones embryonum humanorum* (1799)." In *Geschichte des Ungeborenen: Zur Erfahrungs- und Wissenschaftsgeschichte der Schwangerschaft, 17.–20. Jahrhundert*, edited by Barbara Duden, Jürgen Schlumbohm, and Patrice Veit, pp. 205–35. Göttingen: Vandenhoeck & Ruprecht, 2002.

Epstein, Julia. "The Pregnant Imagination, Women's Bodies, and Fetal Rights." In *Inventing Maternity: Politics, Science, and Literature, 1650–1865*, edited by Susan C. Greenfield and

Carol Barash, pp. 111–37. Lexington: University Press of Kentucky, 1999.

Fassin, Didier, and Richard Rechtman. *The Empire of Trauma: An Inquiry into the Condition of Victimhood.* Trans. Rachel Gomme. Princeton, NJ: Princeton University Press, 2009.

Figlio, Karl. "The Historiography of Scientific Medicine: An Invitation to the Human Sciences." *Comparative Studies in Society and History* 19.3 (1977), pp. 262–86.

Filippini, Nadia Maria. "Die 'erste Geburt': Eine neue Vorstellung vom Fötus und vom Mutterleib (Italien, 18. Jahrhundert)." In *Geschichte des Ungeborenen: Zur Erfahrungs- und Wissenschaftsgeschichte der Schwangerschaft, 17.–20. Jahrhundert*, edited by Barbara Duden, Jürgen Schlumbohm, and Patrice Veit, pp. 99–127. Göttingen: Vandenhoeck & Ruprecht, 2002.

Fischer, Jean-Louis. "La callipédie, ou l'art d'avoir de beaux enfants." *Dix-huitième siècle* 23 (1991), pp. 141–58.

———. "Le concept expérimental dans l'œuvre tératologique d'Étienne Geoffroy Saint-Hilaire." *Revue d'histoire des sciences* 25.4 (1972), pp. 347–64.

———. "Défense et critiques de la thèse 'imaginationiste' à l'époque de Spallanzani." In *Lazzaro Spallanzani e la biologia del settecento*, edited by Giuseppe Montalenti and Paolo Rossi, pp. 413–29. Florence: L. S. Olschki, 1982.

———. "Embryogénése." In *Dictionnaire d'histoire et de philosophie des sciences*, edited by Dominique Lecourt, pp. 392–97. Paris: Presses universitaires de France, 1999.

———. *De la genèse fabuleuse à la morphogénèse des monstres.* Paris: Société française d'histoire des sciences et des techniques, 1986.

———. *Leben und Werk von Camille Dareste, 1822–1899: Schöpfer der experimentellen Teratologie.* Halle an der Saale: Deutsche Akademie der Naturforscher Leopoldina, 1994.

———. "La vie et la carrière d'un biologiste du XIXe siècle, Camille Dareste, 1822–1899." 3 vols. PhD diss., University of Paris 1 E.P.H.E., 1973. Translated as Fischer, Jean-Louis. *Leben und Werk von Camille Dareste, 1822—1899: Schöpfer der experimentellen Teratologie.* Halle an der Saale: Deutsche Akademie der Naturforscher Leopoldina, 1994.

Fischer-Homberger, Esther. *Geschichte der Medizin.* 2nd revised ed. Berlin: Springer, 1977.

———. *Harvey's Troubles with the Egg.* Sheffield, UK: European Association for the History of Medicine and Health Publications, 2001.

———. *Krankheit Frau und andere Arbeiten zur Medizingeschichte der Frau.* Bern: H. Huber, 1979.

———. *Medizin vor Gericht: Zur Sozialgeschichte der Gerichtsmedizin.* Darmstadt: Luchterhand, 1988.

———. *Die traumatische Neurose: Vom somatischen zum sozialen Leiden.* Bern: H. Huber, 1975.

Fleck, Ludwik. *Genesis and Development of a Scientific Fact*. Edited by Thaddeus J. Trenn and Robert K. Merton. Trans. Fred Bradley and Thaddeus J. Trenn. Chicago: University of Chicago Press, [1935] 1979.

Forrester, John. *Dispatches from the Freud Wars: Psychoanalysis and Its Passions*. Cambridge, MA: Harvard University Press, 1997.

Foucault, Michel. *Abnormal: Lectures at the Collège de France, 1974–1975*. Trans. Graham Burchell. London: Verso, 2003.

———. *The Archaeology of Knowledge and the Discourse on Language*. Trans. A. M. Sheridan Smith. New York: Pantheon, [1969] 1972.

———. "The Order of Discourse" (1970). Trans. Ian McLeod. In *Untying the Text: A Post-Structuralist Reader*, edited by Robert Young, pp. 48–78. London: Routledge & Kegan Paul, 1981.

———. *The Order of Things: An Archaeology of the Human Sciences*. Translated from the French. London: Routledge, [1966] 1974.

———. *Security, Territory, Population: Lectures at the Collège de France, 1977–78*. Edited by Michel Sellenart and trans. Graham Burchell. Basingstoke: Palgrave Macmillan, 2009.

Fox Keller, Evelyn. *The Mirage of a Space Between Nature and Nurture*. Durham, NC: Duke University Press, 2010.

———. *Refiguring Life: Metaphors of Twentieth-Century Biology*. New York: Columbia University Press, 1995.

Franklin, Sarah. "Fetal Fascinations: New Dimensions to the Medical-Scientific Construction of Fetal Personhood." In *Off-Centre: Feminism and Cultural Studies*, edited by Sarah Franklin, Celia Lury, and Jackie Stacey, pp. 191–205. London: HarperCollins Academic, 1991.

———. "*In Vitro Anthropos*: New Conception Models for a Recursive Anthropology?" *Cambridge Anthropology* 31.1 (2013), pp. 3–32.

Galibert, Jean-Paul. "Le jeu des temps embryonnaires." In *L'embryon humain à travers l'histoire: Images, savoirs et rites*, edited by Véronique Dasen, pp. 257–65. Gollion: Infolio éditions, 2007.

Gamper, Michael. *Masse lesen, Masse schreiben: Eine Diskurs- und Imaginationsgeschichte der Menschenmenge 1765–1930*. Munich: W. Fink, 2007.

Gaudillière, Jean-Paul. "Le syndrome nataliste: Hérédité, médecine et eugénisme en France et en Grande-Bretagne, 1920–1965." In *L'éternel retour de l'eugénisme*, edited by Jean Gayon and Daniel Jacobi, pp. 177–99. Paris: Presses universitaires de France, 2006.

Gaudillière, Jean-Paul, and Ilana Löwy. "The Hereditary Transmission of Human Pathologies between 1900 and 1940: The Good Reasons Not to Become 'Mendelian.'" In *Heredity Explored: Between Public Domain and Experimental Science, 1850–1930*, edited by Staffan Müller-Wille and Christina Brandt, pp. 311–35. Cambridge, MA: MIT Press, 2016.

——— (ed.). *Heredity and Infection: The History of Disease Transmission.* London: Routledge, 2001.

———. "Introduction: Horizontal and Vertical Transmission of Diseases, The Impossible Separation." In *Heredity and Infection: The History of Disease Transmission*, edited by Jean-Paul Gaudillière and Ilana Löwy, pp. 1–18. London: Routledge, 2001.

Gausemeier, Bernd. "Pedigrees of Madness: The Study of Heredity in Nineteenth and Early Twentieth Century Psychiatry." *History and Philosophy of the Life Sciences* 36.4 (2015), pp. 467–83.

Gélis, Jacques. *L'arbre et le fruit: La naissance dans l'Occident moderne, XVIe–XIXe siècle.* Paris: Fayard, 1984.

———. *La sage-femme ou le médecin: Une nouvelle conception de la vie.* Paris: Fayard, 1988.

Gilbert, Scott (ed.). *A Conceptual History of Modern Embryology.* Baltimore, MD: Johns Hopkins University Press, 1994.

Gluckmann, Peter D., Mark A. Hanson, and Tatjana Buklijaš. "A Conceptual Framework for the Developmental Origins of Health and Disease." *Journal of Developmental Origins of Health and Disease* 1.1 (2010), pp. 6–18.

———. "Maternal and Transgenerational Influences on Human Health." In *Transformations of Lamarckism: From Subtle Fluids to Molecular Biology*, edited by Snait Gissis and Eva Jablonka, pp. 237–50. Cambridge, MA: MIT Press, 2011.

Gould, Steven Jay. *Ontogeny and Phylogeny.* Cambridge, MA: Belknap Press, 1977.

Graf-Nold, Angela. *Der Fall Hermine Hug-Hellmuth: Eine Geschichte der frühen Kinder-Psychoanalyse.* Munich: Verlag Internationale Psychoanalyse, 1988.

Greenfield, Susan C., and Carol Barash (ed.). *Inventing Maternity: Politics, Science, and Literature, 1650–1865.* Lexington: University Press of Kentucky, 1999.

Gross, Michael. "The Lessened Locus of Feelings: A Transformation in French Physiology in the Early Nineteenth Century." *Journal of the History of Biology* 12 (1979), pp. 231–71.

Grubrich-Simitis, Ilse. "Metapsychology and Metabiology." In Sigmund Freud, *A Phylogenetic Fantasy: Overview of the Transference Neuroses.* Edited by Ilse Grubrich-Simitis, pp. 73–107. Trans. Axel Hoffer and Peter T. Hoffer. Cambridge, MA: Belknap Press, 1987.

Haberling, Wilhelm. *Johannes Müller: Das Leben des Rheinischen Naturforschers*. Leipzig: Akademische Verlagsgesellschaft, 1924.

Hacking, Ian. *Historical Ontology*. Cambridge, MA: Harvard University Press, 2002.

Hagner, Michael. *Homo cerebralis: Der Wandel vom Seelenorgan zum Gehirn*. Frankfurt am Main: Suhrkamp, 2008.

———. "Sieben Briefe von Johannes Müller an Karl Ernst von Baer." *Medizinhistorisches Journal* 27 (1992), pp. 138–55.

———. "Vom Naturalienkabinett zur Embryologie: Wandlungen des Monströsen und die Ordnung des Lebens." In *Der falsche Körper: Beiträge zu einer Geschichte der Monstrositäten*, edited by Michael Hagner, pp. 73–107. Göttingen: Wallstein, 1995.

Hanson, Clare. *A Cultural History of Pregnancy: Pregnancy, Medicine and Culture, 1750–2000*. Basingstoke: Palgrave Macmillan, 2004.

Harris, Oliver J. T., and John Robb. "Multiple Ontologies and the Problem of the Body in History." *American Anthropologist* 114.4 (2012), pp. 668–79.

Haynes, Douglas M. "The Human Placenta: Historical Considerations." In *The Human Placenta: Clinical Perspectives*, edited by J. Patrick Lavery, pp. 1–10. Rockville, MD: Aspen, 1987.

Helduser, Urte. *Imaginationen des Monströsen: Wissen, Literatur und Poetik der "Missgeburt," 1600–1835*. Göttingen: Wallstein, 2016.

Helduser, Urte, and Burkhard Dohm (ed.). *Imaginationen des Ungeborenen/Imaginations of the Unborn: Kulturelle Konzepte pränataler Prägung von der Frühen Neuzeit zur Moderne/ Cultural Concepts of Prenatal Imprinting from the Early Modern Period to the Present*. Heidelberg: Universitätsverlag Winter, 2018.

Henare, Amiria, Martin Holbraad, and Sari Wastell. "Introduction: Thinking through Things." In *Thinking Through Things: Theorising Artefacts Ethnographically*, edited by Amiria Henare, Martin Holbraad, and Sari Wastell, pp. 1–31. London: Routledge, 2007.

Hepper, Peter G. "Fetal Psychology: An Embryonic Science." In *Fetal Behaviour: Developmental and Perinatal Aspects*, edited by Jan G. Nijhuis, pp. 129–56. Oxford: Oxford University Press, 1992.

Herschkorn, Paule. "Adolphe Pinard et l'enfant à naître: L'invention de la médecine fœtale." *Devenir* 8.3 (1996), pp. 77–87.

Herschkorn-Barnu, Paule. "Wie der Fötus einen klinischen Status erhielt: Bedingungen und Verfahren der Produktion eines medizinischen Fachwissens, Paris 1832–1848." In *Geschichte des Ungeborenen: Zur Erfahrungs- und Wissenschaftsgeschichte der*

Schwangerschaft, 17.–20. Jahrhundert, edited by Barbara Duden, Jürgen Schlumbohm, and Patrice Veit, pp. 167–203. Göttingen: Vandenhoeck & Ruprecht, 2002.

Holbraad, Martin, and Morten Axel Pedersen. *The Ontological Turn: An Anthropological Exposition*. Cambridge, UK: Cambridge University Press, 2017.

———. "Planet M: The Intense Abstraction of Marilyn Strathern." *Anthropological Theory* 9.4 (2009), pp. 371–94.

Honegger, Claudia. *Die Ordnung der Geschlechter: Die Wissenschaften vom Menschen und das Weib, 1750–1850*. Frankfurt am Main: Campus, 1991.

Hopwood, Nick. "Embryology." In *The Cambridge History of Science*, vol. 6: *The Modern Biological and Earth Sciences*, edited by Peter J. Bowler and John V. Pickstone, pp. 285–315. Cambridge, UK: Cambridge University Press, 2009.

———. "'Giving Body' to Embryos: Modeling, Mechanism, and the Microtome in Late Nineteenth-Century Anatomy." *Isis* 90.3 (1999), pp. 462–96.

———. *Haeckel's Embryos: Images, Evolution, and Fraud*. Chicago: University of Chicago Press, 2015.

———. "The Keywords 'Generation' and 'Reproduction.'" In *Reproduction: Antiquity to the Present Day*, edited by Nick Hopwood, Rebecca Flemming, and Lauren Kassell, pp. 287–304. Cambridge, UK: Cambridge University Press, 2018.

———. "'Not Birth, Marriage or Death, but Gastrulation': The Life of a Quotation in Biology." *The British Journal for the History of Science* 55.1 (2022), pp. 1–26.

———. "Producing Development: The Anatomy of Human Embryos and the Norms of Wilhelm His." *Bulletin of the History of Medicine* 74.1 (2000), pp. 29–79.

Horder, T. J., J. A. Witkowski, and C. C. Wylie (ed.). *A History of Embryology*. Cambridge, UK: Cambridge University Press, 1986.

Huet, Marie-Hélène. *Monstrous Imagination*. Cambridge, MA: Harvard University Press, 1993.

Jablonka, Eva, and Marion J. Lamb. *Evolution in Four Dimensions: Genetic, Epigenetic, Behavioral, and Symbolic Variation in the History of Life*. Cambridge, MA: MIT Press, 2006.

Jackson, Mark. *The Age of Stress: Science and the Search for Stability*. Oxford: Oxford University Press, 2013.

Jacob, François. *The Logic of Life: A History of Heredity*. Trans. Betty E. Spillmann. Princeton, NJ: Princeton University Press, [1973] 1993.

Jahn, Ilse. *Grundzüge der Biologiegeschichte*. Jena: G. Fischer, 1990.

Jordanova, Ludmilla. "Gender, Generation and Science: William Hunter's Obstetrical

Atlas." In *William Hunter and the Eighteenth-Century Medical World*, edited by W. F. Bynum and Roy Porter, pp. 385–412. Cambridge, UK: Cambridge University Press, 1985.

———. "Interrogating the Concept of Reproduction in the Eighteenth Century." In *Conceiving the New World Order: The Global Politics of Reproduction*, edited by Faye Ginsburg and Rayna Rapp, pp. 369–86. Berkeley: University of California Press, 1995.

Kampf, Antje. "Times of Danger: Embryos, Sperm and Precarious Reproduction ca. 1870s–1910s." *History and Philosophy of the Life Sciences* 37.1 (2015), pp. 68–86.

Koselleck, Reinhart. *Futures Past: On the Semantics of Historical Time*. Trans. Keith Tribe. New York: Columbia University Press, [1979] 2004.

Kremer, Richard L. "Physiology." In *The Cambridge History of Science*, vol. 6: *The Modern Biological and Earth Sciences*, edited by Peter J. Bowler and John V. Pickstone, pp. 342–66. Cambridge, UK: Cambridge University Press, 2009.

Latour, Bruno. *We Have Never Been Modern*. Trans. Catherine Porter. London: Harvester Wheatsheaf, 1993.

Leitner, Marina. *Freud, Rank und die Folgen: Ein Schlüsselkonflikt für die Psychoanalyse*. Vienna: Turia & Kant, 1998.

Lenoir, Timothy. *The Strategy of Life: Teleology and Mechanics in Nineteenth-Century German Biology*. Dordrecht: D. Reidel, 1982.

Léonard, Jacques. *La médecine entre les savoirs et les pouvoirs: Histoire intellectuelle et politique de la médecine française au XIXe siècle*. Paris: Aubier-Montaigne, 1981.

Lepenies, Wolf. "Naturgeschichte und Anthropologie im 18. Jahrhundert." *Historische Zeitschrift* 231.1 (1980), pp. 21–42.

Lesch, John E. *Science and Medicine in France: The Emergence of Experimental Physiology, 1790–1855*. Cambridge, MA: Harvard University Press, 1984.

Lesky, Erna. *Die Zeugungs- und Vererbungslehren der Antike und ihr Nachwirken*. Wiesbaden: Franz Steiner, 1950.

Lettow, Susanne. *Biophilosophien: Wissenschaft, Technologie und Geschlecht im philosophischen Diskurs der Gegenwart*. Frankfurt am Main: Campus, 2011.

———. "Generation, Genealogy, and Time: The Concept of Reproduction from *Histoire naturelle* to *Naturphilosophie*." In *Reproduction, Race, and Gender in Philosophy and the Early Life Sciences*, edited by Susanne Lettow, pp. 21–44. Albany: SUNY Press, 2014.

———. "Improving Reproduction: Articulations of Breeding and 'Race-Mixing' in French and German Discourse (1750–1800)." In *The Secrets of Generation: Reproduction in the*

Long Eighteenth Century, edited by Raymond Stephanson and Darren N. Wagner, pp. 120–40. Toronto: University of Toronto Press, 2015.

———. "Population, Race and Gender: On the Genealogy of the Modern Politics of Reproduction." *Distinktion: Scandinavian Journal of Social Theory* 16.3 (2015), pp. 267–82.

——— (ed.). *Reproduction, Race, and Gender in Philosophy and the Early Life Sciences.* Albany: SUNY Press, 2014.

Leys, Ruth. *Trauma: A Genealogy.* Chicago: University of Chicago Press, 2000.

Longo, Lawrence D. *The Rise of Fetal and Neonatal Physiology: Basic Science to Clinical Care.* New York: Springer, 2013.

Longo, Lawrence D., and Lawrence P. Reynolds. "Some Historical Aspects of Understanding Placental Development, Structure and Function." *International Journal of Developmental Biology* 54.2–3 (2010), pp. 237–55.

López-Beltrán, Carlos. "In the Cradle of Heredity: French Physicians and *L'Hérédité Naturelle* in the Early 19th Century." *Journal of the History of Biology* 37.1 (2004), pp. 39–72.

———. "The Medical Origins of Heredity." In *Heredity Produced: At the Crossroads of Biology, Politics, and Culture, 1500–1870*, edited by Staffan Müller-Wille and Hans-Jörg Rheinberger, pp. 105–32. Cambridge, MA: MIT Press, 2007.

Loux, Françoise. *Le jeune enfant et son corps dans la médecine traditionelle.* Paris: Flammarion, 1978.

Löwy, Ilana. *Imperfect Pregnancies: A History of Birth Defects and Prenatal Diagnosis.* Baltimore, MD: Johns Hopkins University Press, 2017.

———. "On Guinea Pigs, Dogs and Men: Anaphylaxis and the Study of Biological Individuality, 1902–1939." *Studies in History and Philosophy of Biological and Biomedical Sciences* 34.3 (2003), pp. 399–423.

Lupton, Deborah. "'Precious Cargo': Foetal Subjects, Risk and Reproductive Citizenship." *Critical Public Health* 22.3 (2012), pp. 329–40.

Maher, JaneMaree. "Visibly Pregnant: Toward a Placental Body." *Feminist Review* 72.1 (2002), pp. 95–107.

Maienschein, Jane. "Heredity/Development in the United States, circa 1900." *History and Philosophy of Life Sciences* 9.1 (1987), pp. 79–93.

———. *Whose View of Life? Embryos, Cloning, and Stem Cells.* Cambridge, MA: Harvard University Press, 2003.

Malich, Lisa. *Die Gefühle der Schwangeren: Eine Geschichte somatischer Emotionalität (1780–2010).* Bielefeld: transcript-Verlag, 2017.

———. "Die hormonelle Natur und ihre Technologien: Zur Hormonisierung der Schwangerschaft im zwanzigsten Jahrhundert." *L'Homme* 25.2 (2014), pp. 69–84.

———. "Zeitpfeile, Zeitfaltungen und Diskursanalyse: Zu Kontinuitäten der Imaginationslehre." *Berichte zur Wissenschaftsgeschichte* 34 (2011), pp. 363–78.

Matthews, J. Rosser. *Quantification and the Quest for Medical Certainty*. Princeton, NJ: Princeton University Press, 1995.

McClive, Cathy. "The Hidden Truths of the Belly: The Uncertainties of Pregnancy in Early Modern Europe." *Social History of Medicine* 15.2 (2002), pp. 209–27.

McLaren, Angus. "Policing Pregnancies: Changes in Nineteenth-Century Criminal and Canon Law." In *The Human Embryo: Aristotle and the Arabic and European Traditions*, edited by G. R. Dunstan, pp. 187–207. Exeter: University of Exeter Press, 1990.

Medvei, Victor C. *The History of Clinical Endocrinology*. Carnforth, UK: Parthenon, 1993.

Meloni, Maurizio. *Impressionable Biologies: From the Archaeology of Plasticity to the Sociology of Epigenetics*. New York: Routledge, 2019.

———. *Political Biology: Science and Social Values in Human Heredity from Eugenics to Epigenetics*. New York: Palgrave Macmillan, 2016.

Mendelsohn, J. Andrew. "Medicine and the Making of Bodily Inequality in Twentieth-Century Europe." In *Heredity and Infection: The History of Disease Transmission*, edited by Jean-Paul Gaudillière and Ilana Löwy, pp. 21–79. London: Routledge, 2001.

Micale, Mark S. "Jean-Martin Charcot and *les névroses traumatiques*: From Medicine to Culture in French Trauma Theory of the Late Nineteenth Century." In *Traumatic Pasts: History, Psychiatry, and Trauma in the Modern Age, 1870–1930*, edited by Mark S. Micale and Paul Lerner, pp. 115–39. Cambridge, UK: Cambridge University Press, 2001.

Micale, Mark S., and Paul Lerner. "Trauma, Psychiatry, and History: A Conceptual and Historiographical Introduction." In *Traumatic Pasts: History, Psychiatry, and Trauma in the Modern Age, 1870–1930*, edited by Mark S. Micale and Paul Lerner, pp. 1–27. Cambridge, UK: Cambridge University Press, 2001.

——— (ed.). *Traumatic Pasts: History, Psychiatry, and Trauma in the Modern Age, 1870–1930*, Cambridge, UK: Cambridge University Press, 2001.

Mol, Annemarie. *The Body Multiple: Ontology in Medical Practice*. Durham, NC: Duke University Press, 2002.

Moravia, Sergio. "From *homme machine* to *homme sensible*: Changing Eighteenth-Century Models of Man's Image." *Journal of the History of Ideas* 39.1 (1978), pp. 45–60.

———. "The Enlightenment and the Sciences of Man." *History of Science* 18.4 (1980), pp. 247–68.

Morel, Marie-France. "Grossesse, foetus et histoire." In *La grossesse, l'enfant virtuel et la parentalité*, edited by Sylvain Missonnier, Bernard Golse, and Michel Soulé, pp. 21–39. Paris: Presses universitaires de France, 2004.

Morgan, Lynn M. "Fetal Relationality in Feminist Philosophy: An Anthropological Critique." *Hypatia* 11.3 (1996), pp. 47–70.

———. *Icons of Life: A Cultural History of Human Embryos*. Berkeley: University of California Press, 2009.

———. "The Potentiality Principle from Aristotle to Abortion." *Current Anthropology* 54.S7 (2013), pp. S15–S25.

Mucchielli, Laurent. "Aux origines de la psychologie universitaire en France (1870–1900): Enjeux intellectuels, contexte politique, réseaux et stratégies d'alliance autour de la *Revue philosophique* de Théodule Ribot." *Annals of Science: The History of Science and Technology* 55.3 (1998), pp. 263–89.

Müller-Wille, Staffan. "Evolutionstheorien vor Darwin." In *Evolution: Ein interdisziplinäres Handbuch*, edited by Philipp Sarasin and Marianne Sommer, pp. 63–78. Stuttgart: Metzler, 2010.

———. "Figures of Inheritance, 1650–1850." In *Heredity Produced: At the Crossroads of Biology, Politics, and Culture, 1500–1870*, edited by Staffan Müller-Wille and Hans-Jörg Rheinberger, pp. 177–204. Cambridge, MA: MIT Press, 2007.

———. "Reproducing Difference: Race and Heredity from a *longue durée* Perspective." In *Reproduction, Race, and Gender in Philosophy and the Early Life Sciences*, edited by Susanne Lettow, pp. 217–35. Albany: SUNY Press, 2014.

———. "Reproducing Species." In *The Secrets of Generation: Reproduction in the Long Eighteenth Century*, edited by Raymond Stephanson and Darren N. Wagner, pp. 37–58. Toronto: University of Toronto Press, 2015.

Müller-Wille, Staffan, and Christina Brandt (ed.). *Heredity Explored: Between Public Domain and Experimental Science, 1850–1930*. Cambridge, MA: MIT Press, 2016.

Müller-Wille, Staffan, and Hans-Jörg Rheinberger. *A Cultural History of Heredity*. Chicago: University of Chicago Press, 2012.

——— (ed.). *Heredity Produced: At the Crossroads of Biology, Politics, and Culture, 1500–1870*. Cambridge, MA: MIT Press, 2007.

———. "Heredity: The Formation of an Epistemic Space." In *Heredity Produced: At the*

Crossroads of Biology, Politics, and Culture, 1500–1870, edited by Staffan Müller-Wille and Hans-Jörg Rheinberger, pp. 3–34. Cambridge, MA: MIT Press, 2007.

Murphy Paul, Annie. *Origins: How the Nine Months before Birth Shape the Rest of Our Lives.* New York: Free Press, 2010.

Needham, Joseph. *Chemical Embryology*, 3 vols. Cambridge, UK: Cambridge University Press, 1931.

———. *A History of Embryology.* Revised ed. Cambridge, UK: Cambridge University Press, 1959.

Newman, Karen. *Fetal Positions: Individualism, Science, Visuality.* Stanford, CA: Stanford University Press, 1996.

Nicolas, Serge. *Histoire de la psychologie française: Naissance d'une nouvelle science.* Paris: In Press, 2002.

Nissen, Gerhardt. *Kulturgeschichte seelischer Störungen bei Kindern und Jugendlichen.* Stuttgart: Klett-Cotta, 2005.

Nijhuis, Jan G. (ed.). *Fetal Behaviour: Developmental and Perinatal Aspects.* Oxford: Oxford University Press, 1992.

Nuño de la Rosa, Laura. "Becoming Organisms: The Organisation of Development and the Development of Organisation." *History and Philosophy of Life Sciences* 32.2–3 (2010), pp. 289–316.

Nye, Robert A. "Love and Reproductive Biology in Fin-de-Siècle France: A Foucauldian Lacuna?" In *Foucault and the Writing of History*, edited by Jan Goldstein, pp. 150–64. Oxford: Blackwell, 1994.

Nyhart, Lynn K. *Biology Takes Form: Animal Morphology and the German Universities, 1800–1900.* Chicago: University of Chicago Press, 1995.

Oakley, Ann. *The Captured Womb: A History of the Medical Care of Pregnant Women.* Oxford: Blackwell, 1984.

Oertzen, Christine von. "Science in the Cradle: Milicent Shinn and Her Home-Based Network of Baby Observers, 1890–1910." *Centaurus* 55.2 (2013), pp. 175–95.

Oppenheimer, Jane M. *Essays in the History of Embryology and Biology.* Cambridge, MA: MIT Press, 1967.

———. "Some Historical Relationships between Teratology and Experimental Embryology." *Bulletin of the History of Medicine* 42.2 (1968), pp. 145–59.

———. "When Sense and Life Begin: Background for a Remark in Aristotle's *Politics* (1335b24)." *Arethusa* 8.2 (1975), pp. 331–43.

Orland, Barbara. "Labor-Reproduktion: Die Identität des Embryo zwischen Natur, Technik und Politik." In *Sexualität als Experiment: Identität, Lust und Reproduktion zwischen Science und Fiction*, edited by Nicolas Pethes and Silke Schicktanz, pp. 311–30. Frankfurt am Main: Campus, 2008.

Otis, Laura. *Müller's Lab*. Oxford: Oxford University Press, 2007.

———. *Organic Memory: History and the Body in the Late Nineteenth and Early Twentieth Centuries*. Lincoln: University of Nebraska Press, 1994.

Ottavi, Dominique. *De Darwin à Piaget: Pour une histoire de la psychologie de l'enfant*. Paris: CNRS Éditions, 2001.

Park, Katharine. *Secrets of Women: Gender, Generation, and the Origins of Human Dissection*. New York: Zone Books, 2006.

Parnes, Ohad S. "On the Shoulders of Generations: The New Epistemology of Heredity in the Nineteenth Century." In *Heredity Produced: At the Crossroads of Biology, Politics, and Culture, 1500–1870*, edited by Staffan Müller-Wille and Hans-Jörg Rheinberger, pp. 315–47. Cambridge, MA: MIT Press, 2007.

Parnes, Ohad, Ulrike Vedder, and Stefan Willer. *Das Konzept der Generation: Eine Wissenschafts- und Kulturgeschichte*. Frankfurt am Main: Suhrkamp, 2008.

Pick, Daniel. *Faces of Degeneration: A European Disorder, c. 1848–c. 1918*. Cambridge, UK: Cambridge University Press, 1989.

Pinell, Patrice. "Degeneration Theory and Heredity Patterns between 1850 and 1900." In *Heredity and Infection: The History of Disease Transmission*, edited by Jean-Paul Gaudillière and Ilana Löwy, pp. 245–59. London: Routledge, 2001.

Piontelli, Alessandra. *Development of Normal Fetal Movements: The First 25 Weeks of Gestation*. Milan: Springer, 2010.

Pomata, Gianna. "Die Geschichte der Frauen zwischen Anthropologie und Biologie." *Feministische Studien* 2.2 (1983), pp. 113–27.

Porqueres i Gené, Enric. "Individu et parenté: Individuation de l'embryon." In *Corps et affects*, edited by Françoise Héritier and Margarita Xanthakou, pp. 139–50. Paris: O. Jacob, 2004.

———. "Personne et parenté." *L'Homme* 210 (2014), pp. 17–42.

Porter, Roy. *The Greatest Benefit to Mankind: A Medical History of Humanity*. New York: W. W. Norton, 1999.

Prochiantz, Alain. *Claude Bernard: La révolution physiologique*. Paris: Presses universitaires de France, 1990.

Prosperi, Adriano. *Infanticide, Secular Justice, and Religious Debate in Early Modern Europe.* Trans. Hilary Siddons. Turnhout: Brepols, 2016.

Reinert, Günther. "Grundzüge einer Geschichte der Human-Entwicklungspsychologie." In *Die Psychologie des 20. Jahrhunderts. Band 1: Die europäische Tradition: Tendenzen, Schulen, Entwicklungslinien*, edited by Heinrich Balmer, pp. 862–96. Zurich: Kindler, 1976.

———. "History of Life-Span Development Psychology." In *Life-Span Development and Behavior*, vol. 2, edited by Paul B. Baltes and Orville G. Brim, pp. 205–54. New York: Academic Press, 1979.

Reiss, H. E. "Historical Insights: John William Ballantyne 1861–1923." *Human Reproduction Update* 5.4 (1999), pp. 386–89.

Renzi, Silvia de. "Resemblance, Paternity, and Imagination in Early Modern Courts." In *Heredity Produced: At the Crossroads of Biology, Politics, and Culture, 1500–1870*, edited by Staffan Müller-Wille and Hans-Jörg Rheinberger, pp. 61–83. Cambridge, MA: MIT Press, 2007.

Rheinberger, Hans-Jörg. *An Epistemology of the Concrete: Twentieth-Century Histories of Life.* Trans. G. M. Goshgarian. Durham, NC: Duke University Press, 2010.

———. *Experiment, Differenz, Schrift: Zur Geschichte epistemischer Dinge.* Marburg an der Lahn: Basilisken-Press, 1992.

———. *Toward a History of Epistemic Things: Synthesizing Proteins in the Test Tube.* Stanford, CA: Stanford University Press, 1997.

Rheinberger, Hans-Jörg, and Michael Hagner. "Experimentalsysteme." In *Die Experimentalisierung des Lebens: Experimentalsysteme in den biologischen Wissenschaften 1850/1950*, edited by Hans-Jörg Rheinberger and Michael Hagner, pp. 7–27. Berlin: Akademie Verlag, 1993.

Rheinberger, Hans-Jörg, and Staffan Müller-Wille. *The Gene: From Genetics to Postgenomics.* Trans. Adam Bostanci. Chicago: University of Chicago Press, 2017.

———. *Vererbung: Geschichte und Kultur eines biologischen Konzepts.* Frankfurt am Main: Fischer, 2009.

Richardson, Sarah S. "Maternal Bodies in the Postgenomic Order: Gender and the Explanatory Landscape of Epigenetics." In Sarah S. Richardson and Hallam Stevens, *Postgenomics: Perspectives on Biology after the Genome*, pp. 210–31. Durham, NC: Duke University Press, 2015.

———. *The Maternal Imprint: The Contested Science of Maternal-Fetal Effects.* Chicago: University of Chicago Press, 2021.

Richardson, Sarah S., and Hallam Stevens (ed.). *Postgenomics: Perspectives on Biology after the Genome.* Durham, NC: Duke University Press, 2015.

Roe, Shirley. *Matter, Life, and Generation: Eighteenth-Century Embryology and the Haller-Wolff Debate.* Cambridge, UK: Cambridge University Press, 1981.

Roger, Jacques. *The Life Sciences in Eighteenth-Century French Thought.* Edited by Keith R. Benson and trans. Robert Ellrich. Stanford, CA: Stanford University Press, [1963] 1998.

Rollet, Catherine. *Les enfants au XIXe siècle.* Paris: Hachette, 2001.

Rose, Nikolas. *Governing the Soul: The Shaping of the Private Self.* London: Free Association Books, [1989] 1999.

Roth, Michael S. *Memory, Trauma, and History: Essays on Living with the Past.* New York: Columbia University Press, 2012.

Rothschuh, Karl E. *Geschichte der Physiologie.* Berlin: Springer, 1953.

Rouch, Hélène. "Le placenta comme tiers." *Langages* 21.85 (1987), pp. 71–79.

Sarasin, Philipp. "Der öffentlich sichtbare Körper: Vom Spektakel der Anatomie zu den *curiosités physiologiques.*" In *Physiologie und industrielle Gesellschaft: Studien zur Verwissenschaftlichung des Körpers im 19. und 20. Jahrhundert*, edited by Philipp Sarasin and Jakob Tanner, pp. 419-52. Frankfurt am Main: Suhrkamp 1998.

———. *Reizbare Maschinen: Eine Geschichte des Körpers 1765–1914.* Frankfurt am Main: Suhrkamp, 2001.

Satzinger, Helga. *Differenz und Vererbung: Geschlechterordnungen in der Genetik und Hormonforschung 1890–1950.* Cologne: Böhlau, 2009.

Schafer, Andrew I. "History of the Physician as Scientist." In *The Vanishing Physician-Scientist?*, edited by Andrew I. Schafer, pp. 17–38. Ithaca, NY: Cornell University Press, 2009.

Scharbert, Gerhard. "Freud and Evolution." *History and Philosophy of the Life Sciences* 31.2 (2009), pp. 295–312.

———. "*Psychologus nemo, nisi Physiologus*: Johannes Müller und die Perspektiven einer médecine philosophique; eine Entdeckung aus dem Universitätsarchiv." *Würzburger medizinhistorische Mitteilungen* 29 (2010), pp. 241–55.

Schiebinger, Londa. *Nature's Body: Gender in the Making of Modern Science.* New Brunswick, NJ: Rutgers University Press, 1993.

Schivelbusch, Wolfgang. *Die Kultur der Niederlage: Der amerikanische Süden 1865, Frankreich 1871, Deutschland 1918.* Berlin: A. Fest, 2001.

Schlumbohm, Jürgen. *Lebendige Phantome: Ein Entbindungshospital und seine Patientinnen,*

1751–1830. Göttingen: Wallstein, 2012.

Schmid, Pia. "Väter und Forscher: Zu Selbstdarstellungen bürgerlicher Männer um 1800 im Medium empirischer Kinderbeobachtungen." *Feministische Studien* 18.2 (2000), pp. 35–48.

Sengoopta, Chandak. *The Most Secret Quintessence of Life: Sex, Glands, and Hormones, 1850–1950*. Chicago: University of Chicago Press, 2006.

Shoja, Mohammadali M., R. Shane Tubbs, Marios Loukas, Ghaffar Shokouhi, and Mohammad R. Ardalan. "Marie-François Xavier Bichat (1771–1802) and His Contributions to the Foundations of Pathological Anatomy and Modern Medicine." *Annals of Anatomy—Anatomischer Anzeiger* 190.5 (2008), pp. 413–20.

Shuttleworth, Sally. *The Mind of the Child: Child Development in Literature, Science, and Medicine, 1840–1900*. Oxford: Oxford University Press, 2010.

Simmel, Georg. "The Problem of Historical Time" (1916). In *Essays on Interpretation in Social Science*. Edited and trans. Guy Oakes, pp. 127–44. Totowa, NJ: Rowman and Littlefield, 1980.

Smith, Justin E. H. "Imagination and the Problem of Heredity in Mechanist Embryology." In *The Problem of Animal Generation in Early Modern Philosophy*, edited by Justin E. H. Smith, pp. 80–99. Cambridge, UK: Cambridge University Press, 2006.

——— (ed.). *The Problem of Animal Generation in Early Modern Philosophy*. Cambridge, UK: Cambridge University Press, 2006.

Stahnisch, Frank W. "François Magendie (1783–1855)." *Journal of Neurology* 256.11 (2009), pp. 1950–52.

———. *Ideas in Action: Der Funktionsbegriff und seine methodologische Rolle im Forschungsprogramm des Experimentalphysiologen François Magendie (1783–1855)*. Münster: LIT, 2003.

Stahuljak, Zrinka. "History as a Medical Category: Heredity, Positivism, and the Study of the Past in Ninenteenth-Century France." *History of the Present* 3.2 (2013), pp. 140–59.

Steedman, Carolyn. *Strange Dislocations: Childhood and the Idea of Human Interiority, 1780–1930*. Cambridge, MA: Harvard University Press, 1995.

Stephanson, Raymond, and Darren N. Wagner (ed.). *The Secrets of Generation: Reproduction in the Long Eighteenth Century*. Toronto: University of Toronto Press, 2015.

Stoff, Heiko. "Der aktuelle Gebrauch der *longue durée* in der Wissenschaftsgeschichte." *Berichte zur Wissenschaftsgeschichte* 32.2 (2009), pp. 144–58.

Strathern, Marilyn. *The Gender of the Gift: Problems with Women and Problems with Society in Melanesia*. Berkeley: University of California Press, 1988.

———. *Kinship, Law and the Unexpected: Relatives Are Always a Surprise*. Cambridge, UK: Cambridge University Press, 2005.

———. *Reproducing the Future: Essays on Anthropology, Kinship and the New Reproductive Technologies*. Manchester: Manchester University Press, 1992.

Struck, Eckhard. "Ignaz Döllinger 1770–1841: Ein Physiologe der Goethe-Zeit und der Entwicklungsgedanke in seinem Leben und Werk." PhD diss., Ludwig Maximilian University, Munich, 1977.

Taithe, Bertrand. *Defeated Flesh: Welfare, Warfare and the Making of Modern France*. Manchester: Manchester University Press, 1999.

Temkin, Owsei. "German Concepts of Ontogeny and History around 1800." *Bulletin of the History of Medicine* 24.3 (1950), pp. 227–46.

Thirard, Jacqueline. "La fondation de la *Revue philosophique*." *Revue philosophique de la France et de l'étranger* 166.4 (1976), pp. 401–13.

Thompson, Charis. *Making Parents: The Ontological Choreography of Reproductive Technologies*. Cambridge, MA: MIT Press, 2005.

Turmel, André. *A Historical Sociology of Childhood: Developmental Thinking, Categorization and Graphic Visualization*. Cambridge, UK: Cambridge University Press, 2008.

Van der Lugt, Maaike. "L'animation de l'embryon humain et le statut de l'enfant à naître dans la pensée médiévale." In *L'Embryon: Formation et animation; Antiquité grecque et latine, tradition hébraïque, chrétienne et islamique*, edited by Luc Brisson, Marie-Hélène Congourdeau, and Jean-Luc Solère, pp. 233–54. Paris: J. Vrin, 2008.

———. *Le ver, le démon et la vierge: Les théories médievales de la génération extraordinaire*. Paris: Les Belles Lettres, 2004.

Vidal, Fernando. "La 'science de l'homme': Désirs d'unité et juxtaposition encyclopédiques." In *L'histoire des sciences de l'homme: Trajectoire, enjeux et questions vives*, edited by Claude Blanckaert, Loïc Blondiaux, Laurent Loty, Marc Renneville, and Nathalie Richard, pp. 61–77. Paris: L'Harmattan, 1999.

Vidal, Fernando, Marino Buscaglia, and J. Jacques Vonèche. "Darwinism and Developmental Psychology." *Journal of the History of the Behavioral Sciences* 19.1 (1983), pp. 81–91.

Vienne, Florence. "Eggs and Sperm as Germ Cells." In *Reproduction: Antiquity to the Present Day*, edited by Nick Hopwood, Rebecca Flemming, and Lauren Kassell, pp. 413–26. Cambridge, UK: Cambridge University Press, 2018.

Vienne, Florence, and Christina Brandt. "Einleitung." In *Wissensobjekt Mensch: Humanwissenschaftliche Praktiken im 20. Jahrhundert*, edited by Florence Vienne and Christina

Brandt, pp. 9–29. Berlin: Kulturverlag Kadmos, 2008.

Viveiros de Castro, Eduardo. "Exchanging Perspectives: The Transformation of Objects into Subjects in Amerindian Ontologies." *Common Knowledge* 10.3 (2004), pp. 463–84.

———. "The Relative Native." *HAU: Journal of Ethnographic Theory* 3.3 (2013), pp. 473–502.

Walentowitz, Saskia. "L'enfant qui n'a pas atteint son lieu: Représentations et soins autour des prématurés chez les Touaregs de l'Azawagh (Niger)." *L'Autre: Cliniques, cultures et sociétés* 5.2 (2004), pp. 227–41.

———. "La vie sociale du foetus: Regards anthropologiques." *Spirales* 36.4 (2005), pp. 125–41.

Watzke, Daniela. "Embryologische Konzepte zur Entstehung von Missbildungen im 18. Jahrhundert." In *Imagination und Sexualität: Pathologien der Einbildungskraft im medizinischen Diskurs der frühen Neuzeit*, edited by Stefanie Zaun, Daniela Watzke, and Jörn Steigerwald, pp. 119–36. Frankfurt am Main: Klostermann, 2004.

Weir, Lorna. *Pregnancy, Risk and Biopolitics: On the Threshold of the Living Subject*. London: Routledge, 2006.

Wellmann, Janina. *The Form of Becoming: Embryology and the Epistemology of Rhythm, 1760–1830*. Trans. Kate Sturge. New York: Zone Books, 2017.

———. "Keine Ikone der Entwicklung: Die *Icones embryonum humanorum* von Samuel Thomas Soemmerring." In *Kulturen des Wissens im 18. Jahrhundert*, edited by Hans Ulrich Schneider, pp. 585–94. Berlin: De Gruyter, 2008.

Wettley, Annemarie. "Zur Problemgeschichte der 'dégénérescence.'" *Sudhoffs Archiv für Geschichte der Medizin und der Naturwissenschaften* 43.3 (1959), pp. 193–212.

Wilkin, Rebecca. "Descartes, Individualism, and the Fetal Subject." *Differences* 19.1 (2008), pp. 96–127.

Williams, Elizabeth A. *The Physical and the Moral: Anthropology, Physiology, and Philosophical Medicine in France, 1750–1850*. Cambridge, UK: Cambridge University Press, 1994.

Wilson, Colette E. *Paris and the Commune, 1871–78: The Politics of Forgetting*. Manchester: Manchester University Press, 2007.

Wilson, Philip K. "Out of Sight, Out of Mind? The Daniel Turner–James Blondel Dispute Over the Power of the Maternal Imagination." *Annals of Science: The History of Science and Technology* 49.1 (1992), pp. 63–85.

Winther, Rasmus. "August Weismann on Germ-Plasm Variation." *Journal of the History of Biology* 34 (2001), pp. 517–55.

Woolgar, Steve, and Javier Lezaun. "The Wrong Bin Bag: A Turn to Ontology in Science and Technology Studies?" *Social Studies of Science* 43.3 (2013), pp. 321–40.

Yehuda, Rachel, Nikolaos P. Daskalakis, Linda M. Bierer, Heather N. Bader, Torsten Klengel, Florian Holsboer, and Elisabeth B. Binder. “Holocaust Exposure Induced Intergenerational Effects on FKBP5 Methylation.” *Biological Psychiatry* 80.5 (2016), pp. 372–80.

Zürcher, Urs. *Monster oder Laune der Natur: Medizin und die Lehre von den Missbildungen 1780–1914*. Frankfurt am Main: Campus, 2004.

List of Illustrations

Index

Zone Books series design by Bruce Mau
Image placement and production by Julie Fry
Typesetting by Meighan Gale
Printed and bound by Maple Press